Fracture—Instability Dynamics, Scaling, and Ductile/Brittle Behavior

MATERIALS RESEARCH SOCIETY
SYMPOSIUM PROCEEDINGS VOLUME 409

Fracture—Instability Dynamics, Scaling, and Ductile/Brittle Behavior

Symposium held November 27-December 1, 1995, Boston, Massachusetts, U.S.A.

EDITORS:

Robin L. Blumberg Selinger

National Institute of Standards and Technology
Gaithersburg, Maryland, U.S.A.

John J. Mecholsky

University of Florida
Gainesville, Florida, U.S.A.

Anders E. Carlsson

Washington University
Saint Louis, Missouri, U.S.A.

Edwin R. Fuller, Jr.

National Institute of Standards and Technology
Gaithersburg, Maryland, U.S.A.

MATERIALS
RESEARCH
SOCIETY

PITTSBURGH, PENNSYLVANIA

CAMBRIDGE
UNIVERSITY PRESS

32 Avenue of the Americas, New York NY 10013-2473, USA

Cambridge University Press is part of the University of Cambridge.

It furthers the University's mission by disseminating knowledge in the pursuit of education, learning and research at the highest international levels of excellence.

www.cambridge.org
Information on this title: www.cambridge.org/9781558993129

CODEN: MRSPDH

Copyright 1996 by Materials Research Society.

A catalogue record for this publication is available from the British Library

Library of Congress Cataloguing in Publication data

Fracture : instability dynamics, scaling, and ductile/brittle behavior : symposium held November 27-December 1, 1995, Boston, Massachusetts, U.S.A. / editors, Robin L. Blumberg Selinger, John J. Mecholsky, Anders E. Carlsson, and Edwin R. Fuller, Jr.
     p. cm.—(Materials Research Society symposium proceedings ; v. 409) Includes bibliographic references and index.
    ISBN 1-55899-312-6 (alk. paper)
    1. Fracture mechanics—Congresses.   I. Blumberg Selinger, Robin L. II. Mecholsky, John J.   III. Carlsson, Anders E.   IV. Fuller, Edwin R., Jr. Series: Materials Research Society symposium proceedings ; v. 409
TA409.F713   1996
620.1′126—dc20   96-10206
                                                      CIP

ISBN 978-1-558-99312-9 Hardback

CONTENTS

*Invited Paper

*Invited Paper

PART IV: <u>FRACTURE IN CERAMICS AND COMPOSITES</u>

*Invited Paper

*Invited Paper

PREFACE

This volume contains the papers that were presented at the 1995 MRS Fall Meeting in Symposium Q, entitled 'Fracture: Instability Dynamics, Scaling, and Ductile/Brittle Behavior.' The purpose of the symposium was to bring together the many communities that investigate the fundamentals of fracture, with special sessions on the ductile/brittle transition, fracture at interfaces, fracture in ceramics and composites, dynamic instabilities in crack propagation, and fractals and scaling in fracture. A full-day joint session was held with Symposium P, 'Materials Theory, Simulations, and Parallel Algorithms.'

What was most striking about the symposium was the rich variety of methods that investigators use to model fracture. At the most detailed level, *ab initio* techniques are used, for instance, to estimate the strength of adhesion at an interface as a function of local structure and composition. At the next level, classical molecular dynamics provide a close up view of a propagating crack in a two- or three-dimensional solid, at least over short length and time scales. Green's function methods provide a way to model the behavior of materials at the atomic scale in the static limit, with zero temperature and strain rate. At the mesoscale, defects such as dislocations may be modelled as point particles (in 2-D) or as line segments (in 3-D) with long-range interactions. The dislocation density may be approximated as a continuous function of position and time, or even treated as an order parameter in a statistical mechanical model. At larger length scales, finite element and continuum models are used to study mixed mode crack propagation, and bond network models are used to represent the behavior of fiber composites.

Experimental papers also emphasized a wide range of techniques for measuring the effects of fractures, ranging from fracture toughness measurements to atomic-force and scanning-electron microscopy. Major topics included impurity embrittlement, interfacial adhesion, the toughness of ceramics and composites, and the morphology of fracture surfaces. In connection with observations of fractal morphology, several papers showed that some fracture surfaces can be modelled as self-affine rather than self-similar; in each case, the surface has scaling properties, but in the former case the scaling perpendicular to the surface is different from that parallel to the surface. Another particularly exciting experimental technique presented was X-ray tomographic microscopy (XTM), which was used to image the interior of a three-dimensional sample undergoing ductile rupture at a material interface.

Thanks are due to Robb Thomson, who assisted in the planning of the symposium; to Heather Mayton, who provided priceless administrative assistance; to the MRS staff, who offered excellent logistical support; and to the 1995 MRS Fall Meeting Chairs, Michael J. Aziz, Leslie J. Struble, and Bernard T. Jonker.

The symposium was generously supported by the Office of Naval Research, and we are grateful to Scientific Officers Steven G. Fishman and Peter Reynolds for their support.

Robin L. Blumberg Selinger
Anders E. Carlsson
John J. Mecholsky
Edwin R. Fuller, Jr.

January 1996

xi

MATERIALS RESEARCH SOCIETY SYMPOSIUM PROCEEDINGS

MATERIALS RESEARCH SOCIETY SYMPOSIUM PROCEEDINGS

Prior Materials Research Society Symposium Proceedings available by contacting Materials Research Society

Part I

Brittle-Ductile Behavior and Crack Tip Processes

MOLECULAR-DYNAMICS SIMULATIONS OF FRACTURE: AN OVERVIEW OF SYSTEM SIZE AND OTHER EFFECTS

B. L. HOLIAN *, S.J. ZHOU *, P.S. LOMDAHL *, N. GRØNBECH-JENSEN *,
D.M. BEAZLEY *, and R. RAVELO **
*Theoretical Division, Los Alamos National Laboratory, Los Alamos, NM 87545
**Department of Physics, University of Texas, El Paso, TX 79968-7046

ABSTRACT

We have studied brittle and ductile behavior and their dependence on system size and interaction potentials, using molecular-dynamics (MD) simulations. By carefully embedding a single sharp crack in two- and three-dimensional crystals, and using a variant of the efficient sound-absorbing reservoir of Holian and Ravelo [Phys. Rev. B **51**, 11275 (1995)], we have been able to probe both the static and dynamic crack regimes. Our treatment of boundary and initial conditions allows us to elucidate early crack propagation mechanisms under delicate overloading, all the way up to the more extreme dynamic crack-propagation regime, for much longer times than has been possible heretofore (before unwanted boundary effects predominate). For example, we have used graphical display of atomic velocities, forces, and potential energies to expose the presence of localized phonon-like modes near the moving crack tip, just prior to dislocation emission and crack-branching events. We find that our careful MD method is able to reproduce the ZCT brittle-ductile criterion for short-range pair potentials [static lattice Green's function calculations of Zhou, Carlsson, and Thomson, Phys. Rev. Letters **72**, 852 (1994)].

We report on progress we have made in large-scale 3D simulations in samples that are thick enough to display realistic behavior at the crack tip, including emission of dislocation loops. Such calculations, using our careful treatment of boundary and initial conditions - especially important in 3D - have the promise of opening up new vistas in fracture research.

INTRODUCTION

Insufficient understanding of the brittle *vs.* ductile behavior of materials has hindered the development of new materials with high strength and toughness. One of the reasons is that these mechanical properties are influenced by material structures at several different levels, including electronic structure, crystal structure, external environment, and imperfections like cracks, dislocations and interfaces. Combining atomistic, microstructural, and continuum theories is an effective way of exploring these mechanical properties. A valuable key to the toughness problem is to distinguish between an intrinsically brittle material and an intrinsically ductile one. In 1974, Rice and Thomson [1] proposed a semi-quantitative criterion for determining the intrinsic ductile

3

vs. brittle behavior of materials in terms of the competition between dislocation emission from crack tips and crack cleavage. Rice has recently improved this model by including some atomistic features into the continuum elastic theory [2].

To gain a comprehensive understanding, however, it is very important to perform large-scale molecular-dynamics (MD) simulations, partly because the strain fields of cracks and dislocations are long-range in character, and partly because moving cracks generate sound waves, which are reflected from (or transmitted through) boundaries. We have developed a massively parallel MD code called SPaSM, which is capable of treating millions of atoms [3,4] and absorbing sound waves and mobile dislocations in boundary regions [5]. With these careful MD simulation techniques, we are able to study many aspects of cracks which have been unobtainable to date. For example, in 2D, we can reproduce the criterion for the brittle-ductile transition for short-range pair potentials [6], and we see dramatic differences between static and dynamic fracture. Moreover, we find that dynamic cracks can significantly facilitate dislocation emission, and that crack branching is initiated by dislocation nucleation. The bond-breaking energy released from a dynamic crack accumulates at the crack tip and strongly excites local phonons on each side of the crack. This is one of the key features of a dynamic crack and has a profound effect on crack instability [7].

Crack-tip fronts in real 3D materials are not always straight; moreover, dislocation nucleation can be a thermally activated process that is dependent on the length scale of the crack front. Ledges and other defects on crack fronts can serve as "easily operated" dislocation nucleation sources [8], which have been experimentally confirmed [9]. Although experimental techniques such as High Resolution Transmission Electron Microscopy are getting more powerful, it is still a challenging task to observe 3D, dynamic, atomistic processes such as dislocation nucleation and emission from a crack tip.

Recently we have performed large-scale 3D MD simulations on a crack in the fcc lattice, similar to the system investigated by deCelis, Argon, and Yip with quasi-3D MD [10]. We have observed blunting half-dislocation loops emitted from both fronts of an embedded crack, at an initial temperature of almost zero. Preliminary results on cracks and dislocations (including jogging dislocations along slip planes that are at an angle to the crack front) are discussed, along with new visualization techniques that are particularly effective in 3D.

RESULTS

Since cracks and mobile dislocations move at appreciable fractions of sound velocities, and since stresses relax by sound waves, it is only natural that dynamic processes accompanying crack propagation impose severe temporal limitations on molecular dynamics simulations, where the inherent length scale is made very small by computer-memory limitations. If, for example, sound waves are permitted to pass through periodic boundaries or reflect from free boundaries, they will eventually return to the region ahead of the crack and impose artificial stresses long before they would have done so in a truly macroscopic sample. For this reason, Holian and Ravelo [5] devised smooth impedance-matched, acoustic-absorbing reservoir regions, where atoms are subjected to artificial viscous damping, which is gradually ramped up from zero at the reservoir-sample interface to a value that critically damps high frequency sound waves. Thus, the whole spectrum of quasiharmonic phonon frequencies of traveling waves is damped out. Other methods of damping that do not use this impedance matching make the reservoir-sample boundary look like a hard wall and reflect a significant fraction of the sound-wave amplitude. A modification of this reservoir approach has been proposed [11], wherein an embedded crack is surrounded by an elliptical "stadium," defined by the function $0 \leq f \leq 1$ ($f = 0$ in the sample region):

$$f(r) = \min\left(1, \max\left(0, \frac{(x/L_x)^2 + (y/L_y)^2 - (a/L_x)^2}{1/4 - (a/L_x)^2}\right)\right) \quad , \tag{1}$$

where L_x and L_y are the widths of the entire computational cell, and a and b are the x- and y-axes of the inner sample region, such that $a/L_x = b/L_y$. The temperature in the reservoir region can then be defined by $kT = \Sigma\ m|u|^2 f / (2\Sigma\ f)$ (u is particle velocity, and the sum extends over all atoms in the system), with viscous damping coefficient $\gamma = 2\omega_E(1 - T_0/T)$ controlling the long-time average of the temperature to be T_0, the initial sample temperature.

In addition to this gentle boundary treatment, the embedded crack itself can be made atomistically sharp, with initial atomic displacements chosen to be nearly the final relaxed values for the critical Griffith strain, according to the continuum elastic solution [11]. Nonlinear relaxation occurs in a tiny region (only two or three atomic spacings away from the crack tip), generating small-amplitude sound waves that are only barely perceptible. Slight additional overstrain can then be achieved by applying a homogeneous strain rate (in the x-direction) $\dot\varepsilon_{xx}$ appropriate to adiabatic expansion; the Hamiltonian equations of motion for the atomic coordinates r and thermal velocities (relative to the imposed strain rate) u are then given by [11]

$$\dot{\mathbf{r}} = \mathbf{u} + \dot\varepsilon_{xx}\, x\, \hat{\mathbf{x}}$$
$$\dot{\mathbf{u}} = \frac{\mathbf{F}}{m} - \dot\varepsilon_{xx}\, u\, \hat{\mathbf{x}} - \gamma\, \mathbf{u}\, f \quad . \tag{2}$$

(Note that viscous damping is *not* applied in the sample, but only in the reservoir.) The result is the gentlest driving possible, with relaxation waves from free surfaces eliminated entirely when periodic boundary conditions are used along with constant boundary velocities $u_p = \pm\dot\varepsilon_{xx}L_x$. Homogeneous strain-rate loading, coupled with the sharp-crack initial condition, as opposed to a

5

blunt notch, allows MD simulations to recover very reproducibly the continuum Griffith criterion for initiation of crack motion. This loading is also preferable to applying constant stress at the boundaries, since the latter invariably introduces perturbing stress waves into the sample region.

When we used these gentle initial and boundary conditions in 2D fracture simulations, we were able to drive a long crack (whose Griffith strain was well below 1%) so delicately that it eventually came to a halt and then receded in length. We were able to determine very precisely the dislocation emission criterion by MD, which compared very favorably with earlier static lattice Green's function calculations by Zhou, Carlsson, and Thomson [6]. We also saw from a series of simulations under a variety of loadings that: (1) dynamic cracks accelerate rapidly to an approximately steady velocity; (2) bond-breaking energy leads to a gradual buildup of a local phonon field near the crack tip; (3) when a critical excitation level is reached, dislocation emission is observed, almost regardless of the interatomic potential (as long as it is longer-ranged than nearest neighbors only); (4) branching is initiated by dislocation emission in almost every case - for the stiffest pair potentials, dislocations cannot even escape the crack; (5) branches zigzag, turning back toward the initial direction of propagation (along the natural cleavage direction; if the strain is contrary to the natural direction, i.e., not along a slip plane in 2D, then the crack behaves in an erratic manner with a diminished rate of dislocation emission). In Figure 1, we show a sequence of snapshots of dynamic crack propagation and branching.

We note that crack velocity is not the only diagnostic that determines when branching occurs; in fact, a crack can travel for some time at a relatively steady velocity before it slows down and then branches. The dynamic buildup of phonon amplitude is also an important part of the dynamic crack propagation process.

In order to visualize dynamic cracks, we have colored the atoms according to their velocities, forces, and potential energies. The first two methods are particularly useful for demonstrating the localized phonon field buildup in a region within a few lattice spacings of the crack tip. Plotting atoms according to their potential energies is especially useful for showing the local structures near the crack tip, with a somewhat more subtle display of the phonon field. For example, with a movie of atom potential energies, we can clearly see crack surfaces and dislocation cores, and the incipient events leading to their formation and emission. (See Figure 1.) In 3D, the effects are even more dramatic, since we can render bulk atoms (with the lowest potential energies) invisible, leaving dislocation loops and crack surfaces exposed to view.

In our recent preliminary 3D simulations, we have used our potential-energy visualization technique to see both blunting and jogging dislocations being emitted from the crack tip, as shown in Figure 2. Almost two million atoms were used in these calculations, and even so, the size was only marginally adequate. Nevertheless, the systems were over an order of magnitude larger in the thickness dimension (along the axis of the elliptical-cylindrical crack) than those of earlier

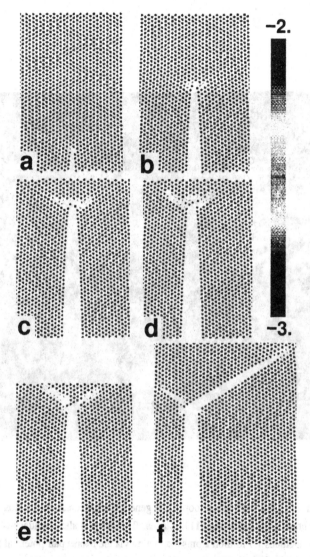

Figure 1. Dislocation emission and crack branching sequence for a 2D dynamic crack (increasing time from a-f, up to 100 vibrational periods). The full computational cell (not shown here in its entirety) includes about 400,000 atoms interacting via the Morse pair potential ($\alpha = 6$). The initial half-crack length is 240 lattice spacings. Particles are colored by a rainbow ranging from deep blue for a minimum potential energy of -3.0 (bulk) to bright red for a maximum value of -2.0 (free surface), indicated by the color bar (shown here in grayscale).

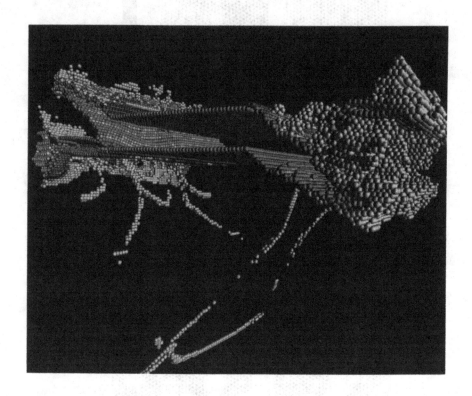

Figure 2.   Blunting half-dislocation loops (3D) generated from crack fronts along {111} planes, and crack branching, also along {111} planes. The full periodic-boundary-condition computational cell includes 1.7 million atoms interacting via the Morse pair potential ($\alpha = 7$). The initial half-crack length is 20 lattice spacings. Particles are colored by potential energy (shown here in grayscale). To visualize dislocation loops, only particles with local potential energy above a certain value are shown. The crack plane is (010) and crack fronts are along the [101] direction.

generations of computers [10]. Generally speaking, as the potential is made softer and longer-ranged, the number of jogging dislocations diminishes, while the number of blunting dislocations increases. Jogging dislocations, originating at the junction of the elliptical-cylinder crack and a free surface, are emitted along slip planes that intersect the crack tip, while blunting dislocations are emitted along slip planes that are coincident with the tip. [The stiffness of the potential is exemplified by the Morse potential form: $\varphi(r) = e^{-2\alpha(r-1)} - 2e^{-\alpha(r-1)}$, where length is measured in units of nearest-neighbor distance r_0 and energy in well-depth ε; the stiffness parameter α is then the ratio of bulk modulus to cohesive energy, and the reduced force constant is $2\alpha^2$.]

Our preliminary 3D calculations utilized the sharp-crack initial conditions and adiabatic expansion boundary conditions, with very low initial temperatures. We saw significant differences between fully periodic boundary conditions and partially free boundary conditions, both in the kinds of dislocations generated and in their shapes.

It will be important in future 3D MD simulations to explore the effects of initial temperature, crack geometry, interaction potential (particularly, embedded-atoms method - EAM - many-body potentials), and system size, as well as utilizing the "stadium" reservoir treatment we have developed in 2D. Already, 3D simulations of crack propagation have revealed a richness of features that could only have been guessed at a few years ago.

CONCLUSIONS

We have studied dynamic cracks using large-scale molecular-dynamics simulations in both 2D and 3D. By setting up initial conditions that are close to those in traditional fracture mechanics treatments, and by absorbing unwanted sound-waves generated at the crack tip, we have been able to compare MD simulations to quasi-static methods. We see differences, however, between the static view and dynamic cracks, namely a gradual buildup of phonon content that is highly correlated in time and intimately associated with the moving crack tip. In dynamic cracks, this buildup then leads to dislocation emission and branching. By coloring atoms according to their potential energy, much of the structure of crack tips and emitted dislocation loops can be seen, even in 3D. We believe that large-scale MD simulations with careful treatments of boundary and initial conditions - especially important in 3D - have the promise of opening up new vistas in fracture research.

ACKNOWLEDGMENTS

It is a pleasure to thank the following people for stimulating discussions of this fracture simulation work: Robb Thomson, Art Voter, Peter Gumbsch, Bill Hoover, Bill Moran, Sid Yip, Vassily Bulatov, Mike Marder, Jim Langer, Alan Needleman, and Rafi Blumenfeld.

REFERENCES

1. J.R. Rice and R. Thomson, Phil. Mag. **29**, 73 (1974).
2. J.R. Rice, J. Mech. Phys. Solids **40**, 239 (1992).
3. P.S. Lomdahl, P. Tamayo, N. Grønbech-Jensen, and D.M. Beazley, Proc. of Supercomputing 93 (IEEE Computer Society Press), 520 (1993).
4. D.M. Beazley and P.S. Lomdahl, Parallel Computing **20**, 173 (1994).
5. B.L. Holian and R. Ravelo, Phys. Rev. B **51**, 11275 (1995).
6. S.J. Zhou, A.E. Carlsson, and R. Thomson, Phys. Rev. Letters **72**, 852 (1994).
7. S.J. Zhou, P.S. Lomdahl, B.L. Holian, and R. Thomson, in *Fractal Analysis and Modelling of Materials: New Directions*, edited by A. Bishop and R. Blumenfeld (World Scientific, 1995, to be published).
8. S.J. Zhou and R. Thomson, J. Mater. Res. **6**, 639 (1991).
9. A. George and G. Michot, Mater. Sci. and Engng. **A164**, 118 (1993).
10. B. deCelis, A.S. Argon, and S. Yip, J. Appl. Phys. **54**, 4864 (1983).
11. S.J. Zhou, P.S. Lomdahl, R. Thomson, and B.L. Holian, "Dynamic Crack Processes via Molecular Dynamics," Phys. Rev. Letters (submitted).

DYNAMICS AND MORPHOLOGY OF CRACKS IN SILICON NITRIDE FILMS: A MOLECULAR DYNAMICS STUDY ON PARALLEL COMPUTERS

AIICHIRO NAKANO, RAJIV K. KALIA, PRIYA VASHISHTA
Concurrent Computing Laboratory for Materials Simulations
Department of Computer Science
Department of Physics and Astronomy
Louisiana State University, Baton Rouge, LA 70803
nakano@bit.csc.lsu.edu
kalia@bit.csc.lsu.edu
priyav@bit.csc.lsu.edu
http://www.cclms.lsu.edu/cclms/

ABSTRACT

Multiresolution molecular dynamics approach on parallel computers has been used to investigate fracture in ceramic materials. In microporous silica, critical behavior at fracture is analyzed in terms of pore percolation and kinetic roughening of fracture surfaces. Crack propagation in amorphous silicon nitride films is investigated, and a correlation between the speed of crack propagation and the morphology of fracture surfaces is observed. In crystalline silicon nitride films, temperature-assisted void formation in front of a crack tip slows down crack propagation.

INTRODUCTION

Porous silica has been the focus of many investigations [1]. This environmentally safe material has numerous technological applications: It is used in thermal insulation of commercial and household refrigerators; in passive solar energy collection devices; in particle detectors; and in catalysis and chemical separation. There is an exciting possibility of utilizing it as an embedding framework in optical switches made of quantum-confined microclusters. Since these applications are due to the remarkable porous structure of the system, it is important to understand the size and spatial distributions of pores and the morphology of pore interfaces.

Silicon nitride has been at the forefront of research for high-temperature, high-strength materials owing to its outstanding properties [2]. The combination of low thermal expansion and high strength makes silicon nitride one of the most thermal-shock-resistant materials currently available. The strong covalent bonding between the atoms results in a superb resistance of mechanical deformation and chemical corrosion. Additionally, silicon nitride based films have found a number of applications in microelectronics technology. For these applications, the mechanism of fracture is one of the most important issues.

The morphology of fracture surfaces has drawn a great deal of attention in recent years [3-7]. It is now well-established that a fracture surface, $z(x,y)$, is a self-affine object in that it remains invariant under the transformation, $(x, y, z) \rightarrow (ax, ay, a^\zeta z)$, where ζ is known as the roughness exponent. A decade ago, Mandelbrot and coworkers reported the first measurements of the roughness exponent for metallic surfaces [3]. Since then there has been a great deal of controversy regarding the value of ζ. Bouchaud, Lapasset, and Planés carried out measurements

11

of ζ for four aluminum alloys with different heat treatments and in each case they obtained ζ = 0.8 [4]. Måløy et al. made measurements on six different brittle materials and found ζ to be 0.87 ± 0.07 [5]. This led them to conjecture that fracture surfaces of brittle and ductile materials had a "universal" roughness exponent, independent of material characteristics and the mode of fracture. Milman et al. questioned the validity of the "universality" of ζ, especially at microscopic length scales, by pointing out that their scanning tunneling microscopy data for MgO, Si, and Cu revealed the roughness exponent to be around 0.6 [6]. Measurements on tungsten and graphite also indicated a low value of ζ ($\approx$ 0.4) [6]. However, recent molecular-dynamics (MD) simulations for such disparate systems as porous silica [7] and two-dimensional Lennard-Jonesium [8] found $\zeta \approx 0.8$.

In this paper we report the results of a MD study of: i) roughening of fracture surfaces in microporous silica; ii) crack propagation and the morphology of fracture surfaces in amorphous silicon nitride films; and iii) temperature effects on crack propagation in crystalline silicon nitride films.

MULTIRESOLUTION MOLECULAR DYNAMICS ALGORITHM

Molecular dynamics (MD) approach provides the phase-space trajectories of particles through the solution of Newton's equations. The compute-intensive part of MD simulations is the calculation of interparticle interactions. We are dealing with materials in which interatomic interactions are characterized by steric repulsion, Coulomb and charge-dipole interactions, and three-body covalent interactions. Highly efficient algorithms have been designed to compute these interactions on parallel machines [9]. The long-range Coulomb interaction is calculated with a divide-and-conquer scheme, called the fast multipole method (FMM), which reduces the computational complexity from $O(N^2)$ to $O(N)$. For short-ranged two- and three-body interactions, we have employed a multiple time-step (MTS) approach in which the force on a particle is subdivided into primary, secondary, and tertiary components (see Fig. 1). A significant reduction in computation is achieved by exploiting different time scales of these force components.

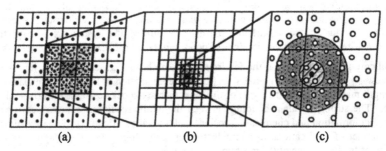

FIG. 1. Schematic representation of the multiresolution algorithm. (a) Periodically repeated images of the original MD box. Replacing each well-separated image by a small number of particles with the same leading multipole expansions reduces the computation enormously while maintaining the necessary accuracy. (b) A hierarchy of cells in the fast multipole method. (c) The near-field forces on a particle are due to primary, secondary, and tertiary neighbor atoms.

Figure 2 shows the performance of this approach on the Touchstone Delta and the IBM SP1 machine at Argonne National Laboratory. For a 4.2 million particle silica glass, the execution time for a single MD time step is only 4.84 seconds on the 512-node Delta machine. The execution time scales linearly with the size of the system and the computation dominates the communication time.

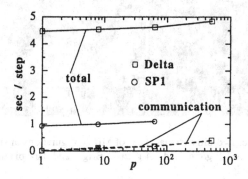

FIG. 2. Execution time per step for multiresolution MD simulations of amorphous SiO$_2$ (solid curves) with 8,232p particles where p is the number of processors. Open squares and open circles are the results on the Delta and SP1, respectively. Dashed curves denote the communication overhead.

FRACTURE OF SILICA GLASS

Structural properties and mechanical failure in porous silica have been investigated with million-particle MD simulations [7]. The results for pore size distribution, internal surface area and surface-to-volume ratio of pores, and fractal dimension are in accordance with structural measurements. We have performed MD calculations on 1.12-million particle amorphous silica systems, investigating the growth of pores with a decrease in the density of the system. As the normal-density glass is uniformly expanded, the pores begin to form when the density of the system is reduced to 1.8 g/cm^3. Further decrease in the density of the system causes an increase in the number of pores and also the pores coalesce to form larger entities (see Fig. 3). There is a dramatic increase in the size of pores when the mass density is reduced to the critical value, $\rho_c = 1.4$ g/cm^3. At the critical density, some pores percolate through the entire system causing fracture. In Fig. 4 we show one of the surfaces of the percolating pore.

The roughness of a fracture surface is calculated from the height-height correlation function, G(σ), where σ is the horizontal distance. The MD results for G(σ) are well-described by the relation, G(σ) ~ σ^ζ with $\zeta = 0.87 \pm 0.02$ for $\sigma < 10$ nm (see Fig. 5). The MD results for the roughness exponent agree with experimental measurements, thus lending further support to experimental claims that the roughness exponent of fracture surfaces is a material-independent quantity. Moreover, the MD results indicate that the universality of the roughness exponent may prevail even at length scales $\leq$ 10 nm.

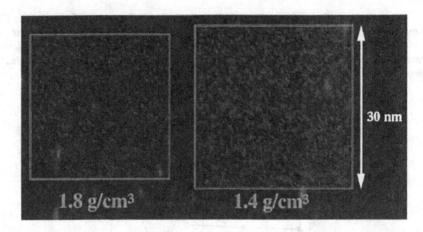

FIG. 3. Snapshots of two-dimensional slices of MD configurations of silica (1.12 million particles) at densities 1.8 g/cm³ and 1.4 g/cm³. Bright and dark pixels represent pores and silica, respectively.

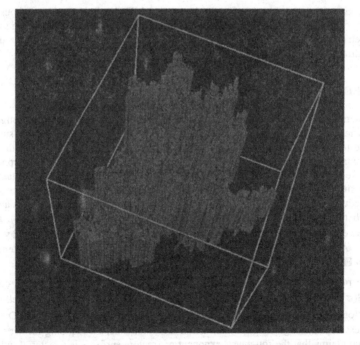

FIG. 4. Snapshot of a fracture surface resulting from a percolating pore in silica glass at a mass density of 1.4 g/cm³.

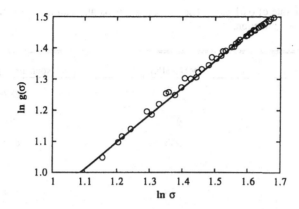

FIG. 5. Height-height correlation function (open circles) versus the in-plane distance, σ, for the fracture surface shown in Fig. 4. The solid line is the best fit, $G(\sigma) \sim \sigma^{\zeta}$ with $\zeta = 0.87 \pm 0.02$ for $\sigma < 10$ nm.

CRACK PROPAGATION IN AMORPHOUS SILICON NITRIDE FILMS

Molecular-dynamics study of fracture in amorphous Si_3N_4 films [10] involved systems with 100,352 atoms (typical dimensions of a film were 220Å × 220Å × 20Å), interacting via a combination of two- and three-body potentials. The two-body potential includes steric repulsion, the effect of charge-transfer via Coulomb interaction, and the large electronic polarizability of anions through the charge-dipole interaction [11]. Three-body interactions with bond-bending and bond-stretching terms were introduced to include covalent effects [11].

To evaluate the quality of the interaction scheme, various structural and dynamical correlations and elastic constants in the bulk Si_3N_4 were computed and the results were compared with available experimental measurements [10, 11]. In the crystalline state, the MD results for interparticle separations (Si-Si, Si-N, and N-N) and bond angle distributions (N-Si-N and Si-N-Si) are in excellent agreement with experiments. Not only that, the MD static structure factor for the glass agrees well with neutron-scattering measurements over the entire range of wave vectors. The specific heat of crystalline α-Si_3N_4, obtained from the phonon density-of-states, is also in excellent agreement with experimental measurements over a wide range of temperatures. The MD results for various elastic modulii are also in reasonable agreement with experimental values [12, 13], see Table I.

We first prepared well-thermalized bulk system by quenching the molten state and then periodic boundary conditions were removed and the systems were relaxed with MD and conjugate-gradient methods. These well-thermalized crystalline and amorphous Si_3N_4 films were subjected to uniaxial tensile loads by displacing atoms uniformly in the leftmost and rightmost layers (thickness ~ 5.5Å each) along the x direction. The strain was applied at a constant rate while maintaining the temperature at 300 K.

To investigate crack propagation, we insert a crack in an uniaxially stretched film (strain ~ 4 %) by removing particles within a region whose projection onto the xy plane is 4 Å × 50 Å, and we use a strain rate of 0.01 ps^{-1}. The crack plane is parallel to the yz plane, and the crack propagates along the y direction.

	MD (GPa)	Experiment (GPa)
β_v^{-1}	287	282[a]
β_a^{-1}	824	847[a]
β_c^{-1}	943	870[a]
E (5°)	419	456[b]
E (64.2°)	375	378[b]
E (83.1°)	386	488[b]

TABLE I. Elastic modulii of α-Si$_3$N$_4$. The MD results for the hydrostatic bulk modulus, β_v^{-1}, the inverse linear compressibilities parallel to the a and c axes, β_a^{-1} and β_c^{-1}, the elastic modulii along the directions with angles 5°, 64.2°, and 83.1° from the c axis, are compared with experiments: [a]Ref. 12; [b]Ref. 13.

Figure 6 shows snapshots of the amorphous Si$_3$N$_4$ film projected onto the xy plane. Initially the crack propagates straight along the y direction, as shown in Fig. 6 (a), for the first 4.5 ps. At 7.9 ps, we observe the formation of voids in front of the crack tip, see Fig. 6 (b). These voids grow and form a secondary crack at t = 10.8 ps, see Fig. 6 (c). Eventually the secondary crack and the initial crack coalesce and the resulting crack surface is very rough, as evident from the snapshot in Fig. 6 (d) at t = 16.4 ps.

We have calculated the height-height correlation function [10],

$$g(y) = \left\langle [h(y+y_0) - h(y_0)]^2 \right\rangle^{1/2},$$

of the crack surface using the height profile, h(y). In the equation, the bracket denotes the average over y_0. For a self-affine surface, the height-height correlation function is expected to obey the scaling relation, $g(y) \sim y^\zeta$ [3-7].

Figure 7 shows a log-log plot of the height-height correlation function for the crack surface in the amorphous film. There are two well-delineated regimes in this figure. For smaller length scales, we observe a rather smooth crack surface with a smaller roughness exponent. By linear fitting, we obtain a roughness exponent of 0.44 ± 0.02 for y < ξ = 25 Å. Beyond this crossover length, ξ, the surface is rougher with a larger exponent, ζ = 0.82 ± 0.02.

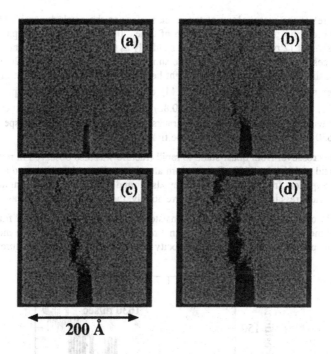

FIG. 6. Snapshots of crack propagation in an amorphous Si_3N_4 film at time (a) 4.5 ps, (b) 7.9 ps, (c) 10.8 ps, and (d) 16.4 ps.

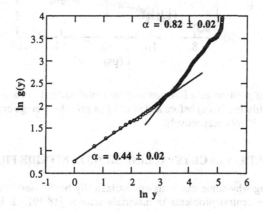

FIG. 7. Log-log plot of the height-height correlation function, g(y) (open circles). The solid curves represent the best fit, $g(y) \sim y^\zeta$, with $\zeta = 0.44 \pm 0.02$ for $y < \xi = 25$ Å, and $\zeta = 0.82 \pm 0.02$ for $y > \xi$.

To relate the morphology of the crack surface to the crack dynamics [14], we show in Fig. 8 the crack tip position as a function of time. Initially the crack propagates slowly and continuously. At 12 ps the crack tip jumps suddenly as the initial and secondary cracks coalesce. The distances between this and each of the subsequent jumps are between 20 and 40 Å, which are close to the crossover length of the height-height correlation function. We thus conclude that the smaller roughness exponent (ζ = 0.44) corresponds to slow crack propagation inside microcracks, and the larger exponent (ζ = 0.82) corresponds to the inter-microcrack propagation associated with the coalescence of microcracks. The average propagation speed in the second regime (1630 m/s) is much larger than in the first regime (640 m/s).

The crossover from quasi-static to rapid fracture was also observed in recent experiments by Bouchaud and Navéos [15]. For titanium aluminum alloys, they observed a crossover of the roughness exponent from 0.45 to 0.84. They also found that the crossover length scale decreases when the local stress intensity factor, or correlatively, the crack velocity increases.

This crossover phenomenon is consistent with a model in which a fracture surface is considered the trace of a line propagating in a random medium [15,16]. The model predicts the roughness exponent to be 0.75 for high-velocity and 0.5 for low-velocity fracture surfaces [17].

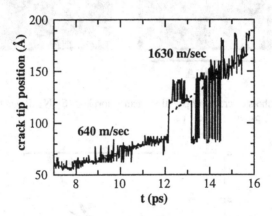

FIG. 8. Crack tip position as a function of time (solid curves) in an amorphous Si_3N_4 film. Linear fits (dashed lines) before and after 12 ps give the average crack tip velocities of 640 m/s and 1630 m/s, respectively.

CRACK PROPAGATION IN CRYSTALLINE SILICON NITRIDE FILMS

Understanding why some materials are intrinsically brittle, and others are intrinsically ductile is one of the central problems in materials science [18,19]. It is also important to understand the effect of temperature on the brittleness of materials. The brittle-to-ductile transition is exhibited by almost all materials. The change in the fracture behavior from brittle cleavage to ductile failure occurs usually in a narrow range of temperature accompanied by a dramatic increase in the fracture toughness [20].

Nucleation and growth of tensile cracks at finite temperatures have been studied in the framework of statistical mechanics [21-23]. Solid under stress is treated as a metastable state, and fracture at a failure threshold corresponds to a metastability limit, or spinodal [22]. Crack near spinodal is a fractal object describable as a percolation cluster; a Griffith-like classical crack [24] is surrounded by a fractal halo of microvoids [21]. Such a nonclassical crack extends through the growth and absorption of voids in front of the crack tip. Recently, the Griffith criterion [24] has been extended to account for the propagation of a self-affine crack [25].

Molecular-dynamics study of fracture in α-crystal Si_3N_4 films involved systems with 100,352 atoms (typical dimensions of a film were $220\text{Å} \times 220\text{Å} \times 20\text{Å}$) at temperatures 300 K and 1,500 K. The film surface is (001) and uniaxial strain is applied in the [210] direction. Cracks propagate in the [010] direction. We insert a crack in an uniaxially stretched film (strain ~ 4 %) by removing particles within a region whose projection onto the xy plane is $4 \text{ Å} \times 50 \text{ Å}$. With this strain condition, the crack starts to propagate without further stretching.

Figure 9 shows snapshots of the α-Si_3N_4 films at temperatures 300 K and 1,500 K, projected onto the (001) plane. At 300 K, the crack propagates straight in a cleavage manner as shown in Fig. 9 (a). At 1,500 K, the crack initially propagates slowly but eventually it stops, see Fig. 9 (b).

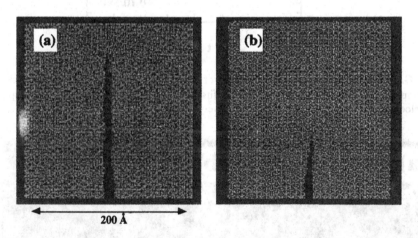

FIG. 9. Snapshots of crack propagation in α-Si_3N_4 film at temperatures (a) 300 K and (b) 1,500 K.

Figure 10 shows the crack tip position as a function of time at the two different temperatures. At 300 K, the crack accelerates continuously to 3,110 m/sec. The Rayleigh wave velocity [24] of α-Si_3N_4 is estimated to be 6,400 m/sec. At 1,500 K, the crack propagates at 940 m/sec before 6 ps, but then the propagation stops.

This difference in crack propagation is a direct consequence of the different atomic structures in front of the crack tip. Figure 11 shows atomic structures at crack tips at the two different temperatures. At 300 K, the crack tip is atomically sharp as shown in Fig. 11 (a). This structure supports the fast, cleavage-like propagation. At 1,500 K, we observe a sequence of microvoids in front of the crack tip. The periodicity of these microvoids is that of the hexagonal symmetry of the (001) surface. At high temperatures, crack propagates through formation and absorption of microvoids. The resulting propagation is slow, and sometimes it cannot be sustained.

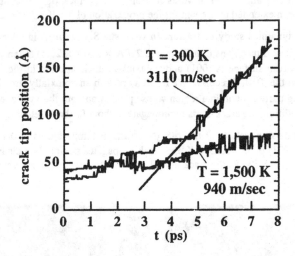

FIG. 10. Crack tip position as a function of time (solid curves) in α-Si$_3$N$_4$ film at temperatures 300 K and 1,500 K. Linear fits (solid lines) give the average crack tip velocities of 3,110 m/s and 940 m/s, respectively.

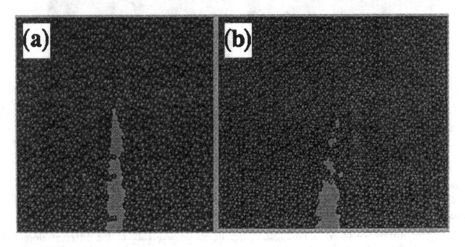

FIG. 11. Atomic structures at crack tips in α-Si$_3$N$_4$ films at temperatures (a) 300 K and (b) 1,500 K.

CONCLUSIONS

We have simulated fracture in ceramic materials using multiresolution molecular dynamics approach on parallel computers. In microporous silica, critical behavior at fracture is analyzed in terms of pore percolation and kinetic roughening of fracture surfaces. In amorphous Si_3N_4 films, correlation between the speed of crack propagation and the morphology of fracture surfaces has been observed. In crystalline Si_3N_4 films, temperature-assisted void formation in front of a crack tip has been shown to slow down crack propagation.

ACKNOWLEDGMENTS

This work was supported by the U.S. Department of Energy, Grant No. DE-FG05-92ER45477, National Science Foundation, Grant No. DMR-9412965, Air Force Office of Scientific Research, Grant No. F 49620-94-1-0444, and USC-LSU Multidisciplinary University Research Initiative, Grant No. F 49620-95-1-0452. A part of these simulations were performed on the 128-node IBM SP computer at Argonne National Laboratory. The computations were also performed on parallel machines in the Concurrent Computing Laboratory for Materials Simulations (CCLMS) at Louisiana State University. The facilities in the CCLMS were acquired with the Equipment Enhancement Grants awarded by the Louisiana Board of Regents through Louisiana Education Quality Support Fund (LEQSF).

REFERENCES

1.   J. Fricke, J. Non-Cryst. Solids **121**, 188 (1990).

2.   J. Mukerji, in *Chemistry of Advanced Materials*, ed. C. N. R. Rao, Blackwell (Oxford, 1993).

3.   B. B. Mandelbrot, D. E. Passoja, and A. J. Paullay, Nature **308**, 721 (1984).

4.   E. Bouchaud, G. Lapasset, and J. Planès, Europhys. Lett. **13**, 73 (1990).

5.   K. J. Måløy, A. Hansen, E. L. Hinrichsen, and S. Roux, Phys. Rev. Lett. **68**, 213 (1992); ibid. **71**, 205 (1993).

6.   V. Y. Milman, R. Blumenfeld, N. A. Stelmashenko, and R. C. Ball, Phys. Rev. Lett. **71**, 204 (1993).

7.   A. Nakano, R. K. Kalia, and P. Vashishta, Phys. Rev. Lett. **73**, 2336 (1994).

8.   F. F. Abraham, D. Brodbeck, R. A. Rafey, and W. E. Rudge, Phys. Rev. Lett. **73**, 272 (1994).

9.   R. K. Kalia, S. W. de Leeuw, A. Nakano, D. L. Greenwell, and P. Vashishta, Comput. Phys. Commun. **74**, 316 (1993); A. Nakano, P. Vashishta, and R. K. Kalia, *ibid.* **77**, 302 (1993); A. Nakano, R. K. Kalia, and P. Vashishta, *ibid.* **83**, 197 (1994).

10.   A. Nakano, R. K. Kalia, and P. Vashishta, Phys. Rev. Lett. **75**, 3138 (1995).

11.   P. Vashishta, R. K. Kalia, and I. Ebbsjö, Phys. Rev. Lett. **75**, 858 (1995); C.-K. Loong, P. Vashishta, R. K. Kalia, and I. Ebbsjö, Europhys. Lett. **31**, 201 (1995).

12.   L. Cartz and J. D. Jorgensen, J. Appl. Phys. **52**, 236 (1981).

13.   A. A. Mukaseev, V. N. Gribkov, B. V. Shchetanov, A. S. Isaikin, and V. A. Silaev, Poroshk. Metall. **12**, 97 (1972).

14. J. Fineberg, S. P. Gross, M. Marder, and H. L. Swinney, Phys. Rev. Lett. **67**, 457 (1991).

15. E. Bouchaud and S. Navéos, J. Phys. I (France) **5**, 547 (1995).

16. J.-P. Bouchard, E. Bouchard, G. Lapasset, and J. Planès, Phys. Rev. Lett. **71**, 2240 (1993)

17. D. Etras and M. Karder, Phys. Rev. Lett. **69**, 889 (1992).

18. J. R. Rice and R. Thomson, Philos. Mag. **29**, 73 (1974).

19. S. J. Zhou, A. E. Carlsson, and R. Thomson, Phys. Rev. Lett. **72**, 852 (1994).

20. M. Khantha, D. P. Pope, and V. Vitek, Phys. Rev. Lett. **73**, 684 (1994).

21. J. B. Rundle and W. Klein, Phys. Rev. Lett. **63**, 171 (1989).

22. R. L. B. Selinger, Z.-G. Wang, W. M. Gelbart, and A. Ben-Shaul, Phys. Rev. A **43**, 4396 (1991).

23. L. Golubovic and S. Feng, Phys. Rev. A **43**, 5223 (1991).

24. B. Lawn, *Fracture of Brittle Solids* (Cambridge University Press, Cambridge, 1993).

25. E. Bouchaud and J.-P. Bouchaud, Phys. Rev. B **50**, 17752 (1994).

CLEAVAGE FRACTURE AND THE BRITTLE-TO-DUCTILE TRANSITION OF TUNGSTEN SINGLE CRYSTALS

J. Riedle*, P. Gumbsch*, H.F. Fischmeister*,V.G. Glebovsky†, and V.N. Semenov†
* Max-Planck-Institut für Metallforschung, Seestr. 92, 70174 Stuttgart, Germany
† Institute of Solid State Physics, Chernogolovka, Russia.

ABSTRACT

The fracture toughness of high purity tungsten single crystals is measured between 77 K and 600 K in 3-point bending. Precracked specimens of 4 different crack systems are used: cracks are injected on both the primary {100} and the secondary {110} cleavage plane and forced to propagate either along the <100> or along the <110> direction.

In the low temperature regime the results allow to discriminate between the intrinsic brittle behavior and the contribution from plasticity. We find that brittle crack propagation is highly anisotropic with respect to the propagation direction. This can be interpreted as a direct consequence of the bond breaking process and is in contrast to the thermodynamic Griffith criterion, which does not discriminate between different crack propagation directions in the same cleavage plane.

At higher temperatures the measured fracture toughness of the (110) crack systems is markedly higher than that of (100) cracks. However, the brittle-to-ductile transition (BDT) temperatures of all 4 crack systems do not vary significantly and are all at about $T_{BDT} = 400 \pm 50$ K. The dependence of the BDT temperature on loading rate gives an apparent activation energy of $Q_{BDT} = 0.2\,eV$.

INTRODUCTION

One of the most interesting and also most challenging problems in fundamental fracture research is to improve our understanding of the so called semi-brittle fracture processes. Semi-brittle fracture is usually found in crystalline materials, which exhibit a brittle to ductile transition (BDT), below the transition temperature. Prominent examples of such materials are the bcc transition metals amongst which single crystalline tungsten has been studied most extensively [1, 2, 3].

Semi-brittle fracture is controlled by a competition of dislocation activity with the brittle cleavage process. Both processes occur simultaneously and a better understanding of these processes will therefore require both, improved knowledge about the perfectly brittle cleavage process and about dislocation activity in the crack tip region. Experimentally, the different contributions to the fracture toughness can best be discriminated by toughness tests on single crystalline material. As an initial effort, we have therefore conducted fracture toughness tests on tungsten single crystals at various temperatures and for two different macroscopic crack propagation directions on each of the two cleavage planes, {100} and {110}. We have tested several crystallographically distinct crack systems, in the following denoted by the specification of the crack plane and the crack front direction, in order to provide experimental data, suitable for a critical assessment of models for the semi-brittle fracture process and the BDT.

EXPERIMENTAL PROCEDURE

Three point bend fracture toughness tests are performed on high purity W single crystal specimens (3 x 6 x 30 mm³). The material [4] and details of the experimental techniques are described elsewhere [5, 6]. Tests were performed between liquid nitrogen tempera-

23

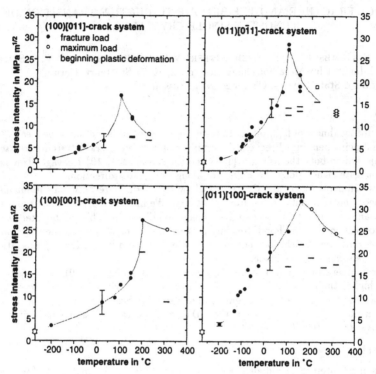

Figure 1: Fracture toughness of tungsten single crystals for the 4 low index crack systems.

ture (77 K) and 650 K. All fracture specimens were precracked at 77 K using the so called bridge-technique [7], where the notched sample is loaded with a supporting steel bar until a (pre-)crack pops in. Precrack length was first measured on the polished side surfaces and specimens are accepted for final testing if the precracks had stopped between 0.45 and 0.7 of the total width of the specimen. For the calculation of the fracture toughness, the length of the precrack was evaluated from the compliance of the specimen and, where possible. also determined *post mortem* on the fracture surfaces. Both values usually agreed very well. Multiple tests (more than 5) have been conducted for all 4 crack systems at room temperature and at 77 K and several individual tests at the other temperatures. Toughness tests were usually performed with a constant loading rate of 0.04 MPa$\sqrt{m}$/s, except for the {011}<0$\bar{1}$1> crack system which was also tested at higher loading rates.

RESULTS

The dependence of the measured fracture toughness on temperature is displayed in figure 1 for all 4 crack systems. The room temperature fracture toughness as well as the 77 K results are also compiled together with the results from atomistic calculations and the measured BDT temperatures in table 1. (Slight differences between the atomistic data given here and those of [8] result from a modified loading procedure as described in [9].) It is evident that both {110} crack systems have a significantly higher room temperature

Table 1: Brittle to ductile transition temperature (BDTT) and fracture toughness of tungsten single crystals for the {100} and {110} cleavage planes with different crack front directions and with results from atomistic modelling. Data at room temperature (RT) and at liquid nitrogen temperature (77K) are mean values and standard deviations from at least 5 individual measurements. Fracture toughness is given in MPam$^{1/2}$.

crack system	BDTT	RT experiment	77K experiment	atomistic modelling
(100)[001]	370 K	8.7±2.5	3.4±0.6	2.05
(100)[011]	470 K	6.2±1.7	2.4±0.4	1.63
(011)[100]	430 K	20.2±5.5	3.8±0.4	2.17
(011)[0$\bar{1}$1]	370 K	12.9±2.1	2.8±0.2	1.56

fracture toughness than the {100} crack systems. With decreasing temperature, however, the crack systems with <$\bar{1}$10> crack fronts, namely the {100}<011> and {110}<$\bar{1}$10> crack systems, are significantly easier to cleave than the crack systems with <100> crack fronts. Both these crack systems are also characterized by a somewhat lower BDT temperature.

Testing at 77 K always resulted in brittle cleavage fracture on the original cleavage plane for all 4 crack systems. The load displacement curves showed perfectly linear elastic behaviour up to the load at which fracture occurred. The fracture surfaces are clean except for a few river lines. On the {100} crack surfaces river lines only occur at or near the location of the stopped precrack, whereas river lines are visible on the whole fracture surface of the {110} specimens. The river lines follow the macroscopic crack propagation direction for the two crack systems with <011> crack fronts. As an example, the river line pattern of a {110}<$\bar{1}$10> specimen is shown in figure 2(a). The river lines deviate significantly from the macroscopic crack propagation direction in the {100}<010> crack system (figure 2(b)) and even tend to align themselves perpendicular to the macroscopic crack propagation direction in the {110}<001> crack system. For the {100} cracks, similar observations have previously been reported from fracture experiments with small, spark induced precracks [1, 3].

At room temperature and above, the load displacement curves revealed increasing deviations from linear elastic behaviour before final fracture. Above the BDT temperature, specimens usually did not fracture but deformed plastically. In all cases the "fracture toughness" K is calculated from the maximum attained load. The fracture surfaces of the specimens tested at room temperature and above are significantly rougher than those of the precracks and of the specimens tested at 77 K. In this temperature regime, both types of {110} cracks display {100} facets in the fracture surface.

The dependence of the fracture toughness on loading rate for the {110}<$\bar{1}$10> crack system is displayed in figure 3. The measured fracture toughness at 77 K does not depend on loading rate. At higher temperatures up to the BDT temperature higher loading rate always results in a lower fracture toughness. The BDT temperature increases significantly with loading rate.

DISCUSSION

Both the qualitative observations and the measured fracture toughness (figure 1) suggest that fracture at 77 K occurs in an almost perfectly brittle way for all 4 crack systems. The measured fracture toughness at 77 K is only about 50% higher than the calculated one (c.f.

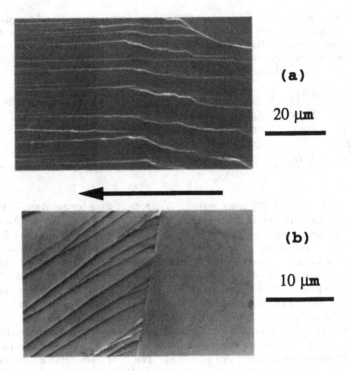

Figure 2: River line patterns on fracture surfaces of the (a) {110}<$\bar{1}$10> crack system and (b) the {100}<010> crack system. Crack propagation direction is from right to the left. The vertical arrest line shows the extent of the initial precrack.

table 1), which may result from (1) a somewhat too low surface energy of the interatomic potential, (2) the still slightly imperfect cleavage surfaces and (3) a possible shielding of the crack by preexisting dislocations. Unfortunately, neither of these influences can be shown to be irrelevant but etching of the fracture surfaces [5] showed a few etch pits near the location of the stopped precrack. This suggests that the restarting of the crack in the final fracture experiment may still have been accompanied by some dislocation activity. Nevertheless, the differences in the measured 77 K fracture toughness between the differently oriented cracks on both cleavage planes appear to be too large to be caused by these effects and one has to conclude that the perfectly brittle fracture process itself must be anisotropic with respect to crack propagation direction.

One arrives at the same conclusion if one compares the temperature dependence of the fracture toughness for the different cleavage systems (figure 1). While both {110} crack systems show significant contributions from dislocation activity at room temperature, the two {100} crack systems show much less of this. By cooling down to 77 K and thereby making dislocation mobility more difficult, remaining dislocation activity should contribute less and less to the measured fracture toughness. However, since the fracture toughness of the {110}<1$\bar{1}$0> crack system drops below the other {110} crack system <u>and</u> the {100}<010>

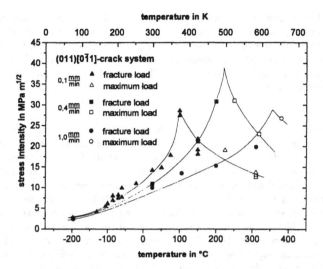

Figure 3: Fracture toughness for the {011}<0$\bar{1}$1> crack system at different loading rates.

crack system as temperature is decreased, one has to conclude that some other effects than dislocation activity must be important for the observed differences.

Further evidence of preferred crack propagation directions can be found, if the river lines are taken as an indication of the local crack propagation direction. The river line patters suggest that crack propagation microscopically follows the <011> directions on the {100} cleavage plane and the <001> direction on the {110} cleavage plane (c.f. figure 2). In both cases, these directions correspond to the crack systems which also give the lowest fracture toughness and the results therefore are internally consistent.

All the qualitative observations and the measurements therefore suggest that cleavage in tungsten is anisotropic with respect to the propagation direction. Since our results seem to rule out dislocation activity as the possible origin for this anisotropy and since tungsten even at 77 K is elastically almost isotropic, we conclude that the brittle fracture process itself must be anisotropic. Such an anisotropy to our knowledge can only be interpreted as an effect of the atomistic nature of the crystal. On the atomic scale the breaking of the bonds will of course depend on the orientation of the bonds with respect to the main loading axis and the lattice trapping effect has indeed been shown to be anisotropic with respect to the orientation of the crack front on a given cleavage plane [8]. Although the lattice trapping effect is not sufficiently well understood to predict its magnitude and the dependence on loading conditions [9] *a priori*, its dependence on crack propagation direction is rather well established [8, 9].

At intermediate temperatures the {100} crack systems generally give somewhat lower fracture toughness than the {110} crack systems. This can be rationalized in terms of the effective loading on the most highly stressed slip systems [10]. However, such static considerations can neither explain the temperature dependence of the fracture toughness nor can they explain the dependence on loading rate. Both these observations, however, are in qualitative agreement with recent modelling of the BDT of Silicon [11, 12, 13], suggesting that a detailed understanding will require the modelling of these fracture processes with a realistic description of the dependence of dislocation velocity on temperature and load. Such theoretical investigations are currently in progress. Without detailed modelling, one may of course directly interpret the BDT as a thermally activated process and interpret the strain rate dependence of the BDT temperature with an Arrhenius law. If the logarithm of the strain rate is plotted against the inverse BDT temperature, an apparent activation enthalpy $Q_{BDT} \approx 0.2eV$ is obtained.

CONCLUSIONS

We have performed fracture toughness tests of tungsten single crystals between 77 K and 650 K for the 4 low index crack systems. The quantitative results as well as the qualitative features on the fracture surfaces indicate that the low temperature (brittle) cleavage process is anisotropic with respect to crack propagation direction on a specific cleavage plane. From the measured temperature dependence of the fracture toughness we can rule out plasticity as the origin of this anisotropy and therefore have to attribute it to the perfectly brittle cleavage process. In agreement with this point of view the results compare extremely well with previous atomistic simulations, which had shown such a cleavage anisotropy and pointed out that the preference of $\{100\}$ cleavage over $\{110\}$ cleavage can be explained as a consequence of this anisotropy. The result of this study therefore indicate that the brittle fracture process should be analyzed from an atomistic rather than a thermodynamic point of view. The dependence of the BDT temperature on loading rate suggests an apparent activation enthalpy of $Q_{BDT} \approx 0.2eV$.

REFERENCES

[1] D. Hull, P. Beardmore, and A. P. Valentine, Phil. Mag. **12**, 1021 (1965).

[2] J. E. Cordwell and D. Hull, Phil. Mag. **26**, 215 (1972).

[3] J. M. Liu and J. C. Bilello, Phil. Mag. **25**, 1453 (1977).

[4] J. Riedle et al., Mat. Letters **20**, 311 (1994).

[5] J. Riedle, Ph.D. thesis, Universität Stuttgart, 1995.

[6] J. Riedle, P. Gumbsch, and H. F. Fischmeister, manuscripts in preparation (unpublished).

[7] G. Bergmann and H. Vehoff, Scripta Metall. Mater. **30**, 969 (1994).

[8] S. Kohlhoff, P. Gumbsch, and H. F. Fischmeister, Phil. Mag. A **64**, 851 (1991).

[9] P. Gumbsch, J. Mat. Res. to be published (1995).

[10] R. A. Ayres and D. F. Stein, Acta metall. **19**, 789 (1971).

[11] M. Brede, Acta metall. mater. **41**, 211 (1993).

[12] P. B. Hirsch and S. G. Roberts, Phil. Mag. **64**, 55 (1991).

[13] S. G. Roberts, A. S. Booth, and P. B. Hirsch, Mater. Sci. Eng. **A176**, 91 (1994).

KINETICS OF DISLOCATION EMISSION FROM CRACK TIPS AND THE BRITTLE TO DUCTILE TRANSITION OF CLEAVAGE FRACTURE.

A.S. Argon *, G. Xu *†, M. Ortiz ‡
* Massachusetts Institute of Technology, Cambridge, MA 02139
† present address: Terra Tek, Inc, 420 Wakara Way, Salt Lake City, Utah 84108
‡ California Institute of Technology, Pasadena, CA 91125

ABSTRACT

Several activation configurations of dislocation embryos emanating from cleavage crack tips at the verge of propagating have been analyzed in detail by the variational boundary integral method, as central elements of the rate controlling process of nucleation governed fracture transitions from brittle cleavage to tough forms, as in the case for BCC transition metals. The configurations include those on inclined planes, oblique planes and crack tip cleavage ledges. Surface ledge production resistance is found to have a very strong embrittling effect. Only nucleation on oblique planes near a free surface and at crack tip cleavage ledges are found to be energetically feasible to explain brittle-to-ductile transition temperatures in the experimentally observed ranges.

INTRODUCTION

Abrupt transitions in fracture between energy absorbing ductile forms and brittle cleavage, with decreasing temperature and increasing strain rate have been, and still continue to be of concern in many structural materials. While the phenomenon has been known in engineering practice since at least the celebrated molasses tank fracture in Boston in 1919 (for a discussion of an historical perspective of non-ship fractures see Shank, [1]), it became of crisis proportions only during World War II through the rash of major fractures of Liberty ships. In the early post-war years the problem received attention through semi-quantitative studies of notch effects and strain rate [2], and through some studies of effects of microstructure and of alloying to suppress the ductile to brittle transition temperature (for a summary see Parker, [3]). In the 50s and early 60s a fundamental recognition was reached that brittle behavior is usually triggered by deformation induced cleavage microcracks introduced into the system by dislocation pile-ups [5,7] by intersection of deformation twins [8] or by cracking of elongated grain boundary carbides [9]. In the range of the transition temperature a substantial concentration of such microcracks are "injected" into the system. Eventual brittle behavior results when one such microcrack succeeds, without being rearrested, to break through grain boundary barriers and propagates long distances [9-11]. The triggering conditions for this final transition in polycrystalline metals were also studied theoretically [12-15].

A fundamental perspective on the class of materials which are capable of fracture transitions starts with Kelly et al [16], and more specifically with Rice and Thomson [17] who have conceived a fundamental behavior pattern for a theoretical criterion establishing intrinsic brittleness vs intrinsic ductile behavior in materials. According to this criterion an atomically sharp crack has means of governing the behavior of a material in the absence of any other form of plastic response of the background, by either nucleating dislocations from its tip or by propagating in a cleavage mode by virtue of the presence of an energy barrier to the emission of such dislocations. In the first instance the material is designated as intrinsically ductile and incapable of exhibiting a fracture transition, while in the second instance it is designated as intrinsically brittle and capable of undergoing a transition from brittle cleavage to ductile forms at a characteristic transition temperature T_{BD}, affected by the rate of loading. Although many experimental studies have demonstrated that mere nucleation of some dislocations from a crack tip does not assure ductile behavior [18,19], the Rice Thomson mechanism comes close to a threshold process that triggers ductile behavior in a class of intrinsically brittle solids with relatively high dislocation mobility such as the BCC transition metals and most alkali halides. However, fully satisfactory experimental confirmation of this limiting response is rare, outside of an elegant experiment by Gilman et al [20] of crack arrest in LiF.

In distinction to the nucleation controlled response is the well established response of Si, and presumably, many other covalent materials and compounds with sluggish dislocation mobility, where the transition between brittle to tough behavior is governed by the mobility of groups of dislocations away from the crack tip [21-25]. In either the nucleation or the mobility controlled dislocation emission scenarios it is now well recognized that, emission of such dislocations from a crack tip occurs preferentially from specific crack tip sites and that assurance for full ductile behavior requires that all parts of the crack front be shielded to prevent continued brittle behavior by local break-out of the cleavage crack from unshielded portions of the crack front [24]. Since even the ductile to brittle transition is triggered by the continued propagation of an "injected" cleavage microcrack, it is now generally accepted that the fundamental fracture transition is governed by the behavior of a cleavage crack. Most modeling studies have been developed around this concept.

The fundamental supposition of the brittle-to-ductile transition models based on the Rice and Thomson scenario [17] is that while background plastic relaxations can serve to suppress the transition temperature, the ultimate arbiter of the transition is the ability, of the crack tip to emit dislocations that can shield the entire crack front and trigger widespread plastic deformation before the crack can propagate by cleavage. However, the approach of the Rice and Thomson model in which the activation configuration consisted of a fully developed dislocation line have proved to seriously over estimate the energy barriers to nucleation of dislocations [26], even after incorporating such refinements as crack tip non-linearity and tension softening [27] across the slip plane. This indicated that the activation configurations must involve imperfect dislocations which must incorporate at least a minimum of atomistic information [26].

The accumulating experimental evidence on Si [21-25] and the insight provided by the most recent modeling studies of Schöck and Püschl, [28], Rice and Beltz [29], Xu et al., [30] suggest that the activation configuration of a dislocation embryo is in the form of a double kink of dislocation core matter. The observation permits the identification of two distinct types of B-D transitions. In the BCC transition metals where barriers to kink mobility along the dislocation are low, the B-D transition is likely to be governed directly by the formation of dislocation embryos at the crack tip, resulting in a nucleation-controlled transition. By contrast, in semi-conductors and compounds the evidence suggests [31-35], and modeling verifies [36], that kink mobility is hindered by substantial energy barriers, rendering the B-D transition controlled by dislocation mobility away from the crack tip.

Ultimately, a full understanding of the B-D transitions must come from atomistic models of the formation and outward propagation of the dislocation embryo at the crack tip. Before such modeling can be attempted, however, much progress can be made by recourse to hybrid continuum-atomistic approaches [28,29,30,37] based on the use of a Peierls interplanar potential [38-40]. In a recent development of this technique by Xu et al [30], an additional surface production resistance was introduced into the interplanar potential, and the appropriate saddle point configurations of the dislocation embryos were determined by recourse to a variational boundary integral method advanced by Xu and Ortiz earlier [41]. Xu et al [30] concluded that the energetics of dislocation embryo formation on inclined slip planes containing the crack tip, against an additional surface production resistance, is quite unfavorable and does not explain the known B-D transition temperatures. It was then conjectured that nucleation may be more favorable on oblique slip planes or, may occur heterogeneously at crack front ledges.

In the present communication we briefly take into account the earlier findings of Xu et al [30] and present new findings on saddle-point configurations and their energies on oblique planes in the interior and near free surfaces, and finally, on cleavage surface ledges at the crack front. In what follows we begin by describing the nucleation controlled alternative mechanisms to be appraised. Following a brief description of the fundamental methodology employed in calculations, results of three nucleation mechanisms are presented. Finally in a discussion, the B-D transition temperatures are estimated from the calculated activation energies, and findings are put in general perspective.

MODES OF DISLOCATION NUCLEATION

Several alternative modes of dislocation nucleation from crack tips have been contemplated in the past. The modes differ mainly in the relative geometry of the slip plane, the crack surface and the crack front. The configurations considered include: nucleation of dislocations on the extension of the crack surface, Fig. 1a; nucleation on an inclined plane containing the crack front, Fig. 1b; nucleation on an oblique plane, Fig. 1c; and nucleation on a cleavage ledge, Fig. 1d.

Of these modes that shown in Fig. 1a is the simplest since it involves no tension across the potential slip plane and no production of a free surface. It has been investigated by Rice, [45] and Rice et al [37] in a two dimensional setting and by Schöck and Püschl [28] in a three dimensional setting of a simple double kink shaped activation configuration. This was also done by a more elegant perturbation analysis by Rice and Beltz [29]. These analyses that have used a conservative tension-shear potential based on a Peierls-Nabarro model to represent the periodic interplanar shear resistance of a slip plane and a tension softening effect have demonstrated that the activation configuration did indeed consist only of dislocation core matter. In 2-D this produces an inelastic crack tip displacement equal to one half of a Burgers displacement and results in a critical energy release rate equal to the unstable stacking energy associated with a half step shear across the slip plane. Rice et al [37] have extended their approach to also deal with the mode of Fig. 1b involving dislocation

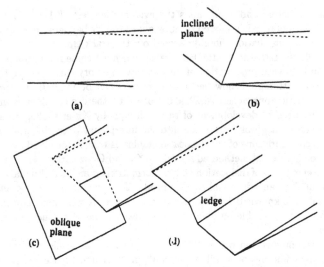

Fig. 1 Alternative modes of dislocation nucleation from a crack tip.

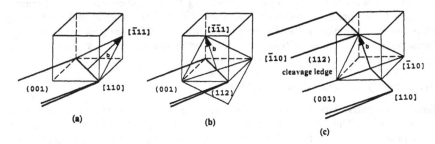

Fig. 2. Modes of dislocation nucleation from crack tips in α-Fe: a) inclined plane; b) oblique plane; c) cleavage ledge.

nucleation on an inclined plane-albeit without the affect of surface ledge production. To deal with such problems more definitively, a surface production resistance was incorporated into their model by Xu et al [30] who have concluded that this effect raises the energy barrier to such a degree that it all but rules out dislocation nucleation on inclined planes.

Experimental observations of Burns and Webb, [18] and considerations of peak stress levels [26] point to the possibility of nucleating dislocations on oblique planes as shown in Fig.1c. However, approximate analysis of this nucleation mode by the Rice and Thomson method have led to very large energy barriers. This conclusion is reinforced by our more accurate analysis presented below. Finally, experimental observations of George and Michot, [25] have indicated that actual dislocation nucleation occurs very frequently on cleavage ledges on the crack front and near free surfaces. We analyze this possibility in Section 4.4 and show that conditions are indeed quite favorable for this mode. Since nucleation controlled fracture transitions are likely to be limited only to BCC transition metals and to some alkali halides, we have performed our analyses for α-Fe. Figures 2a-2c show the specific

32

geometrical settings for Fe that have been considered where the cleavage plane in (001) and the crack front is parallel to the [110] directions which is the prevalent direction of the crack front in Fe giving a minimum of energy release rate [47].

METHOD OF ANALYSIS

The analysis employs a variant of the variational boundary integral method of Xu and Ortiz [41] to encompass problems of dislocation nucleation from atomically sharp cracks. In this model, a slip plane connected to the crack front is viewed as an extension of the crack surfaces with a nonlinear interlayer potential acting across it. Thus, the crack and the slip plane on which the dislocation nucleates are jointly regarded as a three-dimensional crack system embedded in the linear elastic solid. The interlayer potential acting across the slip plane is modeled by combining the universal binding energy relation of Rose et al [48] with a skewed shear resistance profile [40]. The interplanar displacements and the crack opening displacements are represented by a continuous distribution of curved dislocations. This approach introduces no artificial discontinuity between the elastic crack opening and inelastic interplanar slip and separation. The technique has been described in complete detail by Xu and Ortiz [41] and its application to the present study by Xu et al, [30,47]. In this section, we briefly outline those aspects of the method which are pertinent to the treatment of the special activation configurations described in the previous section.

We consider a semi-infinite cleavage crack and a slip plane intersecting the crack front as shown in Fig. 1b. The crack/slip plane system is loaded remotely by a K-field. The crystallographic slip plane is chosen to be the most advantageous for slip. As the driving force increases, an embryonic dislocation forms progressively until it reaches an unstable equilibrium configuration. The load corresponding to this unstable configuration is defined as the critical driving force for nucleation. The embryonic dislocation profile is characterized as a distribution of interplanar inelastic displacements, defined by Rice [45] as:

$$\delta = \Delta - \Delta^e \tag{1}$$

where Δ and Δ^e denote the total and the elastic interplanar displacements respectively. The opening displacements u of the crack surface, including the inelastic displacements along the slip plane, can be written as:

$$u = \bar{u} + \delta \tag{2}$$

where $\bar{u}$ is the displacement of a standard K-field for a reference semi-infinite crack. The term $\bar{u}$ matches the behavior of the opening displacements of the crack far away from the crack tip and serves as a boundary condition of the system. Consequently, the additional term δ, modifying the former, and which is the primary unknown in the analysis, is expected to decay rapidly to zero with distance away from the crack tip in either direction. In this manner, δ can be considered in a finite domain around the crack front on the crack surface and on the slip plane connected to the crack.

Following the procedure introduced by Xu et al, [30,47] the potential energy of the whole system can be written in the form.

$$\Pi[\overline{u} + \delta] = W[\overline{u} + \delta] + V[\delta] = W_1[\overline{u} + W_1[\delta] + W_2[\overline{u}, \delta] + V[\delta], \tag{3}$$

where we identify $W_1[\overline{u}]$ as the elastic strain energy of the system, free of inelastic modifications, $W_1[\delta] + V[\delta]$ as the self energy of the system of inelastic modifications consisting of the distributed dislocations and the interplanar interaction energy on the slip plane, and $W_2[\overline{u}, \delta]$ is the interaction energy of the initial unmodified system with the second system of modifications. Of these energies the ones of relevance in the variational approach are those that depend on the unknown inelastic modification δ. Their specific forms have been given elsewhere [30, 47].

The potential energy $V[\delta]$ of the interplanar inelastic deformation on the slip plane is a key ingredient of the method and is given as:

$$V[\delta] = \int_{\hat{S}_s} \Phi[\delta] dS, \tag{4}$$

where $\Phi[\delta]$ is the interplanar tension/shear potential, defined per unit area of the slip plane $\hat{S}_s$. It adopts the constrained displacement hypothesis of Rice [45] and Sun et al [49], in which the interplanar shear displacement Δ_r is constrained along the Burgers vector direction. The shear and tension separation resistances, τ and σ respectively, follow as functions of the inelastic shear displacements δ_r and tensile separation displacement δ_θ on the slip plane. The associated traction-displacement relation has been modeled by Rice et al [37] and was modified somewhat by Xu et al [30] to incorporate an element of skewness and a surface production resistance. The resulting forms have been presented in detail elsewhere [30].

The unknown displacements follow by rendering the potential energy $\Pi[\overline{u} + \delta]$ stationary. This is achieved by discretizing the integral equation with six noded elements distributed on the crack surface. The non-linear equations are solved by a Newton-Raphson iteration. The saddle point configurations are activated by introducing a small perturbation into the system at the bifurcation point, based on the solution of a first-order eigen-value problem if necessary. Solutions are obtained by recourse to interplanar displacement control achieved through the introduction of Lagrange multipliers [30,47].

The material constants for α-Fe used in the present calculations are given in Table I.

Table I. Material Properties for α-Fe [47]

slip system	T (°K)	μ (10^5MPa)	c	$\gamma_{us}^{(u)}$ (Jm^{-2})	$2\gamma_s$
$(1/2)[111](1\overline{1}0)$	4.2	0.756	3.125	0.517	3.33
$(1/2)[\overline{1}\overline{1}1](112)$	4.2	0.756	3.125	0.581	3.80

where

μ	shear modulus
c	uniaxial strain elastic modulus
$\gamma_{us}^{(u)}$	unrelaxed unstable stacking energy
γ_s	surface energy

RESULTS

Nucleation of Dislocations from a Shear Crack in Mode II

Because of its simplicity and the ease of systematically incorporating important effects, we consider first the nucleation analysis in Mode II as indicated in Fig. 1a. The crack surface coincides with the (001) plane, the crack front with the [110] direction and the slip plane with the (1$\bar{1}$2) plane, Fig.2a. The crack/slip plane system is subjected to pure mode II loading. Two representative saddle point configurations of the dislocation embryo corresponding to normalized load levels G_{II}/G_{IIcd} =0.75 and 0.50 are shown in Figs 3 and 4, respectively. In the figures $x_1 = 0$ indicates the initial crack front and the level contours represent the shear displacements between the crack faces (for $x_1 < 0$), and the inelastic displacements across the slip plane (for $x_1 > 0$). The double-kink nature of the saddle point configuration is clearly visible. The dependence of the activation energy on the crack driving force is plotted in Fig. 5. The crack driving force G_{II} has been scaled with the athermal driving force G_{IIcd} for the 2-D release of a dislocation line from the entire crack front, where G_{IIcd} is the unrelaxed unstable stacking energy $\gamma_{us}^{(u)}$.

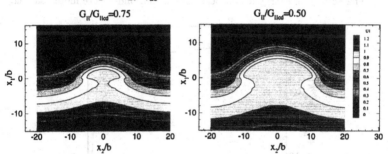

Figs. 3 & 4. Saddle-point configurations at a crack tip under Mode II loading for $G_{II}/G_{IIcd} = 0.75$, and 0.5.

Nucleation of Dislocations on Inclined Planes in Mode I.

To obtain from the results of the section above the condition for nucleating dislocations on inclined planes two modifications need to be considered.

First, on any inclined slip plane passing through a Mode I crack the nucleation of a dislocation will occur under a mixture of effective Modes II and Mode I where the local effective Modes I and II are related to the Mode I acting across the crack by [50],

$$K_I^{eff} = K_I cos^3(\theta/2) \quad \therefore \quad K_{II}^{eff} cos^2(\theta/2)sin(\theta/2) \qquad (5a,b)$$

Instead of solving the specific problem of Fig. 2b at hand, we examine the general modifying effect on K_{II} of a K_I^{eff} acting across the slip planes, that is required to achieve an athermal 2-D release of a straight dislocation on this plane. This result is given in Fig. 6 [30] and demonstrates that while the effect of a K_I^{eff} acting across the slip plane does indeed produce some reduction in the required K_{II}^{eff}, this reduction is relatively minor.

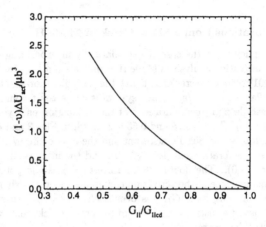

Fig. 5. The activation energy for dislocation emission under pure mode loading.

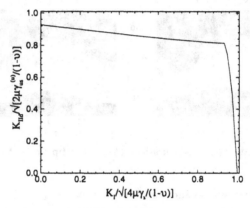

Fig. 6. Dependence of critical K_{II} for dislocation emission on K_I present on the $\frac{1}{2}[\bar{1}\bar{1}1](112)$ system of α-Fe.

Second, on any inclined slip plane the nucleation of a dislocation will produce a free surface ledge after the dislocation is emitted and has been sent into the background. During the actual nucleation process when the incipient surface is not yet relaxed the effect must be considered as an additional surface production resistance to be added to the shear resistance of the tension shear potential [30, 51] to at least the first row of atoms that will be bared as a free surface. The result of a specific analysis of this effect is shown in Fig. 7 again for the release of a straight dislocation from the crack tip onto an inclined plane making an angle of 45° with the crack plane. The dependence of the critical energy release rate G_{cd}, normalized with the crack driving force G_{IC} on the parameter $q = \gamma_{us}/2\gamma_s$ is shown for a case with no surface production resistance ($\lambda = 0$) and for a surface production resistance affecting substantially only the first row of atoms ($\lambda = 1$). The slanted dash-dotted line shows a somewhat more approximate estimate of Rice for no surface production resistance [45], while the horizontal dashed line gives the condition where cleavage is obtained on the slip plane. $G_{cd}/G_{IC} < 1.0$ indicates nucleation of a dislocation prior to propagation of the crack. The solutions including the surface production resistance indicates that for all values of q crack propagation on the cleavage plane will precede athermal release of a dislocation on the 45° inclined plane. Values of q range from 0.081 for Al to 0.091 for Ni (both for Shockley partial dislocations) to 0.217 for α-Fe and 0.376 for Si (Shockley partial) [30]. Regardless of the values of q, since all cases lie above G_{cd}/G_{IC} of 1.0, it must be concluded that homogenous nucleation of dislocations on inclined planes is not possible prior to crack propagation for any material, regardless of whether it is intrinsically brittle or ductile [30].

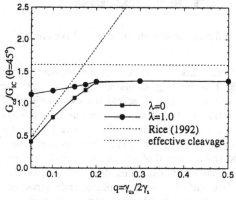

Fig. 7. Dependence of the critical energy release rate G_{cd}/G_{IC} on the factor $q = \gamma_{us}/2\gamma_s$ for dislocation nucleation at 45°.

Nucleation of Dislocations on the Oblique Plane.

As remarked above, dislocation nucleation on oblique planes has been put forth as a likely mode. However, the capability required for the analysis of this mechanism had been heretofore unavailable. Approximate analyses based on the Rice and Thomson method have led to estimates of the B-D transition temperatures several orders of magnitude higher than what is experimentally observed, prompting suggestions that nucleation should involve fractional dislocation [26].

Here we provide a direct analysis of the formation of dislocation embryos on oblique planes in α-Fe in the configuration of Fig. 2c as the most favorable for such nucleation. In the analysis, we have artificially constrained the cleavage plane from propagating to enable the computation of the critical driving force for dislocation nucleation. If the resulting athermal critical driving force is less than G_{IC} for cleavage, the crystal is intrincsically ductile. Contrariwise, if the critical driving force is greater than that for cleavage, the crystal is cleavable or intrinsically brittle. However, on a crack, on the verge of crack propagation by cleavage, a dislocation can still be nucleated through thermal activation. The calculation of the activation energy for such thermally assisted dislocation nucleation was performed for Fe in the configuration of Fig. 2c [47].

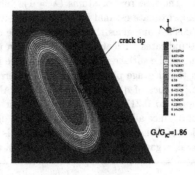

Fig. 8. Saddle-point configuration of a dislocation embryo emitted from a crack tip on an oblique plane.

A typical saddle point configuration of an embryonic dislocation loop emanating from the crack tip is shown in Fig. 8. The resulting calculated dependence of the activation energy on the crack driving force, near the athermal threshold is shown in Fig. 9. The critical driving force at the athermal threshold, and the attendant activation energies are so high that they render the nucleation mechanism most unlikely. The activation energy at $G_I/G_{IC} = 1.0$ can be estimated from the extrapolated curve, and forms an adequate basis for reaching a firm negative conclusion on the likelihood of nucleating a dislocation in the interior by this mechanism. A very different conclusion can be reached, however, for this mechanism where the crack front reaches a free surface. Here, no plane strain stress constraint exists and the resolved shear stresses on the oblique planes become much higher. An estimate of this is readily obtained by re-scaling the driving forces in Fig. 9 in proportion to the resolved shear stresses on the oblique slip planes near the surface vs those in the interior as (for the geometry of Fig. 2b):

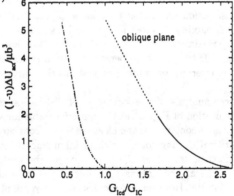

Fig. 9. Activation energies for dislocation nucleation on an oblique plane. Dash-dotted curve for such a plane near the surface.

38

$$\frac{G_{Is}}{G_{Ii}} = \left(\frac{3-4\nu}{3}\right)^2 \tag{6}$$

where G_{Is} and G_{Ii} are the respective energy release rates for initiation near the surface and in the interior. The result for Fe with $\nu = 0.291$ is shown as the dash-dotted curve in Fig. 9, which now suggests almost spontaneous nucleation near the surface. This, however, is not true since in this instance some surface ledge needs to be produced which will make the nucleation more difficult, but presumably still much easier than in the interior.

Nucleation of Dislocations on a Cleavage Ledge

Cleavage surfaces in metallic crystals invariably contain ledges parallel to the direction of crack propagation. These are likely to form when the direction of the principal tension driving the crack deviates slightly on a local scale, requiring the crack to make small adjustments along its front. This unavoidable micro-roughness of the cleavage surface depends on the crystallography of the cleavage planes and the crack propagation direction, as well as on temperature. The height of the observed ledges can range from several atomic spacings to microns. Numerous observations [24,25,42] have revealed that dislocation nucleation at a crack front is a relatively rare phenomenon associated with such crack front heterogeneities [52]. In what follows we analyse this mechanism as it is likely to operate in α-Fe.

Consider a cleavage crack propagating under Mode I loading, and containing ledges of a width of roughly a hundred atomic spacings distributed along its front. The presence of a considerable local Mode III stress intensity factor acting on the ledge, the fact that no free surface is produced, and that the nucleated embryo will be nearly screw in nature are all expected to promote nucleation.

The distribution of stress intensity factors on the crack front can be readily calculated by the boundary element method [41]. On the ledge, the dominant stress intensity factors are $K_I^{ledge} \sim 0.81 K_{IC}$ and $K_{III}^{ledge} \sim 0.35 K_{IC}$ at the verge of brittle propagation.

The athermal condition for nucleating a dislocation in Mode III loading, as it exists on the ledge, was determined by Rice [45], as:

$$K_{IIIcd} = \sqrt{(1-\nu)\gamma_{us}/2\gamma_s} K_{IC} = 0.357 K_{IC} > 0.350 K_{IC} \tag{7}$$

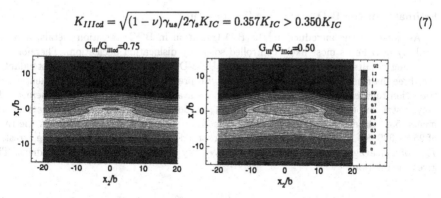

Figs. 10 & 11. Saddle-point configurations of a dislocation embryo on a cleavage ledge at $G_{III}/G_{IIIcd} = 0.75$, and 0.5.

Although, this suggests that dislocation formation on the ledges should be close to spontaneous according to this simple estimate, we merely conclude that it should be in the range to initiate thermally assisted embryo formation. To analyse such thermally assisted embryo formation we consider a saddle point analysis of a crack in pure Mode III, to apply to the problem at hand, of the cleavage ledge loaded in Mode III. Figures 10 and 11 show the saddle point configuration under normalized load levels of G_{III}/G_{IIIcd} of 0.75 and 0.5 respectively. The resulting dependence of the activation energies for such configurations on actual normalized crack driving forces G_I/G_{IC} are shown in Fig. 12, for a case with no K_I component acting across the ledge ($\triangledown$) and the case with the appropriate K_I component of $0.81K_{IC}$ (o) acting across the ledge as it should be the case of the geometry of Fig. 2c. The differences are small.

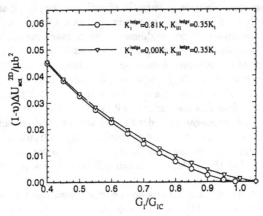

Fig. 12. Activation energies for dislocation nucleation on a cleavage ledge.

DISCUSSION

Estimates of the B-D Transition Temperature

As noted in the introduction, the B-D transition in BCC transition metals, and particularly in α-Fe, is most likely controlled solely by dislocation nucleation. Therefore, the preceding results can be used to estimate the B-D transition temperatures attendant to the three main nucleation modes considered. No precise experimental measurements of the transition temperature of single crystal α-Fe are available. The transition temperature for polycrystal low carbon steel is about $250°K$, as determined from Charpy impact experiments [53]. Based on this, we take the transition temperature for α-Fe to be in the range of $250 - 300°K$. A B-D transition scenario proposed by Argon [26] consists of the arrest, at T_{BD}, of a cleavage crack propagating with a velocity v against a temperature gradient. This gives the relation,

$$T_{BD} = \left[\frac{ln(c/v)}{\alpha} + \eta\frac{T_o}{T_m}\right]^{-1} T_o \qquad (8)$$

where $T_o \equiv \mu b^3/k(1 - \nu) \sim 1.2 \times 10^5 K$; the melting temperature $T_m = 1809K$ for α-Fe; $\alpha = (1 - \nu)\Delta U_{act}/\mu b^3$ is the normalized activation energy; c is speed of sound; $v \sim 1cm/s$ is a typical crack propagation velocity, giving $\ln(c/v) \sim 10$; $\eta \sim 0.5$ is a coefficient describing the temperature dependence of the shear modulus which, to a first approximation, is presumed of the form $\mu = \mu_o(1 - \eta(T/T_m))$.

The activation energy at the critical driving force for cleavage , i.e., at $G_I/G_{IC} = 1$, determines the transition temperature through Eqn. 8. This relation is plotted in Fig. 13 together with reasonable ranges of activation energies for nucleation on inclined planes, oblique planes and on ledges. The results for the oblique plane near a free surface should be close to that for cleavage ledges. Also shown in the figure is the value of the transition temperature for polycrystalline Fe and its melting temperature. It is evident that only nucleation on cleavage ledges and on oblique planes near a free surface result in transition temperatures that approach the expected value for α-Fe. The dislocation loops which eventually shield the entire crack are apparently emitted from ledges distributed along the crack front as was noted by George and Michot [25].

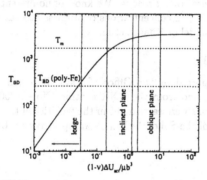

Fig. 13. Estimates of B-D transition temperatures in α-Fe vis-a-vis the three nucleation configurations considered.

Other Possibilities

The fracture transition modeled in the present study is meant for dislocation-nucleation-controlled- transitions, characteristic of BCC transition metals, and not for transitions controlled by the mobility of dislocations away from the crack tip [44] which is typical of semiconductors and compounds. The distinguishing characteristic between these two behaviors is the mobility of kinks on dislocations, which is high in BCC transition metals, and very low in Si and other compounds for which good information on dislocation mobility exists [31-35].

An additional effect which can influence fracture behavior is crack-tip shielding by general 'background' plasticity. A particularly elegant and compelling analysis of this mechanism was advanced by Freund and Hutchinson [54], who have demonstrated that brittle-like fracture should take place at high crack propagation velocities with progressively diminishing inelastic response. However, this transition is smooth and spread out, and far from being abrupt.

The importance of background plasticity effects has been demonstrated experimentally by Hirsch et al [24] who have shown that the sharp B-D transition in dislocation-free Si becomes diffuse and moves to somewhat lower temperatures-when the crystals are initially dislocated by a pre-deformation step. The effect of background plasticity can therefore be regarded to modulate the B-D transition, with the ultimate controlling mechanism still residing in crack-tip initiated processes.

In closing, we note that the ability of dislocation nucleation at the crack tip to account for the exceedingly sharp transitions observed in some materials has been questioned by Khanta et al [55, 56] who have advocated a critical phenomena approach, akin to defect- mediated melting. However, the preponderance of the observational evidence appears to support the crack-tip dislocation nucleation mechanism. Indeed, the detailed and meticulous direct X-ray imaging experiments of George and Michot [25] of the stages of evolution of the crack tip plastic response, starting from nucleation at crack tip heterogeneities and followed by the very rapid spread and multiplication of dislocation length from such sources, is a convincing direct demonstration of the vast numbers of degrees of freedom available to dislocations for populating the highly stresses crack tip zone. We know of no present experimental evidence for the large thermal equilibrium concentrations of stiffness-attenuating dislocation dipoles that are predicted by the model of Khanta et al.

ACKNOWLEDGMENTS

This research was supported by the Office of Naval Research (ONR) under contract N00014-92-J-4022 at MIT, and from the ONR contract No. N00014-90-J1758 at Brown University. The computations were carried out in the facilities of the Mechanics of Materials Group at MIT, and those of the Solid Mechanics Group at Brown University.

REFERENCES

1.   M.E. Shank, *Mech. Eng.*, *76*, 23 (1954).

2.   E. Orowan, in "Repts. Prog. Physics", vol 12, p.185 (1949).

3.   E.R. Parker, "Brittle Behavior of Engineering Structures", J. Wiley, New York (1957).

4.   C. Zener, in "Fracturing of Metals", ASM, Metals Park Ohio, p.3 (1949).

5.   A.N. Stroh, *Proc. Roy. Soc.*, *A223*, 404, (1954).

6.   A.N. Stroh, *Proc. Roy. Soc.*, *A232*, 548 (1955).

7.   D. Hull, *Acta Metall*, *8*, 11 (1960).

8.   C. McMahon, "Micromechanisms of Cleavage Fracture in Polycrystalline Iron", ScD Thesis, M.I.T., Cambridge, MA (1963).

9.   G.T. Hahn, B.L. Averbach, W.S. Owen, and M. Cohen, in "Fracture", edited by B.L. Averbach et al, MIT Press, Cambridge, MA, 91 (1959).

10. J.F. Knott and A.H. Cottrell, *J. Iron Steel Inst.*, *201*, 249 (1963).

11. M. Cohen and M.R. Vukcevich, in "Physics of Strength and Plasticity", edited by A.S. Argon, MIT Press, Cambridge, MA p.295 (1969).

12. A.N. Stroh, *Adv. Phys.*, 6, 418 (1957).

13. R.O. Ritchie, J.F. Knott and J.R. Rice, *J. Mech. Phys. Solids*, 21, 395 (1973).

14. T. Lin, A.G. Evans and R.O. Ritchie, *J. Mech. Phys. Solids*, 34, 477 (1986).

15. T. Lin, A.G. Evans and R.O. Ritchie, *Met. Trans.*, 18A, 641 (1987).

16. A. Kelly, W.R. Tyson and A.H. Cottrell, *Phil. Mag.*, 15, 567 (1967).

17. J.R. Rice and R. Thomson, *Phil. Mag.*, 29, 73 (1974).

18. S.J. Burns and W.W. Webb, *J. Appl. Phys.*, 41, 2078 (1970).

19. S.J. Burns and W.W. Webb, *J. Appl. Phys.*, 41, 2086 (1970).

20. J.J. Gilman, C. Knudsen and W.P. Walsh, *J. Appl. Phys.*, 29, 600 (1958).

21. C. StJohn, *Phil. Mag.*, 32, 1193 (1975).

22. M. Brede and P. Haasen, *Acta Metall*, 36, 2003 (1988).

23. P.B. Hirsch, J. Samuels, and S.G. Rloberts, *Proc. Roy. Soc.*, A421, 25 (1989).

24. P.B. Hirsch, S.G. Roberts, J. Samuels and P.D. Warner, in "Advances in Fracture Research" edited by K. Salama et al, Pergamon, Oxford vol 1, p.139 (1989).

25. A. George and G. Michot, *Mater. Sci. Engng.*, A164, 118 (1993).

26. A.S. Argon, *Acta Metall.*, 35, 185 (1987).

27. K.S. Cheung, A.S. Argon and S. Yip, *J. Appl. Phys.*, 69, 2088 (1991).

28. G. Schöck and W. Püschl, *Phil. Mag.*, A64, 931 (1991).

29. J.R. Rice and G.E. Beltz, *J. Mech. Phys. Solids*, 42, 333 (1994).

30. G. Xu, A.S. Argon and M. Ortiz, *Phil. Mag.*, 72, 415 (1995).

31. K. Sumino, in "Structure and Properties of Dislocations in Semiconductors", edited by S.G. Roberts et al, Inst. Phys., Bristol, England, p.245 (1989).

32. K. Maeda and Y. Yamashita, same as Ref. 31, p.269 (1989).

33. I. Yonenaga, U. Oriose and K. Sumino, *J. Mater. Res.*, 2, 252 (1987).

34. I. Yonenaga, K. Sumino, G. Izawa, H. Watanabe and J. Matsui, *J. Mater. Res.*, 4, 361 (1989).

35. I. Yonenaga, and K. Sumino, *J. Mater. Res.*, 4, 355 (1989).

36. V.V. Bulatov, S. Yip, and A.S. Argon, *Phil. Mag.*, 72, 452 (1995).

37. J.R. Rice, G.E. Beltz, and Y. Sun, in ``Topics in Fracture and Fatigue", edited by A.S. Argon, Springer, Berlin, p.1 (1992).

38. R.E. Peierls, *Proc. Phys. Soc.*, *A52*, 34 (1940).

39. F.R.N. Nabarro, *Proc. Phys. Soc.*, *A59*, 256 (1947).

40. A.J. Foreman, M.A. Jaswon, and J.K. Wood, *Proc. Phys. Soc.*, *A64*, 156 (1951).

41. G. Xu and M. Ortiz, *Intern. J. Num. Methods Engng.*, *36*, 3675 (1993).

42. Chiao, Y-H, and D.R. Clarke, *Acta Metall.*, *47*, 203 (1989).

43. J. Samuels and S.G. Roberts, *Proc. Roy. Soc.*, *A421*, 1 (1989).

44. M. Brede, *Acta Metall., et Mater.*, *41*, 211 (1993).

45. J.R. Rice, *J. Mech. Phys. Solids*, *40*, 235 (1992).

46. A.S. Argon and D. Deng, unpublished research, available on request.

47. G. Xu, A.S. Argon and M. Ortiz, submitted to Phil. Mag.

48. J.H. Rose, J. Ferrante and J.R. Smith, *Phys. Rev. Letters.*, *47*, 675 (1981).

49. Y. Sun, G.E. Beltz and J.R. Rice, *Mater. Sci. Engng.*, *A170*, 67 (1993).

50. B. Cotterell and J.R. Rice, *Intern. J. Fract.*, *16*, 155 (1980).

51. E. Kaxiras and Y. Juan, private communication, to be published.

52. S.J. Zhou and R. Thomson, *J. Mater. Res.*, *6*, 639 (1991).

53. F.A. McClintock and A.S. Argon ``Mechanical Behavior of Material", Addison Wesley, Reading MA (1966).

54. L.B. Freund and J.W. Hutchinson, *J. Mech.Phys.Solids*, *33*, 169 (1985).

55. M. Khanta, D.P. Pope and V. Vitek, *Phys. Rev. Letters*, *73*, 684 (1994).

56. M. Khanta, D.P. Pope and V. Vitek, *Scripta Metall et Mater.*, *31*, 1349 (1994).

PLAN-VIEW TRANSMISSION ELECTRON MICROSCOPY OF CRACK TIPS IN BULK MATERIALS

H.Saka*, G.Nagaya*,T.Sakuishi*,S.Abe*,A.Muroga**

*Department of Quantum Engineering, Nagoya University,Nagoya, 464-01, Japan
**Aichi Steel Works Ltd.,Tokai,476, Japan

ABSTRACT

A focused ion beam (FIB) system has been applied to prepare a thin foil specimen of Si, MgO and alumina which contained cracks in the plane of foil. It was possible to observe a much larger area at and near a crack tip than has been hitherto possible. FIB was also applied to observation of microstructure near a crack tip evolved during severe rolling contact fatigue in a steel.

INTRODUCTION

The fracture of materials is of great practical as well as academic importance. The behaviour of a crack tip propagating under the action of an applied stress plays a very important role in determining whether fracture is ductile or brittle[1]. Many attempts have been made to observe a crack tip at a resolution high enough for detailed information to be obtained; they include optical microscopy[2,3], X ray topography [4,5]and TEM[6,7]. Of these techniques, the resolution of optical microscopy and X-ray topography is too poor. By contrast, the resolution of TEM is high enough for detailed information to be obtained. However, in order to carry out TEM observation, a thin foil specimen must be prepared, and it is usually extremely difficult to prepare such a specimen from regions containing crack tips.

Recently a focused ion beam (FIB) technique has been developed and applied to the preparation of thin foil specimens for TEM observation[8,9,10]. Briefly, in this technique a focused beam of gallium ions is scanned across a bulk sample and in doing so the sample is fabricated into a thin foil. It is possible to prepare the thin foil from pre-selected regions within an accuracy better than 100nm. In the present study the FIB technique is applied to prepare foil specimens of Si, MgO and alumina which allow the plan-view TEM observation of crack tips. FIB was also applied to observation of microstructure near a crack tip evolved during severe rolling contact fatigue in a steel.

EXPERIMENTAL PROCEDURES

Single crystal of Si, MgO and alumina were used. Vickers indentations were performed in air, at

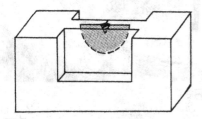

Fig.1 Geometry of a specimen.

Fig.2 Indent on a (001) Si.

Mat. Res. Soc. Symp. Proc. Vol. 409 ⁰1996 Materials Research Society

room temperature, using a 50g load and a 15s dwell–time. Indentation resulted in radial cracks which are approximately perpendicular to each other. After indentation, the specimen was transferred into a FIB system. The system used in this study is a Hitachi FB–2000 system. Two trenches were milled in such a way that a thin wall is left behind between the two trenches, the wall being thin enough to be transparent to electrons when tilted by 90 degrees(Fig.1). The specimen was examined in the high–voltage electron microscope of Nagoya University, Hitachi HU–1000D, operated at an accelerating voltage of 1000kV. This accelerating voltage was used in order to observe as thick a specimen as possible.

RESULTS AND DISCUSSION

Si

Fig. 2 shows an optical micrograph of cracks formed at an indent on a(001) surface of Si. The cracks were parallel to <110>directions in this case. Fig.3 shows an electron micrograph of the area of the indent(denoted by I). The following features are evident.
1) Within the indent the crystal is deformed severely and some dislocations(denoted by D) have glided out of the indent.
2) Near the indent, cracks piercing the foil are observed(denoted by Cr). There are two types of

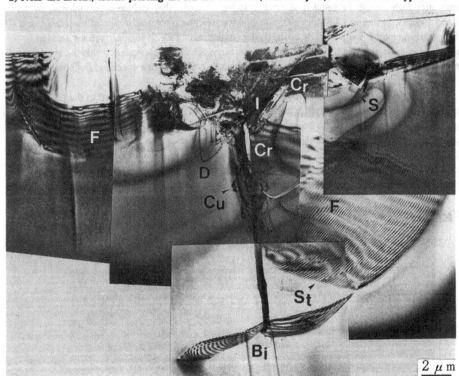

Fig. 3

46

such cracks. One is almost parallel to the foil surface, i.e., (001) plane. Occasionally cracks of this type are bent sharply as is shown by S. They have a tendency to lie on the {111} plane. The other type is perpendicular to the foil surface and lies on the {110} plane, but , in depth, cracks of this type again have a tendency to lie approximately parallel to the foil surface. It is noted that the bend is very smooth in this case.

3) On both sides of the indent and underneath the indent, fringe contrasts can be seen(denoted by F). It is natural to consider the areas with fringe contrast as containing cracks.The fringes can be explained by moire fringes which result from the crystal being separated into two by the crack. Close examination reveals that the cracks are not simply planar but have features like steps(denoted by S_t) and cusps(denoted by C_u).

4) Bifurcation of a crack is also observed at B_i.

Fig. 4 shows another crack which was almost parallel to the foil surfaces. In the lower right corner moire fringes are observed, while in the upper left such fringes are not present. No defects such as dislocations are observed ahead of the tip but, in the wake of the crack, butterfly−like contrasts are observed in the vicinity of the crack tip. The directions of the line of no contrast were always perpendicular to the diffraction vectors used. Thus, it is supposed that the contrast results from spherical strain centres[11]). Since the purity of Si used in this study is very high, the centres are unlikely to be clusters of impurities; they must be some kind of defect associated with the propagation of the crack tip. At the moment, the nature of these defects is not understood.

MgO

Fig. 5 shows microstructure near an indent on a (001) surface of MgO. In this case, many dislocations are introduced by indentation. This is in good agreement with that MgO is plastic

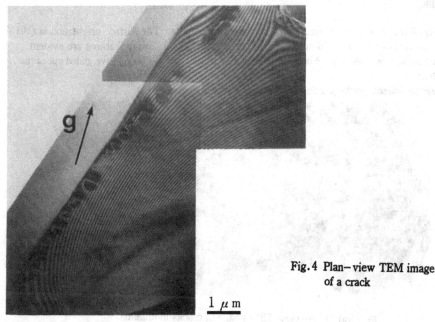

g

Fig. 4 Plan−view TEM image
of a crack

1 μ m

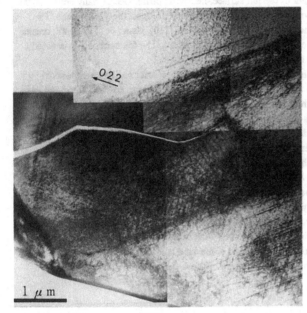

Fig. 5 Dislocation structure near an indent in MgO.

022

1 μm

even at room temperature [12].

<u>Alumina</u>

Fig. 6(a) shows microstructure near an indent in alumina. The surface orientation is $(10\bar{1}0)$ and the cracks were approximately parallel to $<0\bar{3}1>$. The following features are evident.
1) Within the indent the crystal is deformed severely and many dislocations have glided out of the indent.
2) Near the indent, cracks piercing the foil are observed(denoted by C_p).

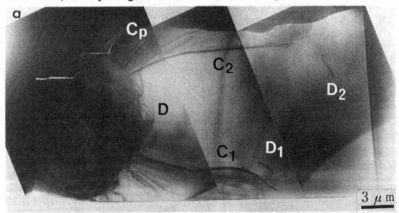

Fig. 6(a) Plan–view TEM image of cracks in alumina.

Fig.6(b)(c)

2 μ m

Fig.7 Optical micrograph of
brightly etched regions
(BER).

20 μ m

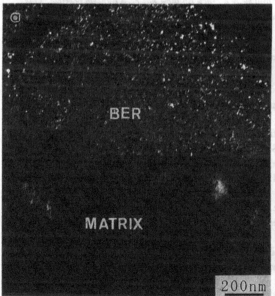

BER

MATRIX

200nm

Fig.8(a) Dark-field electron micrograph of
a BER and matrix.

(b) Diffraction pattern from BER.
(c) Diffraction pattern from matrix.

3) There are two cracks(C_1 and C_2) nearly parallel to the foil surface. From the tip of C_1, at least two dislocations(D_1, D_2) seemed to have been emitted. These dislocations disappeared in $\mathbf{g}=0\bar{1}3$(Fig. 6(c)). This can be explained by assuming that the $\mathbf{b}$ of these dislocations are in the basal plane.

4) Near the tip and in the wake of C_1 strong line contrasts are observed(Fig. 6(b)). It is inferred that these contrasts are formed as a result of the lattice distortion associated with the opening of the crack and steps on the crack surfaces.

Rolling contact fatigue crack in steel

It is well known that a steel which is subjected to fatigue under severe conditions shows peculiar microstructures. Here, a quench–tempered JIS SUJ2 steel was subjected to a rolling contact fatigue with a high speed, a high load and a severe vibration. During the fatigue a peculiar microstructure, which is called brightly etched regions(BER) was formed[13]. Fig. 7 shows an optical micrograph of such BER. The microstructure of BER was examined combining FIB with HVEM. Fig. 8(a) shows a dark–field electron micrograph of both the BER and the matrix. It is clear that the structure of the BER is much finer than the matrix. Fig. 8(b)(c) show the electron diffraction patterns from BER and the mattrix, respectively. The diffraction pattern from the matrix shows a network of spots which is typical to s single crystal. In contrast, the diffraction pattern from the BER is composed of rings typical to polycrystals.

In conclusion, using a FIB combined with a high–voltage electron microscope it has been made possible to observe a much wider area at and near a crack tip than has been hitherto possible. This technique will open a new way to the study of the fracture of materials.

REFERENCES

[1] Smith,E.,1985, in *"Dislocations and Properties of Real Materials"*,(The Inst. Metals, London), pp.205.

[2] Samuels,J. and Roberts,S.G.,1989,Proc.Roy.Soc.London,**A421**,1.

[3] Cook,R.F., and Pharr,G.M.,1990,J.Am.Ceram.Soc.,**73**,787.

[4] St.John,C.,1975,Phil.Mag.,**32**,1193.

[5] Michot ,G.and George,A.,1986,Scripta metall.,**20**,1495.

[6] Majumdar,B.S. and Burns,S.J.,1980,Scripta metall.,**14**,653.

[7] Farber,B.Y.,Chiarelli,A.S., and Heuer,A.H.,1994,Phil.Mag.,A,**70**,201.

[8] Park,K.,1990,MRS Symp. Proc.,Vol.199,*"Specimen Preparation for Transmission Electron Microscopy of Materials, II"*, edited by R.Anderson, (MRS, Pittsburgh),271.

[9] Young,R.J.,Kirk,E.C.G.,Williams,D.A., and Ahmed,H. 1990,MRS Symp. Proc.,Vol.199,*"Specimen Preparation for Transmission Electron Microscopy of Materials, II"*, edited by R.Anderson, (MRS, Pittsburgh),205.

[10] Saka,H.,Kuroda,K.,Hong,M.H.,Kamino,T.,Yaguchi,T.,Tsuboi,H.,Ishitani,T.,Koike,H., Shibuya,A., and Adachi,Y.,1994, *Proc. 13th International Congress on Electron Microscopy*,edited by B.Jouffrey and C.Colliex, Vol.1,1009.

[11] Ashby,M.F.,and Brown,L.M.,1963,Phil.Mag.,**8**,1083.

[12] Appel,F., Messerschmidt,U.,and and Kuna,M.,1979,Phys.Stat.Solidi(a),**55**,529.

[13] Muroga, A and Saka, H. , Scripta metall. et materi. , 1995, **33**, 151.

COOPERATIVE DISLOCATION GENERATION AND
THE BRITTLE-TO-DUCTILE TRANSITION

M. KHANTHA, D. P. POPE and V. VITEK
Department of Materials Science and Engineering
University of Pennsylvania, Philadelphia, PA 19104-6272.

ABSTRACT

The characteristic features of the brittle-to-ductile transition are explained using a model of cooperative dislocation generation. In two dimensions, the onset of the ductile behavior corresponds to a thermally-driven, stress-assisted dissociation of many atomic-size dislocation dipoles in the vicinity of the crack tip above a critical temperature T_c. The instability is caused by thermally induced screening of dislocation interactions as in the Kosterlitz-Thouless phase transition. However, the critical temperature is well below the melting temperature in the presence of a stress. The nature of dislocation dynamics in the vicinity of the crack tip is also described and its role in the onset of the cooperative instability is examined. The origin of the correlation between the strain-rate dependence of the transition temperature and the temperature dependence of dislocation mobility is explained.

INTRODUCTION

The brittle-to-ductile transition (BDT) is ubiquitous in crystalline materials with the possible exception of f.c.c. metals (for reviews, see [1-5]). Two characteristic features associated with this transition are qualitatively similar in most materials:

1.   A dramatic increase in the fracture toughness occurs in a narrow temperature range around the brittle-to-ductile transition temperature (BDTT). For example, the toughness of steels in the brittle region (approximately 0-200 K) is typically 30-60 MPam$^{1/2}$ while it increases to over 250 MPam$^{1/2}$ in a range of 50 degrees around the BDTT (200-270 K) [1]. The toughness of an intermetallic compound such as NiAl is in the range 5-10 MPam$^{1/2}$ in the brittle region (0-400 K) and increases approximately four-fold to 20-30 MPam$^{1/2}$ in a range of 10-30 degrees around the BDTT (400-500K) [3, 4]. The toughness of a semiconductor like Si is in the range 1-2 MPam$^{1/2}$ in the brittle region (up to 800K) and increases to 7-8 MPam$^{1/2}$ in a narrow transition region of less than 10 degrees around the BDTT (800-1000 K) [6-8]. There is strong evidence that extensive dislocation activity begins abruptly near the crack tip at the BDTT in Si [6-16] while there is no evidence of dislocation generation even 5° below the transition [6]. This confirms the notion that massive dislocation activity is responsible for the rapid increase in fracture toughness and the change in the fracture mode from brittle to a plasticity-controlled failure mode above the BDTT. The similarity in the variation of toughness around the BDTT in different materials suggests a common underlying mechanism for the transition in all materials.

2.   The transition temperature, hereafter referred to as T_c, is usually a function of the applied rate of stress intensity, $\dot{K}$, and exhibits the following behavior:

$$\dot{K} \propto \exp(-U_s/k_B T_c) \tag{1a}$$

where k_B is the Boltzmann constant and U_s represents the activation energy associated with the process. This Arrhenius-type strain-rate dependence is observed in steels [17], semiconductors

[6] and intermetallics [4]. In most materials the variation of the dislocation mobility (or velocity) (v) with temperature (T) and stress (τ) can be expressed by an empirical relation [18]

$$v = A\,\tau^m \exp(-U_m/k_B T) = v_0 \tau^m \qquad (1b)$$

where A is a constant, m is the power-law stress exponent which usually does not vary with temperature, U_m is the activation energy for dislocation motion and v_0 is equal to $A\exp(-U_m/k_B T)$. In Si [6-8] and Ge [19], it was found that the value of U_s measured from the strain-rate dependence of T_c (Equation 1a) was almost exactly equal to the activation energy for dislocation motion (U_m) measured from the variation of v with T (Equation 1b). There is an indirect evidence that this correlation may also be true in intermetallics [4]. The value of U_s measured from the BDT experiments is in a reasonable agreement with the estimate of the activation barrier for dislocation motion even in steels [17]). This suggests that the correlation between dislocation mobility and the strain-rate dependence of T_c is a distinct, and perhaps, universal feature of the BDT.

A general model of cooperative dislocation generation which explains the dual characteristics of the BDT in all materials was put forward recently [20-22]. In this paper we explain the underlying mechanism of cooperative dislocation generation in detail. We also describe the effect of dislocation dynamics and its influence on the strain-rate dependence of the BDTT. The major differences between the present approach and other models [7, 23-27] are also summarized.

THE MECHANISM OF COOPERATIVE DISLOCATION GENERATION

A cooperative dislocation generation mechanism was first proposed by Kosterlitz and Thouless (hereafter referred to as K-T) [28] in the context of phase transitions in two dimensional (2D) solids. According to the K-T model [28-32], the melting of 2D crystals in the absence of applied loads can be described as a cooperative dissociation of many dislocation dipoles above a critical temperature. It is caused by thermally-induced screened dislocation interactions and the instability occurs close to the melting temperature. Our model [20, 21] is based on the K-T concept of screened dislocation interactions [32] but describes the collective dissociation of dipoles in a loaded 2D solid at finite temperatures. The presence of an applied stress lowers the formation energy of dipoles and enables the instability to occur well below the melting temperature.

The K-T mechanism, based on fundamental principles of statistical mechanics, is well known in the context of 2D systems. It is much less known that the mechanism underlying the cooperative instability is also applicable to 3D systems [1]. The K-T model describes the entropy-driven cooperative instability of topological defects which possess finite energies of a logarithmic form. These include dislocation dipoles [2] in 2D with energies proportional to lnr, where r refers to the dipole size, dislocation loops in 3D with energies proportional to rlnr, where r is the radius of the loop, vortex dipoles and loops in superfluids, etc. The instability is thermally driven and occurs at a critical temperature when the entropic and the self-energy contributions to the free energy of a collection of topological defects become equal. Since the entropy is always a logarithmic function, such a balance is especially possible when the self-energy is also of a logarithmic form. The free energy of the system becomes negative above the critical temperature and causes spontaneous cooperative dissociation/expansion of many dipoles/loops.

It is well known that the entropy of a single dislocation dipole or loop is negligible compared to its self-energy [35]. It appears therefore that the entropic contribution of a collection of dipoles/loops can become equal to the self-energy contribution only at unrealistically high temperatures. This is indeed true for non-interacting dipoles/loops pre-existing in a crystal. Even in the presence of an applied stress, the entropy of a single dipole/loop is negligible compared to its enthalpy and the activation barrier for unstable dissociation/expansion is very large for stresses smaller than the theoretical shear stress.

---

[1] The 3D applications of the K-T transition were first developed in the context of superfluidity [33, 34].
[2] In 2D, the dipole refers to a planar projection of a dislocation loop composed of 'point' dislocations with equal and opposite Burgers vectors.

While this much was already known, Kosterlitz and Thouless showed that a collection of dipoles/loops behaves in a different manner at finite temperatures compared to a single dipole/loop at very low temperatures. This difference arises due to the following reasons: At finite temperatures, even in the absence of applied loads, spontaneous strain fluctuations can occur in a crystal resulting in the formation of small atomic-size <u>bound</u> dislocation dipoles or loops. This is possible because such configurations have finite energies of the order of 1eV and can therefore form by a thermally activated process. K-T then examined the effect of such configurations using the 'fluctuation-dissipation' (F-D) theorem, which links all linear response functions of a system to the spectrum of spontaneous fluctuations (in the absence of the external perturbation) in appropriate physical variables [36, 37]. For example, according to the F-D theorem, the compliance tensor of a solid which represents the response to an applied stress can be expressed in terms of the fluctuations of the total strain. The formation of dipoles/loops introduces a small amount of plastic strain which increases the total strain in the system compared to its value in the absence of defects. This, in turn, increases the effective compliance of the system or equivalently, decreases the effective shear modulus of the system. Thus, the spontaneous formation of atomic-size dipoles/loops causes a change in the effective response of the system. There is however, no observable change in the mechanical properties since the dipoles/loops are smaller than the critical size needed for spontaneous dissociation/expansion under an applied shear stress.

The self-energy of a dipole/loop is proportional to the shear modulus of the solid. Hence, a lowering of the effective modulus due to the presence of dipoles/loops lowers the energy of formation of subsequent dipoles/loops. Consequently, the probability of forming dipoles/loops in the vicinity of other existing dipoles/loops increases, which in turn, causes a further increase in the total strain and a lowering of the effective shear modulus. Concurrently, the entropy of the dipoles/loops also increases with both increase in temperature and increase in density of dipoles resulting in lower total free energy. K-T formulated this (positive) feedback mechanism between the reduction in the self-energy contribution and the increase in the entropic contribution by introducing a plastic polarizability, ε, in analogy with the polarization of electric charges in a dielectric medium [28-32]. Just as the Colomb interaction of charges becomes screened in a dielectric medium, the plastic polarizability lowers the free energy by screening the $(1/r)$ interaction between dislocations comprising a dipole, to the form $(1/\varepsilon r)$. Using the F-D theorem and the tools of equilibrium statistical mechanics, the polarizability ε can be calculated as a function of temperature.

The behavior of ε as a function of temperature is an important indicator of whether the free energy of the system of dipoles/loops is positive or negative. ε, which is a response function, can be expressed in terms of the plastic strain-strain correlation function using the F-D theorem and can be evaluated using the tools of statistical mechanical ensemble averages. As long as the dipoles/loops remain bound, unable to expand freely in the presence of a stress, the polarizability remains finite whereas when the free energy becomes negative causing the spontaneous collective unstable dissociation/expansion of many dipoles/loops, the plastic strain and the polarizability diverge. The temperature above which ε diverges is identified as the 'critical temperature' of the cooperative instability. In the absence of stress, the critical temperature is close to the melting temperature.

Based on this qualitative description, it is easy to see why the presence of an applied stress lowers the critical temperature well below melting. In a loaded crystal, the formation energy of atomic-size dipoles/loops is smaller compared to an unloaded crystal due to the contribution from the work done by the external loads to the enthalpy. Hence the probability of nucleating dipoles/loops is correspondingly higher. The total free energy of the system is also lower due to the additional entropic contribution from the enhanced probability of dipole/loop formation. Consequently, the balance between the enthalpy and entropy of a collection of dipoles/loops occurs at a lower temperature in the presence of a stress than in the case of an unloaded crystal. If the applied stress is large enough such that spontaneous unstable dissociation/expansion of a single dipole/loop becomes possible at 0K, then the free energy is negative at all non-zero temperatures. Thus, the stress level corresponding to the limit of zero transition temperature is the critical stress for the homogenous nucleation of a single dislocation. In this limit, due to the absence of thermal fluctuations, there is no thermally-induced screening of dislocation interactions.

In the case of dislocation loops in 3D, the self energy is proportional to rlnr (as opposed to lnr in the case of dipoles) and it is not immediately apparent that the entropy, which is purely a logarithmic function, can become equal to the self energy of a collection of loops. However, it turns out that the plastic polarizability and the entropy of the loops is significant to cause a cooperative instability of a collection of interacting loops even in an unloaded crystal [3]. For the sake of simplicity, we restrict our attention to 2D and give a brief quantitative description of the cooperative dissociation of dipoles in a loaded solid at finite temperatures. The details are described in [20, 21].

COOPERATIVE DISSOCIATION OF DIPOLES IN A 2D LOADED SOLID

Consider a loaded elastic solid such that a shear stress σ_{xy} acts on a potential slip plane. The self-energy of a dipole (composed of dislocations with Burgers vectors $b = \pm be_x$ where e_x is the unit vector in the x-direction) of size r in the presence of the stress is given by

$$E^d(r) = \frac{K_0 b^2}{4\pi}\left\{\ln\frac{|r|}{r_0} - \frac{1}{2}\cos 2\theta\right\} + 2E_c - \sigma br\cos\theta, \quad K_0 = \frac{4\mu_0 B_0}{(\mu_0 + B_0)} = \frac{2\mu_0}{(1-v)} \quad (2)$$

where μ_0, B_0 are the shear and bulk modulus, respectively, defined in the absence of dislocations, v represents the Poisson ratio, θ is the angle between b and r, r_0 is an inner elastic cut-off radius and E_c is the core energy of a dislocation.

At a fixed temperature, the probability of formation of a dipole depends only on its energy and thus dipoles of various sizes can be present in the solid. Consider a test dipole of size r formed in the presence of other dipoles. At any temperature, dipoles much larger than the test dipole have only a small probability of existence because their energies are appreciably higher. Thus, the dominant contribution to screening of dislocation interactions of a test dipole of size r comes from dipoles which are smaller than r. In effect, the plastic polarizability, ε, representing the screening is scale-dependent and is a function of the test dipole size r. The screened Hamiltonian of the test dipole can be written in the form

$$H_{scr}^d(r) = V(r) + 2E_c - \sigma br, \quad V(r) = \frac{K_0 b^2}{4\pi}\int_{r_0}^{r}\left(dr'/r'\,\varepsilon(r')\right) \quad (3)$$

In the above expression, we have neglected the angular interaction, $(K_0 b^2/8\pi)\cos(2\theta)$, of the dislocations comprising the dipole (see equation 2) and the angular variation in energy due to the orientation of the dipole with respect to the shear stress (i.e., put $\cos\theta = 1$ in Equation 2) in order to simplify the algebra. The polarizability can be expressed in terms of the plastic strain-strain correlation function of all possible dipole configurations of size less than r, taking into account the density of dipoles using a grand canonical ensemble. Using the standard tools of statistical mechanics, the following self-consistent integral equation for the scale-dependent polarizability, e(r), can be obtained after some algebra [20, 21, 28-32]:

$$\varepsilon(r) = 1 + 4\pi\int_{r_0}^{r}\int_{0}^{2\pi}\alpha(r')\,n(r',\theta)\,r'\,dr'\,d\theta,$$

$$\alpha(r) = \frac{1}{16\pi}K_0\,\beta\,b^2 r^2\,; \quad n(r,\theta) = \left(1/r_0^4\right)\exp\left[-\beta H_{scr}^d(r)\right]; \quad \beta = (k_B T)^{-1} \quad (4)$$

where $\alpha(r)$ is the polarization of a single dipole of separation r in the limit of zero stress as in the K-T theory and n(r) is the number of dipoles of separation r at finite σ and T. In Equation 4, the upper limit of integration for calculating the polarizability is set equal to r and represents the dominant contribution to the screening. The neglect of dipoles of size bigger than r does

---

[3] The analog of the K-T transition for dislocation loops in 3D solids in the absence of applied loads was developed by Lund et al. [38, 39].

not introduce large errors because ultimately the instability of a test dipole is calculated for very large r corresponding to the test dipole existing in the thermodynamic limit of an infinite system. This procedure is necessary in all statistical mechanical calculations because quantities such as the entropy are well-defined only in the thermodynamic limit [36, 37].

As described previously [20, 21], the behavior of ε in the thermodynamic limit can be examined by introducing the following transformations which enable the integral equation to be expressed as a set of coupled differential equations. Let us define

$$\ell = \ln(r/r_0) \quad ; \quad h(\ell) = \beta q^2 / \pi \varepsilon(r) , \quad y(\ell) = (r/r_0)^2 \exp((-2E_c - V(r) + \sigma br)\beta / 2) \quad (5)$$

The function $y(\ell)$ is proportional to the probability of dipoles of size ℓ at a given T and σ while $h(\ell)$ represents the screening due to dipoles of separations less than $r_0 \exp(\ell)$, where ℓ represents the basic length scale in the problem. Differentiating the expression for $\varepsilon(r)$, and using Equation (5), we obtain (to first order in y)

$$\frac{dh^{-1}(\ell)}{d\ell} = 4\pi^3 y^2 \quad ; \quad \frac{dy}{d\ell} = \left(2 - \pi h + \frac{\beta \sigma br}{2}\right) y \quad (6)$$

subject to the boundary conditions

$$h(\ell = 0) = \beta q^2 / \pi \quad ; \quad y(\ell = 0) = \exp\left[(-2E_c + \sigma br_0) \beta / 2\right] \quad (7)$$

The integral equation is thus transformed to a set of coupled non-linear equations which describe the scaling behavior of screened dislocation interactions.

The transition temperature (T_c) is the critical point (unstable fixed point) of the differential equations which demarcates the temperature regime above which $\varepsilon(r)$ diverges for a given value of σ. It can be evaluated in two ways.

First, the coupled system of non-linear equations (Equation 6) can be iterated successively starting from a given initial condition (equation 7) and this procedure corresponds to calculating $\varepsilon(r)$ for progressively larger systems. However, in the presence of an applied stress, the dipole Hamiltonian is divergent in the thermodynamic limit of an infinite system and it is this feature which causes the coupled system of equations to be non-autonomous. In order to distinguish the cooperative instability induced by screening from the inherent divergence of the dipole Hamiltonian, it is necessary to approximate the σbr term in all the above equations by $\sigma br'$ where r' is a constant. In [20, 21], r' was identified with r_0 where r_0 is the inner elastic cut-off radius. The repeated iterations then show that for all $T \le T_c$, $\varepsilon(r)$ tends to zero for large r while for $T > T_c$, $\varepsilon(r)$ diverges for $r \ge r_c$. Just above T_c, r_c is very large and approaches ∞, while for $T >> T_c$, r_c approaches atomic dimensions.

The second procedure for calculating T_c is to carry out a linear stability analysis around the critical point of the coupled system of equations. The critical point is given by the following implicit equation:

$$T_c = \frac{1}{2k_B} \left(\frac{K_0 b^2}{8\pi \varepsilon(r_c)} - \frac{\sigma br_0}{2} \right) \quad (8a)$$

The stress-dependent term in Equation 6 has again been approximated by σbr_0 to avoid the divergence of the dipole Hamiltonian in the thermodynamic limit and to make the set of coupled equations autonomous; r_c corresponds to the length scale ℓ_c at which the dislocation interactions become completely screened causing a change in the nature of trajectories. This behavior is shown in Figure 1 below. The linear stability analysis around the critical point [20, 21] enables one to write T_c as the solution of the following non-linear equation

$$(\beta_c)^{-1} = k_B T_c = \frac{\mu_0 b^3 \left(1 - 2\pi \exp\left(-\beta_c E_c b + \sigma b^2 \beta_c r_0/2\right)\right)}{8\pi(1-\nu)\left(1 + \sigma b^2 \beta_c r_0/4\right)} \qquad (8b)$$

The stresses, elastic moduli and all other parameters in the above equation have three-dimensional units so that material constants can be used directly in the evaluation of the transition temperature. This has been done by using the Burgers vector as a scaling parameter to express the 2D stresses and moduli in equations 6 and 7 with dimensions N/m in terms of the respective 3D quantities with dimensions N/m^2. (It is important to note that this procedure does not correspond to an analysis of the instability in 3D because the latter involves dislocation loops and not dipoles as we have considered above.) It can be verified that the values of T_c calculated using the two procedures are virtually identical.

The nature of the trajectories below and above T_c can be seen by solving the coupled differential equations numerically. Let us use the material constants of silicon for purposes of illustration and transform all quantities in equations 6 and 7 suitably to 3D units as indicated above. We set b = 0.218 nm (corresponding to the partial dislocation of type (1/6)<1 1 2> in Si), the elastic cut-off distance $r_0 = 0.3776$ nm (corresponding to the Burgers vector of the total dislocation (1/2)<1 1 0>, $\mu_0 = 60.5$ GPa (the shear modulus on the (111) slip plane), $\nu = 0.215$ (the Poisson ratio), $\sigma = 3$ GPa and $\sigma b r$ is set equal to $\sigma b r_0$ at all temperatures. The dislocation core energy is written arbitrarily in the form $E_c = (\mu b^2/2\pi(1-\nu))$ ln (b/ζ) where ζ is the dislocation core width which by definition is always less than the Burgers vector. Here, we choose $\zeta = 0.1$ nm.

Figure 1 shows the evolution of $h^{-1}(\ell)$ and $y(\ell)$ under repeated scale transformations corresponding to increasing values of ℓ (indicated by arrows). At T = 1100K and 1130K, $y(\ell)$ decreases continuously starting from the initial value and tends to zero for large ℓ while $h^{-1}(\ell)$ increases and approaches an asymptotic value. This behavior persists until T = 1164 K and the probability of dipoles of very large separations is virtually zero even in an infinite system.

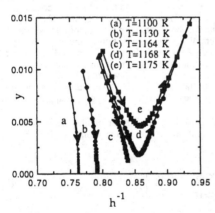

Figure 1. Variation of $y(\ell)$ versus $h^{-1}(\ell)$ for five different initial conditions.

Starting at a temperature a few degrees higher, T = 1168K, the qualitative behavior is completely different. $y(\ell)$ decreases initially as a function of ℓ, reaches a minimum and then $y(\ell)$ increases rapidly to unity. (Figure 1 only shows the increase over a small range of y in order to see the behavior near the critical trajectory clearly.) $h^{-1}(\ell)$ increases gradually at the beginning but as ℓ increases beyond a certain value corresponding to $r=r_c$, $h^{-1}(\ell)$ diverges which implies that for large ℓ or r, $\varepsilon(r)$ is divergent. At still higher temperatures, the approach of $y(\ell)$ to unity and the corresponding divergence of $h^{-1}(\ell)$ is more rapid. For the given choice of parameters, the T=1164 K trajectory corresponds to the critical 'separatrix' of the system of equations.

The separatrix demarcates the following two temperature regimes in the thermodynamic limit of an infinite system: The temperature up to T_c, where $\varepsilon(r)$ tends to unity and the probability of dipoles of infinite separation (namely, dissociated dipoles) is zero, and the temperature above T_c where $\varepsilon(r)$ tends to infinity and the probability of dissociated dipoles is unity.

In the case of a crystal containing a loaded crack, the stress varies spatially as opposed to the uniformly loaded crystal considered so far. However, due to the relatively slow ($1/\sqrt{r}$)

variation of a crack stress field, the expression for T_c obtained above can be considered a reasonable approximation for cooperative dislocation generation in a macroscopic region around the crack tip. The massive dislocation activity that ensues above T_c (see Figure 1) effectively blunts the crack and prevents its propagation above this temperature. The dislocations generated by the cooperative process are capable of extensive glide and thus help dissipate large amounts of energy which can result in a dramatic increase in the toughness. The prediction of BDTT (using the constant stress approximation) for other material parameters is summarized in Table 1.

TABLE 1. The BDTT for different materials

El	μ_{slip} (GPa)	v	σ (GPa)	b (nm)	ζ (nm)	r_0 (nm)	T_c (K)
Si	60.5	0.215	1.0	0.218	0.10	0.378	1679
			2.0	0.218	0.10	0.378	1421
			3.0	0.218	0.10	0.378	1164
			3.0	0.218	0.06	0.378	1319
TiAl	70.0	0.26	3.0	0.224	0.15	0.224	1259

For TiAl, the Burgers vector corresponds to (1/2)<110> and the shear modulus corresponds to the value on the cube plane. The elastic inner cut-off radius is chosen as the Burgers vector of a lattice dislocation in all cases. The calculated values of T_c are in reasonable agreement with observations. The BDTT for TiAl is approximately 700°C [40].

In summary, the mechanism that gives rise to the cooperative instability is general and applicable to all solids. The starting point in the cooperative process is the formation of atomic-size dipoles/loops by spontaneous fluctuations at finite temperatures. This is not in principal different from the starting configuration in the models of BDT based on thermally activated dislocation generation [23, 25-27]. The difference between the present approach and the other models lies in the subsequent evolution of the dipoles/loops until their unstable dissociation/expansion. While the present model considers the evolution of *many* dipoles/loops at finite temperatures and takes into account the statistical mechanics based coupling between fluctuations and response, the independent evolution of *individual* dipoles/loops is considered via a thermally activated process in the earlier models. The latter process is always possible and can also occur in the cooperative instability model. However, a thermally-driven instability becomes probable under certain conditions when dipoles/loops which are only slightly bigger than atomic dimensions can dissociate in a cooperative manner without reaching the critical sizes needed for single non-interacting dipoles/loops to become unstable under similar conditions of applied loads. This is primarily due to the screening of interactions in a system containing many dipoles/loops as opposed to the behavior of a single non-interacting dipole or loop. If the temperature or stress is such that a complete screening of interactions does not occur, as in the range below T_c, then the collective instability cannot take place and independent thermally activated process of dissociation/expansion is the only available alternative for dislocation generation near the crack tip. This process is expected to give a smooth increase in the toughness with increasing temperature but cannot give rise to sudden massive ductility in a narrow temperature range. The latter implies a very large increase in the density of mobile dislocations. The cooperative Kosterlitz-Thouless mechanism is one possible way in which a large increase in dislocation density can suddenly occur above a certain temperature.

DISLOCATION DYNAMICS IN THE VICINITY OF A LOADED CRACK

We have so far considered a static description of the BDT. As discussed in the Introduction, there is an unambiguous evidence that the strain-rate dependence of the transition temperature is of an Arrhenius form similar to the variation of dislocation mobility with temperature [4, 6, 17]. This characteristic correlation, observed irrespective of the magnitude of dislocation mobility, can be explained in the following manner [22]: We first demonstrate that dislocation dynamics in the vicinity of a loaded crack is of a "similarity" form when the mobility of dislocations is of the power-law Arrhenius type given in Equation 1b. This special type of dynamics exists under very general conditions and is applicable to a number of situations in which the crack stress field dominates. If we assume the same dynamical solution for the evolution of dipoles in the vicinity of a crack, then substituting the similarity form in the criterion for the cooperative instability provides a straightforward explanation for the observed strain-rate dependence. In particular, it shows why the activation energy associated with the strain-rate dependence of T_c is almost equal to the activation energy for dislocation motion in many materials.

Consider the dynamics of a group of interacting dislocations in the vicinity of a Mode III loaded crack. The stress acting on a dislocation with the Burgers vector b at the position x_i is [41]

$$\tau(x_i) = \frac{K}{\sqrt{2\pi x_i}} - \frac{\mu b}{4\pi x_i} + \frac{\mu b}{2\pi} \sum_{j \neq i} \left(\frac{x_j}{x_i}\right)^{\frac{1}{2}} \frac{1}{(x_i - x_j)}, \tag{9}$$

where K is the crack tip stress intensity factor and μ, the shear modulus on the slip plane. The terms on the right represent respectively, the stress due to the crack, the contribution of the 'image' dislocation associated with the dislocation at x_i and the stress induced by other dislocations and their images (the image dislocations ensure traction-free boundary conditions on the crack surface). The image and dislocation interaction terms are important when the dislocations are close to the crack tip while the crack stress field (represented by the first term on the right) dominates at large distances. Let us assume that dislocations are positioned initially such that the crack tip is fully shielded. The resulting non-singular stress field is of the form

$$\tau(x_i) = \frac{\mu b}{4\pi x_i} + \frac{\mu b}{2\pi} \sum_{j \neq i} \left(\frac{x_i}{x_j}\right)^{\frac{1}{2}} \frac{1}{(x_i - x_j)} \tag{10}$$

The complete shielding implies the following relation between x_i's and K:

$$K = \sum_j \frac{\mu b}{\sqrt{2\pi x_j}} \tag{11}$$

Let us assume that the velocity of each dislocation is given by the empirical power law dislocation mobility given in Equation 1b. Substituting Equation 10 for τ in Equation 1b we obtain the following equations of motion:

$$\left(\frac{1}{v_0} \frac{dx_i}{dt}\right)^{1/m} = \tau(x_i) = \frac{\mu b}{4\pi x_i} + \frac{\mu b}{2\pi} \sum_{j \neq i} \left(\frac{x_i}{x_j}\right)^{\frac{1}{2}} \frac{1}{(x_i - x_j)} \tag{12}$$

Here v_0 is proportional to $(\exp(-U_m/k_B T)$ where U_m is the activation energy for dislocation motion. The above equations have simple "similarity" solutions of the form

$$x_i(t) = g(t) X_i, \tag{13}$$

where g(t) is a dimensionless function which describes the time dependence and is the same for all dislocations, and X_i is independent of time and depends only on the position of all the

58

dislocations [4]. The function $g(t)$ can be determined by solving a first order differential equation obtained by substituting the above solution in Equation 12. In the case of a constant applied stress intensity (K), the similarity solutions are of the form

$$g(t) = \left\{(m+1)\,\alpha^m\,v_0 t\right\}^{1/(m+1)} + c \qquad (K = \text{constant}) \qquad (14)$$

where the constant of integration can be chosen from the initial position and α^m is a constant with dimensions μ^m/b. When the applied stress intensity is not a constant but varies linearly with time it can be shown that dislocation dynamics is again of the similarity form given by

$$g(t) = \left\{\alpha^m\,v_0 t\right\}^{2(m+1)/(m+2)} + c \qquad (K = \beta t) \qquad (15)$$

with only the power law exponent being different from equation 14. This solution can be derived in a straightforward manner by expressing μb in Equation 12 in terms of $K(t)$ using the shielding condition given in Equation 11.

It can be shown that similarity solutions exist even when dislocation motion is considered only under the influence of the crack stress field. When the applied stress intensity varies linearly with time, one can show that the similarity solution is of the same form as in equation 15. The latter represents the solution for dislocations moving in a non-singular stress field (see equation 10). It is not surprising that the same type of solution is obtained for the singular crack stress field because the two fields are almost indistinguishable far away from the crack tip. The dislocation dynamics described by Equations 1b and 9 were simulated by Hirsch et al. [24] in the context of a dislocation mobility based model for the BDT. It was found that when the dislocations move away from the tip, the dynamics is of the similarity form given in equation 15.

The similarity solutions characterize an important feature of dislocation dynamics in the vicinity of a loaded crack. The solutions show that the dynamics is of the same form for all dislocations irrespective of their precise positions from the crack tip. This 'similarity' dynamics is established on a time scale which is large enough that transient effects can be ignored. The transients are important only when dislocations are very close to the crack tip which is the case for small time scales.

APPLICATION OF SIMILARITY DYNAMICS TO THE STRAIN-RATE DEPENDENCE

Several models have been proposed [7, 11, 13, 24, 42] which consider dislocation mobility to be the rate-controlling aspect of the BDT while dislocation nucleation in the vicinity of the crack is believed to be a relatively easy process. These models were originally developed to explain the BDT in Si and in particular the strain-rate dependence of T_c. The models involve numerical simulation of dislocation dynamics in the vicinity of the crack tip assuming the empirical thermally-activated power-law form for the dislocation mobility (equation 1b). The results obtained by numerical simulation clearly display the similarity nature of dislocation dynamics after the initial transient dynamics [11, 24]. A quantitative comparison of the BDTT predicted by the mobility-based models and the present cooperative approach is however not possible. The former descriptions rely on the precise positions of initial dislocation sources, the crack geometry and the dislocation emission criterion used in the model while the present model is phenomenological and only depends on the parameters that define the energy of a dipole in a linear elastic description. The dislocation mobility based models can generally reproduce the observed strain-rate dependence of the BDTT. We now

---

[4] When dislocations begin to move, new dislocations need to be generated in order to continuously shield the crack. The new dislocations also exhibit similarity motion of the same type as the existing ones and therefore we do not discuss the generation aspect explicitly.

show how the qualitative strain-rate dependence can be deduced analytically using the cooperative instability model and the similarity nature of dislocation dynamics.

The cooperative instability involves the thermal and stress induced evolution of (atomic-size) bound dislocation dipoles as opposed to macroscopic mobile dislocations. We assume that the dynamical evolution of thermally induced dislocation dipoles in the vicinity of the crack tip can also be described by similarity solutions as in the case of mobile interacting dislocations [22]. This assumption may be questioned but it seems reasonable since the similarity solutions are very general and thus applicable to a wide variety of cases. Furthermore, all dislocations display the similarity dynamics irrespective of their position from the tip. In the cooperative instability model, the dipoles are formed in regions where the stress is considerably high, and thus, in the case of a crystal containing a loaded crack, dipoles are formed in the high stress region which extends from the tip to many atomic distances away. The average size of the dipole in the presence of a stress is larger than its value in the absence of applied loads and we assume that the stress-assisted dynamical size of all dipoles to be given by the similarity solutions. The instability occurs when many small dipoles are equivalent to a test dipole of size r_c that is much larger than atomic dimensions and for which dislocation interactions become completely screened (see equation 8a). In effect, we can regard the dynamical evolution of the test dipole from an initial size, r_0, to a critical size r_c at T_c to be described by similarity solutions of the form

$$r_c(t) = \left\{ \alpha \exp(-U_m/k_B T_c) t \right\}^\gamma \qquad (16)$$

where α and γ are functions of the velocity power-law exponent m. Equation 16 is analogous to the solutions in Equation 14 and 15 with the temperature T in v_0 being set equal to T_c. If r_c is approximately a constant for all strain rates, it is straightforward to deduce the correlation between the strain-rate dependence of T_c and the variation of dislocation mobility with temperature. In a constant $\dot{K}$ test, the time t can be written as $K/\dot{K}$ and taking logarithms on both sides of equation 16, we obtain

$$\ln(\dot{K}) = -\frac{U_m}{k_B T_c} + \text{constant} \qquad (17)$$

The strain-rate dependence of T_c thus exhibits an Arrhenius behavior with the same activation energy as for dislocation motion. It is not surprising that this correlation is observed in many materials because it is only based on the general nature of dislocation dynamics in a crack stress field.

CONCLUSIONS

A comprehensive analysis of the BDT has been presented in two parts. Until recently, the BDT has been described as a stress-induced phenomenon occurring in the vicinity of the crack tip and effects of temperature were introduced only via thermally activated process of either dislocation generation or dislocation motion. We have described a model in which the spontaneous generation of dislocations corresponds to a collective instability driven primarily by thermal fluctuations and assisted by the applied stress. Such a transition is caused by the thermally-induced screening of dislocation interactions above a certain critical temperature. The cooperative nature of the instability helps to explain the rapid increase in dislocation density observed suddenly at a certain temperature. While many details of the crack geometry that play a role in the BDT have been ignored in this simple 2D model, it is well known that even single crystals not containing a pre-existing crack are often brittle at low temperatures and undergo a transition to ductile behavior at high temperatures. The present approach gives a new perspective for studying the onset of ductile behavior not only in terms of crack tip activity but also by simply investigating the temperature dependence of the onset of plastic flow in crystals not containing cracks. Perhaps, a convincing proof of the applicability of the present model to the broader description of tensile ductility may be found in the measurements of yield and flow behavior as a function of temperature in whiskers. In the case of Si whiskers [43], it was found that the onset of plastic flow occurred suddenly above a certain temperature, completely analogous to the BDT observed in bulk single crystals with a crack.

The strain-rate dependence of the BDT is related to the similarity dislocation dynamics near the crack tip which occurs in many materials under general conditions of crack geometry. It is the combination of the similarity dynamics and the scale-invariance of dislocation interactions at a critical temperature that leads to a simple correlation between the strain-rate dependence of T_c and the temperature dependence of dislocation mobility. The present approach demonstrates that both dislocation generation and dislocation dynamics play important roles in the BDT but in distinct aspects of the problem.

ACKNOWLEDGMENTS

This research was supported by the U.S. Air Force Office of Scientific Research grant No. 95-1-0143.

REFERENCES

1. J. F. Knott, *Fundamentals of Fracture Mechanics (Revised)*, Butterworths: London (1979).
2. R. W. Hertzberg, *Deformation and Fracture Mechanics of Engineering Materials*, Wiley: New York (1989).
3. H. Vehoff, *Ordered Intermetallics - Physical Metallurgy and Mechanical Behavior* (edited by C. T. Liu, R. W. Cahn and G. Sauthoff), Kluwer, NATO ASI Series E: Vol. 213, p. 299 (1992).
4. H. Vehoff, *High-Temperature Ordered Intermetallic Alloys V* (edited by I. Baker, R. Darolia, J. D. Whittenberger and M. H. Yoo), Materials Research Society, Vol. 288, p.71 (1993).
5. R. Thomson, *Materials Science and Engineering A* **176**, 1 (1994).
6. C. St. John, *Philos. Mag.* **32**, 1193 (1975).
7. M. Brede and P. Haasen, *Acta Metall.* **36**, 2003 (1988).
8. J. Samuels and S. G. Roberts, *Proc. R. Soc. Lond.* A **421**, 1 (1989).
9. G. Michot and A. George, *Scr. Metall.* **20**, 1485 (1986).
10. M. Brede, K. J. Hsia and A. S. Argon, *J. Appl. Phys.* **70**, 758 (1991).
11. P. B. Hirsch and S. G. Roberts, *Philos. Mag. A* **64**, 55 (1991).
12. W. Zielinski, M. J. Lii and W. W. Gerberich, *Acta Metall. Mater.* **40**, 2861 (1992).
13. M. Brede, *Acta Metall. Mater.* **41**, 211 (1993).
14. A. George and G. Michot, *Materials Science and Engineering A* **164**, 118 (1993).
15. G. Michot, A. L. de Oliveira and A. George, *Materials Science and Engineering A* **176**, 99 (1994).
16. K. J. Hsia and A. S. Argon, *Mat. Sci. & Eng.* **A176**, 111 (1994).
17. N. Urabe and H. Ichinose, *Trans. ISIJ* **18**, 279 (1978).
18. U. F. Kocks, A. S. Argon and M. F. Ashby, *Progress in Materials Science* **19**, 1 (1975).
19. F. C. Serbena and S. G. Roberts, *Acta Metall. Mater.* **42**, 2505 (1994).
20. M. Khantha, D. P. Pope and V. Vitek, *Phys. Rev. Lett.* **73**, 684 (1994).
21. M. Khantha, D. P. Pope and V. Vitek, *Scr. Metall. Mater.* **31**, 1349 (1994).
22. M. Khantha, *Scr. Metall. Mater.* **31**, 1355 (1994).
23. J. R. Rice and R. Thomson, *Philos. Mag.* **29**, 73 (1974).
24. P. B. Hirsch, S. G. Roberts and J. Samuels, *Proc. R. Soc. Lond.* A **421**, 25 (1989).
25. G. Schoeck, *Philos. Mag. A* **63**, 111 (1991).
26. J. R. Rice, *J. Mech. Phys. Solids* **40**, 239 (1992).
27. J. R. Rice and G. E. Beltz, *J. Mech. Phys. Solids* **42**, 333 (1994).
28. J. M. Kosterlitz and D. J. Thouless, *J. Phys. C:Solid State Phys.* **6**, 1181 (1973).

29. D. R. Nelson, *Phys. Rev. B* **18**, 2318 (1976).
30. D. R. Nelson and B. I. Halperin, *Phys. Rev. B* **19**, 2457 (1979).
31. A. P. Young, *Phys. Rev. B* **19**, 1855 (1979).
32. A. P. Young, *NATO Advanced Study Institute on Ordering in Strongly Fluctuating Condensed Matter Systems* (edited by T. Riste), Plenum, p. 271 (1980).
33. G. A. Williams, *Phys. Rev. Lett.* **59**, 1926 (1987).
34. S. R. Shenoy, *Phys. Rev. B* **40**, 5056 (1989).
35. F. R. N. Nabarro, *Theory of Crystal Dislocations*, Dover: New York (1967).
36. L. E. Reichl, *A Modern Course in Statistical Physics*, Univ. of Texas Press: Austin (1980).
37. K. Lindenberg and B. J. West, *The Non-equilibrium Statistical Mechanics of Open and Closed Systems*, VCH Publishers: New York (1990).
38. F. Lund, A. Reisenegger and C. Utreras, *Phys. Rev. B* **41**, 155 (1990).
39. F. Lund, *Phys. Rev. Lett.* **69**, 3084 (1992).
40. H. A. Lipsitt, D. Shechtman and E. Schafrik, *Metall. Trans. A* **6A**, 1975 (1975).
41. R. Thomson, *Solid St. Phys.* **39**, 1 (1986).
42. V. R. Nitzsche and K. J. Hsia, *Mat. Sci. & Engg.* **A176**, 155 (1994).
43. G. L. Pearson, W. T. Read Jr. and W. L. Feldmann, *Acta Metall.* **5**, 181 (1957).

EFFECTS OF EXTERNALLY GENERATED DISLOCATIONS ON BRITTLENESS/DUCTILITY OF CRYSTALS

SINISA DJ. MESAROVIC
Harvard Univ., Div. of Appl. Sciences, Cambridge, MA 02138, mesarovic@husm.harvard.edu

ABSTRACT

Direct interactions of externally generated dislocations with a moving, non-emitting crack tip, have been investigated. Dislocations of the appropriate Burgers vector, initially residing in a strip ahead of the incoming crack tip, are funneled toward the tip, with their motion restricted to the slip plane. Once drawn into the vicinity of the tip, these dislocations inevitably cause local, atomic scale blunting of the tip. The resulting crack trapping process is modelled, resulting in the temperature, crack velocity, and dislocation density dependent, tip toughness.

INTRODUCTION

While the analyses of competition between dislocation nucleation at the tip and cleavage [1, 2], have been largely successful in explaining the intrinsic cleavability of crystals, the phenomenon of brittle to ductile transition with temperature has defied definite and general explanation. As temperature increases, the fracture mechanism of intrinsically cleavable crystals changes from cleavage to ductile mechanisms. Two basic types of transition have been observed: a sharp transition (Si [3]) and a gradual transition (Mo [4], W [5, 6]). Iron alloys may exhibit both types of transition [7-9], depending on experimental conditions and alloying elements.

The models which account for the effects of externally generated dislocations and their mobility generally do so through the continuum theories of plasticity [10-13]. The macroscopic toughness (far field energy release rate required to sustain crack growth), Γ_{far}, is a function of the tip toughness, Γ_{tip}, and the flow stress, σ_{flow}, which depends on the loading rate, i.e., on the crack tip velocity, v_{crack}, temperature, T, and the hardening level, symbolically represented by a scalar dislocation density, ρ:

$$\Gamma_{far} = f\left(\Gamma_{tip}, \sigma_{flow}(v_{crack}, T, \rho)\right) \tag{1}$$

Effects of inertia may be formally included by insisting on explicit dependence of f on v_{crack}. The effect of externally generated dislocations is seen through the difference $(\Gamma_{far} - \Gamma_{tip})$ and is due to the dissipation through dislocation motion and storage. Γ_{tip} is often tacitly assumed to be the Griffith, surface energy based, toughness, 2γ. f is an increasing function of Γ_{tip} and a decreasing function of σ_{flow}. The usual conclusion is that predeformation embrittles the crystal.

There has been a number of reports in the literature [e.g., 10, 14] about toughening by prior deformation. The Liu and Shen [6] experiments on dynamic cleavage of tungsten single crystals seem to be the only ones with clear interpretation. For the same impact energy, their predeformed specimens show lower crack velocities, and higher dynamic toughness, contrary to the predictions of continuum models (1). For the same specimen, the toughness is higher at lower velocities by an amount not explainable by inertial effects and this effect is stronger for predeformed specimens. The river lines on the fracture surfaces generally become apparent at lower velocities and appear to be denser for the predeformed specimen. The trend in etch pit density on fracture surfaces was observed, showing an increase in density for lower velocities. Hull and Beardmore [15] also analyzed the fracture surfaces of tungsten single crystals and found that the density of river lines increased with temperature and measured (far field) toughness.

Mat. Res. Soc. Symp. Proc. Vol. 409 © 1996 Materials Research Society

A moving crack strongly attracts certain types of dislocations. Once drawn into the vicinity of the tip, these dislocations inevitably cause local, atomic scale blunting of the tip, either by direct absorption or by nucleation of another dislocation. Crack trapping by local blunting is modelled phenomenologically, resulting in an increase of the effective tip toughness:

$$\Gamma_{tip} = g(2\gamma, v_{crack}, T, \rho) \tag{2}$$

The energy dissipation, $(\Gamma_{tip} - 2\gamma)$ is the result of the trapping mechanism which requires increase in the rate of energy supply for breaking or bypassing the obstacles created by local blunting. In that sense, the model developed here serves as a temperature and crack velocity dependent input to the above mentioned continuum models (1). The effective tip toughness, Γ_{tip}, in (2), is an increasing function of dislocation density, so that from (1) and (2), one cannot conclude, in general, whether the far field toughness, Γ_{far}, will increase or decrease when dislocation density is increased.

BEHAVIOR OF DISLOCATIONS IN THE PRESENCE OF A MOVING CRACK

Two-dimensional geometry of a bcc crystal is shown in *Fig. 1*. Consider the slip system denoted "A": Dislocations denoted *positive* are those which would be nucleated at the tip and emitted in the upper half plane, if the tip was emitting. The same sign convention is adopted for both, Burgers vectors and dislocation velocities.

A quasi-statically moving crack is considered and the motion of dislocations is limited to glide. Dislocation-dislocation interactions are neglected and only the effect of the moving crack on each individual dislocation is considered.

The resolved shear stress, at a position (r, θ) in polar coordinates, measured from the tip of the crack propagating with the tip energy release rate, $G_{tip} = \Gamma_{tip}$, is given by [16]:

$$\frac{\tau}{\mu} = sign(b) \cdot \sqrt{\frac{\Gamma_{tip}}{\mu|b|4\pi(1-v)}} \cdot \sqrt{\frac{|b|}{r}} \cdot F_k(\theta, \alpha) - \frac{1}{2\pi(1-v)} \cdot \frac{|b|}{r} \cdot F_i(\theta, \alpha) \tag{3}$$

where b is the Burgers vector (in this case, scalar with a sign), μ and v are shear modulus and Poisson ratio, and F_k and F_i are non-dimensional functions of α and θ. The first term is due to the crack tip field and the second is the image force term due to the presence of free surfaces.

The dislocation velocity - stress law is governed by two physical mechanisms [17]. For low stresses/velocities the motion takes place by thermally activated glide over Peierls barriers, while at high stresses/velocities, the phonon and electron drag dominate:

$$\tau = \tau_{Peierls}(v, T) + \tau_{drag}(v, T) \tag{4}$$

Simple and convenient expressions are chosen, which represent the basic features of dislocation behavior. The material parameters used are those for α–iron [17,18]. The resulting velocity-stress law [19] is plotted in *Fig. 2* for several temperatures.

The equations of motion of a single dislocation, confined to its glide plane, written in the moving coordinate system (x, y) attached to the crack tip, are

$$\dot{x} = v(\tau(x, y)) \cdot \cos\alpha - v_{crack} \quad \text{and} \quad \dot{y} = v(\tau(x, y)) \cdot \sin\alpha \tag{5}$$

The results of numerical integration are shown in *Fig. 3*. All the considered dislocations, which

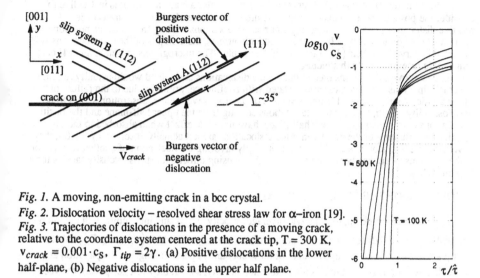

Fig. 1. A moving, non-emitting crack in a bcc crystal.

Fig. 2. Dislocation velocity – resolved shear stress law for α–iron [19].

Fig. 3. Trajectories of dislocations in the presence of a moving crack, relative to the coordinate system centered at the crack tip, T = 300 K, $v_{crack} = 0.001 \cdot c_S$, $\Gamma_{tip} = 2\gamma$. (a) Positive dislocations in the lower half-plane, (b) Negative dislocations in the upper half plane.

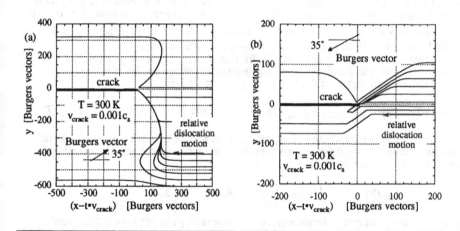

initially reside in a strip ahead of the crack, are funneled towards the crack tip as it approaches. The width, $d(T, v_{crack}, \Gamma_{tip})$, of this influence zone depends on the crack velocity, dislocation mobility (i.e., temperature) and the tip toughness. At relevant distances, the contribution of the image force is negligible compared to the crack tip field and hence, the distance d scales with the square root of the tip toughness (cf. 3):

$$d(T, v_{crack}, \Gamma_{tip}) = d_0(T, v_{crack})\sqrt{\Gamma_{tip}/2\gamma} \; ; \quad d_0(T, v_{crack}) = d(T, v_{crack}, 2\gamma) \qquad (6)$$

The function $d_0(T, v_{crack})$ is shown in Fig. 4. Calculations are given in [19].

Since many experimental observations are made after a crack arrests, it is instructive to consider the possible near-tip dislocation structure for that case. In *Fig. 3(a)*, trajectories of positive dislocations coalesce into a single curve. Dislocations starting from various positions are simultaneously on that curve. If the crack suddenly arrests, the etch pits of these dislocations will outline this curve, and their {112} traces will be visible. The micrographs in Tetelman [14] bear remarkable resemblance to this picture.

The above calculations predict that all the dislocations considered would actually run into the crack tip. This, however, cannot be stated with confidence, primarily due to the failure of linear elasticity within several Burgers vectors of the tip. Nevertheless, absorption seems to be a real possibility. Both, the negative dislocations coming from the upper half-plane and the positive dislocations coming from the lower half-plane, have the same effect when absorbed to the tip: atomic scale blunting of the tip. Even if the dislocations are not actually absorbed by the tip, they nevertheless cause atomic scale blunting by strongly favoring the nucleation of another dislocation at the tip. This effect has been quantified in [19], by using directional stress intensity factors for a slip plane [2].

MODELLING OF THE CRACK TRAPPING

Consider a crack tip, moving with average velocity v_{crack} and interacting with incoming externally generated dislocations. Dislocations from the strip of width $d(T, v_{crack}, \Gamma_{tip})$ ahead of the tip are relevant. The average number of these encounters per unit length of propagation at a point along the tip is ρd, where ρ is the dislocation density, so that the average spacing between arrests at a point along the tip is $1/\rho d$. These events are local since dislocations usually appear in loops or in segments pinned at the ends. The average length of an arrested segment of the crack tip may be taken to be proportional to the average dislocation spacing: $\ell = 1/(\psi\sqrt{\rho})$, where ψ varies between 0.1 and 10 depending on the details of dislocation arrangements.

At any time during the propagation, the fraction of arrested parts, Φ, is given by the product of the average length of an arrested segment and the average number of arrests per unit length of propagation, so that, recalling the scaling relation for d (6):

$$\Phi = \ell\rho d = \left(\sqrt{\rho}/\psi\right)\cdot d_o\left(T, v_{crack}\right)\cdot\sqrt{\Gamma_{tip}/2\gamma}, \quad \text{if } \ell\rho d < 1 \text{ and } \Phi = 1, \text{ if } \ell\rho d \geq 1. \tag{7}$$

Crack trapping effects have been modelled mostly for applications to composite materials. Mower and Argon [20] found that the best fit to experimental data was obtained by a combination of models available in literature:

$$\Gamma_{tip}/2\gamma = 1 - \Phi\cdot\left(1 - (A + B\Phi)^2\right), \quad A = 2.1, \quad B = 2.4 \tag{8}$$

This relation was developed under assumption of impenetrable particles (obstacles) and hence, depends only on the fraction of the arrested tip. Using (7) and (8) one computes the toughening ratio, shown in *Fig. 5*. The tip toughness, Γ_{tip}, depends strongly on dislocation density, and, at higher temperatures, on the crack velocity. The strong temperature dependence of Γ_{tip} at the velocity dependent transition temperature, is preceded by a slow increase in Γ_{tip} with temperature, over about 200 K.

DISCUSSION

The scenarios for gradual and abrupt brittle to ductile transition are discussed in [19] within a framework based on the present model for tip toughness, and a continuum plasticity model [12] for the functional relation between the far field (measured) toughness, and the tip toughness.

Fig. 4. The width of the influence zone, d, for $\Gamma_{tip} = 2\gamma$.

Fig. 5. The tip toughening ratio as a function of crack velocity and temperature, $\rho/\psi^2 = 10^{12}$ m^{-2}.

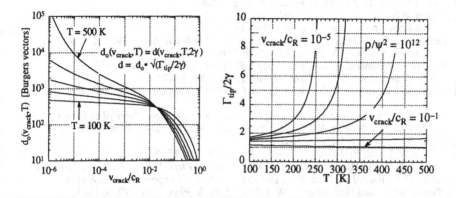

Generally, the conditions which increase σ_{flow}, such as high crack velocities, precipitation and solution hardening, and neutron irradiation, would favor abrupt transitions.

At 77 K, iron and its alloys exhibit a toughness, Γ_{far}, which is 2-3 orders of magnitude higher than the Griffith value, 2γ. For example, $\Gamma_{far}/2\gamma \approx 800$ in iron-silicon [8]. Tungsten, at this temperature, exhibits $\Gamma_{far}/2\gamma \approx 1.4$ [5]. In [19], the Beltz *et al.* [12] continuum plasticity results, based on the elastic enclave concept, are used to examine whether this enormous difference in low temperature behavior in iron-silicon and tungsten could be explained by the continuum plastic dissipation in (1). It is shown that a significant increase in Γ_{tip} (over 2γ) in iron-silicon is necessary to explain this behavior, in agreement with the Curry and Knott [21] estimate of Γ_{tip} in mild steel. It is then argued that the frequency of dislocation - crack tip interactions is higher in iron-silicon than in tungsten, i.e., that the present model describes the mechanism responsible for the high toughness in iron-silicon at low temperatures.

The crack trapping model is based on a number of assumptions. It is incomplete, in the sense that for any quantitative prediction, it must be supplemented by a continuum plasticity model, such as [12]. Nevertheless, the model demonstrates that the toughness of a crystal, and hence the nature of brittle to ductile transition, must be regarded as the result of an interplay between processes occurring at two length scales: the dislocation – crack tip interaction scale, and, the continuum plasticity scale.

The river lines on fracture surfaces of metallic single crystals become stronger and denser at lower velocities, higher temperatures and higher measured toughnesses. Motivated by their observations of fracture surfaces of tungsten single crystals, Hull and Beardmore [15] concluded that the river lines are the result of "plastic relaxation around the tip of a propagating crack" and that "in most metals such relaxation is difficult to avoid even in fractures at very low temperatures." It is possible that a careful examination of the topology of these markings, as well as the correlations between their density and experimental conditions, may provide an unambiguous explanation of their origin.

ACKNOWLEDGEMENTS: This work was supported by the University Research Initiative Program, based on ONR/DARPA grant, and by the Office of Naval Research. The author gratefully acknowledges discussions with J. R. Rice, A. S. Argon, R. Thomson and J.-S. Wang.

REFERENCES

[1] Rice, J.R. and Thomson, R.M. (1974) *Phil. Mag.* **29**, pp. 73-97.

[2] Rice, J.R. (1992) *J.Mech. Phys. Solids* **40**, pp. 239-271.

[3] Hirsch, P.B. (1995) In *Plastic Deformation of Ceramics*. Bradt, R.C., Brookes, C.A., and Routbert, J.L., Eds. Plenum, New York.

[4] Beardmore, P. and Hull, D. (1967) In *Refractory Metals and Alloys IV*, AIME Symp., Amer. Inst. Mining & Metall. Eng., New York.

[5] Hull, D., Beardmore, P., and Valentine, A.P. (1965) *Phil. Mag.* **12**, p. 1021.

[6] Liu, J.M. and Shen, B.W. (1984) *Metall. Trans.* **15A**, p. 1247.

[7] Pickington, R. and Hull, D. (1968) In *Conf. on Fract. Toughness*, Iron and Steel Inst. **20**, p. 5.

[8] Wang, J.S. (1994) *Proc. 5th Int. Conf. in Hydrogen Effects on Material Behavior*, Moran, WY. Moody, W.R., and Thompson, A.W., Eds., TMS, Warrendale, PA.

[9] Leslie, W.C. (1972) *Metall. Trans.* **3**, p. 5.

[10] Ashby, M.F. and Embury, J.D. (1985) *Scripta Metall.*, **19**, 557-562.

[11] Freund, L.B. and Hutchinson, J.W. (1985) *J. Mech. Phys. Solids* **33** (2), pp. 169-191.

[12] Beltz, G.E., Rice, J.R., Shih, C.F., and Xia, L. (1995) To appear in *Acta Metall.*

[13] Tvergaard, V. and Hutchinson, J.W. (1992) *J.Mech. Phys. Solids* **40**, p. 1377.

[14] Tetelman, A.S. (1964) *Acta Metall.* **12**, p. 993.

[15] Hull, D., and Beardmore, P. (1966) *Int. J. Fract. Mech.* **2**(1), pp. 468-486.

[16] Lin, I.-H., and Thomson, R. (1986) *Acta Metall.* **34** (5), pp. 187-206.

[17] Frost, H.J. and Ashby, M.F. (1982) *Deformation-Mechanism Maps*, Pergamon Press, New York.

[18] Hirth J.P. and Lothe, J. (1982) *Theory of Dislocations*, John Wiley & Sons, New York.

[19] Mesarovic, S.Dj. (1996) Submitted to *J.Mech. Phys. Solids.*

[20] Mower, T.M. and Argon, A.S. (1995) *Mechanics of Materials* **19**, pp. 343-364.

[21] Curry, D.A. and Knott, J.F. (1978) *Metal Science* **12**, pp. 511-514.

Stress-assisted Kosterlitz-Thouless Dislocation Nucleation: the Statistical Physics of Plastic Flow

Robin L. Blumberg Selinger

Materials Science and Engineering Laboratory, National Institute of Standards and Technology, Gaithersburg, MD 20899; Dept. of Physics, Catholic University of America, Washington, DC 20064-0001

The stress-assisted Kosterlitz-Thouless (K-T) transition was proposed by Khantha, Pope and Vitek [1] as a possible mechanism for the abrupt transition from brittle to ductile behavior of single crystals as a function of temperature. We argue that in a two-dimensional crystal under elevated temperature and applied shear stress, the microscopic mechanism of the stress-assisted K-T transition is the nucleation of dislocation dipole pairs with a preferred Burgers vector orientation. This defect population gives rise to an anisotropic hexatic phase with quasi-long range order along one axis and short-range order along the orthogonal axis. The laboratory signature of such a phase is scattering peaks that are "streaked," i.e. strongly broadened along a single direction. We discuss experimental data for an analogous two-dimensional system which displays order of this type. Potential implications for the brittle-ductile transition in three dimensions are outlined.

In 1994, Khantha, Pope and Vitek [1] first proposed that the transition from brittle to ductile behavior in crystals under increasing temperature is due to the cooperative nucleation of dislocations via a stress-assisted Kosterlitz-Thouless (K-T) transition. Their motivation to pursue such a theory was the observation that thermal activation alone cannot account for the extremely sharp temperature dependence of the toughness of single crystal silicon. Through a renormalization group calculation they demonstrated that the application of an applied stress to a two dimensional crystal reduces the K-T transition temperature. They went on to conjecture that in three-dimensions, this same effect could lead to the cooperative nucleation of dislocation loops in a region of elevated stress near a crack tip, at temperatures well below the normal melting temperature. This conjecture has been controversial in both the statistical physics and materials communities, in part because it has been commonly believed that the K-T transition in three dimensions is pre-empted by first-order melting.

According to the Kosterlitz-Thouless-Halperin-Nelson-Young [4] theory of second-order melting, a two dimensional crystal under increasing temperature goes through a second-order melting transition in which pairs of dislocation dipoles of all orientations proliferate, destroying quasi-long range translational order but preserving long-range orientational order. This peculiar phase–not crystalline, but more ordered than a liquid–has been dubbed a hexatic. If the hexatic phase is heated, at a yet higher temperature first-order melting occurs via the nucleation of disclinations, and long-range orientational order is destroyed, leaving an ordinary liquid phase. Crystal-hexatic-liquid phase transitions have been observed in both quasi-two-dimensional experimental systems and in computer simulations of two-dimensional materials.

Khantha et al [1] explored the question of how an externally imposed stress on a two-dimensional crystal would affect the 2nd-order melting transition. They determined that the result would be that the transition temperature T_{K-T} would be reduced by the application of stress.

Here we explore the question of the nature of the resulting phase; in addition to changing the transition temperature, the applied stress can also change the nature of the phase transition itself. This is a question that was not so far addressed by Khantha et al. Our proposal is that a non-hydrostatic applied stress will break the symmetry in the energy of dipole pairs as a function of the orientation of the Burgers vector. Dipole pairs with the lower energy orientation will be nucleated by the K-T mechanism at a somewhat lower transition temperature. Slightly above that transition temperature the defect population will be dominated by these low-energy dipole pairs.

This defect population is not consistent with an isotropic hexatic phase of the kind discussed above. A hexatic has dislocations of all possible orientations which destroy quasi-long-range translational order in all directions. If dislocations with one Burgers vector orientation predominate, then they will destroy quasi-long-range order *along one axis only*. The resulting phase will have short-range translational order along this axis while retaining quasi-long-range order along the orthogonal axis. This phase could be called an "anisotropic hexatic" and would be characterized by two distinct correlation lengths. Alternatively it could be called a "two-dimensional smectic" in analogy to a liquid crystal phase of similar symmetry. Such a phase was predicted already in another context by Ostlund and Halperin [5]. They discussed the nature of the K-T transition in the case that the elastic moduli of a two-dimensional crystal are anisotropic.

In Figs. 1 and 2 we present an elementary demonstration of the way dislocations of restricted Burgers vector orientation alter translational order. In Fig. 1(a) we show a triangular lattice with five dislocation pairs included, all with their Burgers vector parallel to the y axis. Fig. 1(c) shows that if the picture in 1(a) is distorted via elongation along the x axis, one may see that the crystalline rows are uninterrupted by the dislocations, so that translational order in the x direction is to lowest order unaffected. Fig. 1(c) shows the same picture distorted via elongation along the y axis, allowing one to see that the diagonal rows are interrupted by the presence of dislocations, interrupting the coherency of the lattice in the y-direction.

Figure 2 shows the Fourier transform intensity as a function of k_x, k_y for the crystal shown in Fig. 1(a). The peaks are somewhat broadened in both x and y directions due to the effects of the finite size of the crystalline sample. However the peaks are broadened much more dramatically along the y direction. This "streaking" of the crystal peaks along one special direction is produced by the dislocations and is in fact the signature of the anisotropic hexatic phase. The widths of the peaks along the x and y axes are proportional to $1/\lambda_x$ and $1/\lambda_y$, respectively, where the latter are the two correlation lengths that characterize the anisotropic hexatic phase. In this case it appears that λ_x is large while λ_y is small.

A crystal containing dislocations may flow under an applied stress. The visco-elastic response of the 2-d solid under stress will depend on the nucleation rate for dislocations and other potential defects such as voids, and on the mobility of defects once they are nucleated. A theory based on such ideas is problematic to formulate for several reasons. First, the dependence of dislocation mobility on applied stress and temperature is not well known.

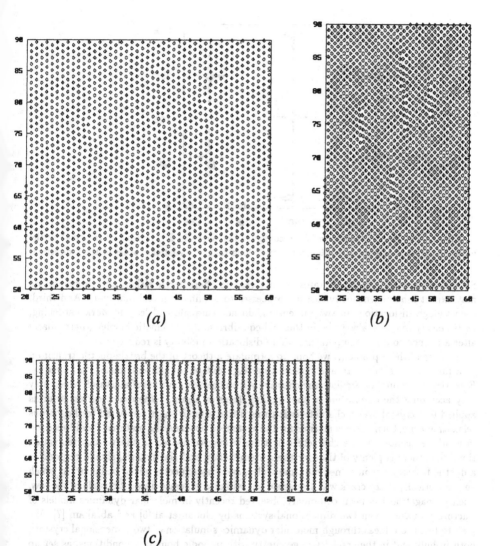

FIG. 1. (a) A triangular lattice with five dislocation pairs, all with Burgers vector parallel to the y-axis. Their effect on coherence of the lattice can be more easily seen if the lattice is elongated (b) along the y-direction, or (c) along the x-direction.

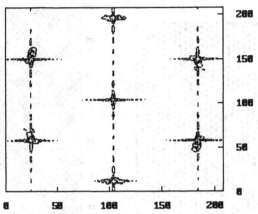

FIG. 2. The Fourier transform of the configuration shown in Figure 1(a). "Streaking" of the peaks along the y-axis indicates the reduction of long-range translational order in the y-direction.

Second, dislocation nucleation rates are not easy to characterize either, and in general we may expect inhomogeneous nucleation processes to dominate in real materials. And third, even though dislocations in two dimensions do not entangle and lead to work hardening, their density may not always be in thermal equilibrium due to kinetic barriers, particularly after a quench to a low temperature where dislocation mobility is reduced.

In spite of these problems we hope to formulate a theory of the brittle-ductile transition as a function of strain rate and pre-existing (e.g. out of equilibrium) dislocation density. Essentially we hope to calculate the effective viscous relaxation time of a crystal, which may vary according the orientation of the applied stress or strain. Where a stress or strain is applied to a crystal over a characteristic time, if that time is short compared to the viscous relaxation time, then there is no time for the system to flow and the material should respond in a brittle fashion. If the characteristic time is long compared to the viscous relaxation time, then there is plenty of time for the system to flow and the material should respond in a ductile fashion. For intermediate time scales, one may expect a mixed behavior, in which an atomistically sharp crack will propagate but will also generate dislocations along the way. Such propagation has been commonly observed recently in molecular dynamics models of fracture propagation in two-dimensional systems by Holian et al [6] and abraham [7]. We plan to test these ideas through molecular dynamics simulation of two dimensional crystals, both in bulk and in the crack/strip geometry with periodic boundary conditions, as set up by Peter Gumbsch (see his contribution in this volume.)

Now we turn to a discussion of an analog system in two dimensions. In a recent paper [8],

Viswanathan et al investigated phase transitions in a Langmuir-Blodgett film consisting of a small number of mono-molecular layers of rod-shaped molecules lying on a solid substrate. The structure of the top layer is probed via atomic force microscope (AFM). When the number of layers is large, the system is crystalline; within each layer the rods lie in a stretched triangular lattice, where the elongation along one axis arises because the rods are uniformly tilted along one nearest-neighbor direction. The energy of a dislocation starting at the substrate below and reaching up to terminate on the top layer is proportional to the number of layers in the film. Thus, when the number of layers is large, the energy of these defects is high compared to the temperature and they are few in number. As the number of layers is reduced, the energy of dislocations goes down and the system undergoes a K-T transition into a hexatic phase.

Real-space AFM images of the film show that the hexatic phase appears highly anisotropic, with a striking directionality that is easily visible to the eye. Viswanathan et al calculated the correlation function and found short-range translational order along one axis. While they do not comment on it, their analysis also shows that there is much longer-range order along the orthogonal direction. Here, as in the discussion above, we have a system with two different correlation lengths, one of which is as short as a few lattice spacings, while the other appears at last as long as 20 lattice spacings, perhaps longer.

The reason for this anisotropy is that the underlying stretch distortion in the crystal lattice, which arose from the molecular tilt, breaks the symmetry in the energy of dislocations as a function of Burgers vector orientation. The result is that as the number of layers decreases, the lower-energy dislocations nucleate first, giving rise to this anisotropic hexatic phase. A Fourier transform of the real-space AFM images indeed shows that the crystal peaks have been strongly streaked along one lattice direction. We interpret this data to indicate that the defect population in this system is dominated by dislocations of restricted Burgers vector orientation.

Acknowledgments: The author wishes to thank Robb Thomson, M. Khantha, Joe Zasadzinski, and Karl Sieradzki for fruitful discussions.

[1] M. Khantha, D. P. Pope, and V. Vitek, Phys. Rev. Lett. 73, 684 (1994).
[2] M. Khantha, D. P. Pope, and V. Vitek, Scripta Met 31, 1349 (1994).
[3] M. Khantha, Scripta Met 31, 1355 (1994).
[4] J. M. Kosterlitz and D. J. Thouless, J. Phys. C: Solid State Phys. 6, 1181 (1973); see also David Nelson in Domb and Green, vol. 7.
[5] Ostlund and Halperin, Phys. Rev. B 23, 335 (1981).
[6] B. L. Holian and R. Ravelo, Phys. Rev. B 51, 11275 (1995).
[7] F. Abraham, D. Brodbeck, and R. A. Rafey, Phys. Rev. Lett. 73, 272 (1994).
[8] R. Viswanathan, L. L. Madsen, J. A. Zasadzinski, and D. K. Schwartz, Science 269, 51 (1995).

MOLECULAR STATICS SIMULATION OF CRACK PROPAGATION IN α-FE USING EAM POTENTIALS

Vijay Shastry and Diana Farkas
Department of Materials Science and Engineering,
Virginia Tech., Blacksburg VA 24061.

ABSTRACT

The behavior of mode I cracks in α-Fe is investigated using molecular statics methods with EAM potentials. A double ended crack of finite size embedded in a cylindrical simulation cell and fixed boundary conditions are prescribed along the periphery of the cell, whereas periodic boundary conditions are imposed parallel to the crack front. The displacement field of the finite crack is represented by that of an equivalent pileup of opening dislocations distributed in a manner consistent with the anisotropy of the crystal and traction free conditions of the crack faces. The crack lies on the {110} plane and the crack front is located either along ⟨100⟩, ⟨110⟩ or ⟨111⟩ directions. The crack tip response is rationalized in terms of the surface energy (γ_s) of the cleavage plane and the unstable stacking energies (γ_{us}) of the slip planes emanating from the crack front.

INTRODUCTION

Recent atomistic investigations using embedded atom (EAM) empirical potentials of the fracture properties of iron suggest that brittle cleavage is the preferred mode of failure at low temperatures, *i.e.* at temperatures of 100K or lower. These investigations have modeled the response of mode I cracks embedded in homogeneous single crystals of iron using different EAM potentials to describe interatomic interactions. For instance, Cheung *et. al* [1] and deCelis *et. al* [2] have used molecular dynamics to study the behavior of crack tips with Voter FeA [3] and Johnson potentials [4], respectively. Cheung *et. al* show that a brittle to ductile transition occurs between 100 K and 200 K for various crack tip geometries. At low temperatures, brittle crack propagation is observed to occur either on {100} or {110} planes. Profuse dislocation emission occurs above the DBT temperature and ⟨111⟩(011) is identified as one of the active slip systems. deCelis *et. al* also find brittle cleavage to occur in α-Fe at 0 K. However, the preferred cleavage plane is observed to be {100}. Kohlhoff *et. al* [5] have used a combined finite element and atomistic model with Finnis-Sinclair [6] and Johnson potentials [4] to show that cracks propagate in a brittle manner in α-Fe at 0 K. Cleavage is observed either on {110} or {100} planes.

In practice, definite predictions of the macroscopic fracture behavior in Fe must be made after evaluating the response of cracks with different crack tip-slip plane geometries. In this paper, cracks on {110} planes but with geometries other than those considered previously are modeled. Dislocation emission is observed when either {112} or {110} slip planes with large Schmid factor emanate from the crack front. The Simonelli potential [7] is used to describe the interatomic interaction in α-Fe.

BOUNDARY CONDITIONS

The 2D displacement field of a finite mode I crack is calculated by linearly superposing the displacement fields of model opening Volterra dislocations in an equivalent double pileup pinned at the two tips. The displacement fields of individual Volterra dislocations are obtained from the sextic theory of Stroh described by Hirth and Lothe [8]. The crack plane is located midway between two adjacent atomic planes which are then displaced symmetrically in opposite directions, by amounts prescribed by the opening profile of pileup. This method of constructing the crack preserves the bonds that link atoms located on either side of the crack plane. In the EAM simulations, the elastic crack displacement field is approximated by that of a crack with traction free faces.

The dislocation distribution in the equivalent pileup is calculated by evaluating the condition of continuity of normal tractions at discrete locations along the crack. For a crack stretching from $-\ell/2 \leq x \leq \ell/2$ and loaded by a uniform external stress σ^{ext} normal to the crack plane, this condition is

$$\sigma^{ext} + \frac{\mu^{eff}}{2\pi} \int_{-\ell/2}^{\ell/2} \frac{\phi(t)}{(x-t)} dt = 0 , \ (-\ell/2 \leq x \leq \ell/2). \tag{1}$$

Shear tractions are always zero along the plane of the mode I crack. $\phi(x) = d[\Delta u_y(x)]/dx$ is the dislocation distribution function, where $\Delta u_y(x)$ is the crack opening at location x. μ^{eff}, the effective shear modulus for the cleavage plane is elegantly phrased in terms of elastic constants (C_{ij}), when the coordinate axes fixed on a crack in a cubic crystal, lie parallel to directions with even-fold symmetry,

$$\mu^{eff} = \frac{\overline{C}_{11} - C_{12}}{(C_{22}/C_{11})^{1/4}\overline{C}_{11}} \left[C_{22} + C_{12} \left(\frac{C_{22}}{C_{11}} \right)^{1/2} \right] \left[\frac{\overline{C}_{11}C_{66}}{\overline{C}_{11}^2 - C_{12}^2 + 2C_{66}(\overline{C}_{11} - C_{12})} \right]^{1/2} \tag{2}$$

Here, C_{ij} are transformed to the coordinate system fixed on the crack. The basis vector $\hat{y}$ of the coordinate system lies along the normal to the crack plane, whereas $\hat{z}$ contains the crack front and $\hat{x} = \hat{y} \times \hat{z}$. $\overline{C}_{11} = [C_{11}C_{12}]^{1/2}$. The untransformed elastic constants corresponding to the Simonelli potential [7] for α-Fe are $C_{11} = 1.51 ev/\mathring{A}^3$, $C_{12} = 0.91 ev/\mathring{A}^3$ and $C_{44} = 0.7 ev/\mathring{A}^3$.

Eqn. (1) is solved using a method developed by Erdogan and Gupta [9] at discrete points, $x_i = \ell/2 \cos(\pi i/N)$, for $\phi(x)$ which is singular at $x = \pm\ell/2$. N, the total number of discrete points, equals 160. The Burgers vectors of dislocations located at x_i, are obtained by integrating $\phi(x)$ over a length interval assigned to x_i.

Initially, atoms in the simulation cell are displaced by magnitudes equal to sum of the 2-D elastic displacements due to the crack (loaded under plain strain conditions) and that due to to the uniform external stress. Rigid boundary conditions are imposed along the periphery of the cylindrical simulation cell, while periodic conditions exist along its axis. The resulting atomic configuration is relaxed using a conjugate gradient scheme. The relaxation is stopped when the norm of the magnitude of the atomic forces falls below a critical value, usually 10^{-10}. A simulation cell with radius $160\mathring{A}$ has been determined to be adequate for determining the regime of crack loadings in which cleavage or emission occurs.

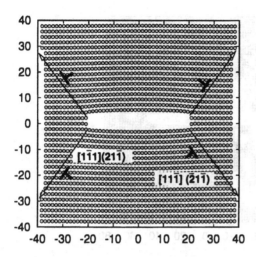

Figure 1: The relaxed atomic configuration associated with with [011](01$\bar{1}$) crack, loaded by an external stress, $\sigma^{ext} = 0.06ev/\AA^3$. Dislocation emission occurs from both crack tips symmetrically on ⟨211⟩ slip planes located on either side of the crack plane.

RESULTS

Figure 1 shows the response of the [011](01$\bar{1}$) crack, loaded by an external stress equal to $0.06ev/\AA^3$. The crack has half length, $\ell/2 = 22.94\AA$. The Griffith value of external loading for this crack is approximately $0.062ev/\AA^3$. The crack tips are observed to be blunted due to emission of $\frac{1}{2}\langle111\rangle$ dislocations on {211} slip planes which emerge from the crack front symmetrically on either side of the crack plane. The dislocations are observed to have moved away far from the crack tips and are located near the periphery of simulation cell, where further motion is restricted by rigid boundary conditions imposed there.

The [100](01$\bar{1}$) crack cleaves when loaded under uniform external stress $\sigma^{ext} = 0.07ev/\AA^3$. Fig. 2 shows an intermediate and unstable configuration of the growing crack. The crack propagates symmetrically in both positive and negative x directions. The original half length of the crack is, $\ell/2 = 22.30\AA$ and its corresponding Griffith loading is $0.058ev/\AA^3$. The crack is observed to be stable in the range of intermediate loadings, $\sigma^{ext} = 0.06 - 0.07\ ev/\AA^3$. At lower values of external stress, $\sigma^{ext} < 0.06\ ev/\AA^3$, the crack retreats and collapse completely under the influence of bonds.

Figure 3 shows the relaxed configuration of a [1$\bar{1}$1](110) crack which has emitted $\frac{1}{2}\langle111\rangle$ dislocations on {110} slip planes when subjected to an external stress, $\sigma^{ext} = 0.06ev/\AA^3$. Emission occurs on {110} slip planes which are located on one side of the crack plane. Here, the nucleating dislocations do not have a pure edge character, rather the Burgers vectors are inclined at an angle of 70.5° with the dislocation line.

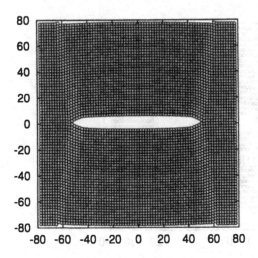

Figure 2: An unstable atomic configuration associated with with a growing $[100](01\bar{1})$ crack, loaded by a supercritical level of external stress, $\sigma^{ext} = 0.07ev/\AA^3$. The original length of the crack is $44.6\AA$.

DISCUSSION

The results show evidence of a potentially ductile response for cracks embedded in α-Fe modeled using the Simonelli potential at 0 K. The critical loading for cleavage observed here is within 10-15 % of the corresponding Griffith value for a traction free crack. The observed agreement with Griffith loading indicates that the effect of bonding between crack faces does not significantly affect the critical loading to cleave for the cases considered here. Further, it also validates the accuracy of boundary conditions used in modelling.

Dislocation emission is observed when the crack fronts lies along the [100] or [011] directions. The $[011](01\bar{1})$ crack emits pure edge dislocations with $\frac{1}{2}\langle 111 \rangle$ Burgers vectors on $\{211\}$ slip planes, inclined at 54.73° with the crack plane. However, the $[1\bar{1}1](110)$ crack emits dislocations with $\frac{1}{2}\langle 111 \rangle$ Burgers vectors on $\{110\}$ planes, inclined at 60° with the crack plane. The emitted dislocations are found to move away from the crack tip and are observed to be piled up against the boundary.

Table I compares the range of loadings (σ^{crit}) in which emission or cleavage was observed in the EAM simulation with the critical value of loading predicted by Rice [10] to emit (σ_{Ie}) and the Griffith value for cleavage (σ_{IC}). The Rice approach determines an approximate value of loading to emit by omitting considerations of ledge formation on an inclined plane when a dislocation is emitted on that plane. Here, in calculating the Rice value of loading to emit, the shear modulus (μ) is approximated by the Voigt average of elastic moduli, 0.54 $ev/\AA^3$ and the Poisson's ratio is approximated by the ratio, $\lambda/[2(\lambda+\mu)] = (2C_{iijj} - C_{ijij})/(2.0 * C_{ijij} + C_{iijj} + 2.0C_{iijj} - C_{ijij}) \approx 0.29$. The Griffith values for cleavage are calculated using formulae given by Sih and Liebowitz [11]. The

Crack plane /Front	Slip system	γ_{us}	γ_s		σ_{IC}	σ_{Ie} Rice	σ^{crit} EAM	γ_{us}/γ_s Crossover	γ_{us}/γ_s EAM
[011]($01\bar{1}$)	$\frac{1}{2}(111)\{112\}$	0.054	0.089	E	0.062	0.093	0.05-0.06	0.26	0.64
[100]($01\bar{1}$)	$\frac{1}{2}(111)\{110\}$	0.046	0.089	C	0.058	0.11	0.06-0.07	0.25	0.52
[$1\bar{1}1$](110)	$\frac{1}{2}(111)\{110\}$	0.046	0.089	E	0.063	0.089	0.05-0.06	0.28	0.52

Table I: The range of values of external loading for which a crack tip instability – emission or cleavage – is observed in the simulations, are compared to the Rice value of loading to emit and the Griffith value to cleave the crack. The letter "E" denotes crack tip emission while "C" denotes cleavage. γ_s and γ_{us} have units, $ev/\mathring{A}^2$, whereas the loadings (σ^{crit}, σ_{Ic}, σ_{Ie}) are expressed in $ev/\mathring{A}^3$.

values of γ_{us} and γ_s of the slip and cleavage planes, respectively, are obtained from the constitutive properties predicted by the Simonelli potential. While the critical loading to cleave is in good agreement with the Griffith value, the emission occurs at loadings which are about 50% lower than the values predicted by Rice. Further, emission occurs when the external loading is lower than the Griffith value by about 10%. This indicates that for the cases considered, the crack does not recede when the external loading is slightly lower than the Griffith value. Spontaneous dislocation emission can only occur from a crack which is equilibrated by external loading. At lower levels of loading, the crack recedes and dislocation emission cannot occur.

Table I also compares the Rice prediction of the ductility ratio, γ_{us}/γ_s, for crossover with the corresponding values predicted by the Simonelli potential. In general, dislocation emission is expected to be the preferred crack tip response when γ_{us}/γ_s is lower than the critical value for ductile-brittle crossover, while cleavage is predicted when this critical value is exceeded. However in the EAM simulations, emission is observed even when the ductility ratio exceeds the critical Rice value.

Zhou, Carlsson and Thomson (ZCT) [12] have suggested that ledge formation increases the difficulty in emission and predict the critical loading for emission to be higher than the Rice value. Therefore, ZCT predict the ductile-brittle crossover line to be shifted to lower values of γ_{us}. Recently however, Xu at. al [13] have used Peierls calculations to demonstrate that when γ_{us} is large, incorporating ledge effects does not effect emission and the critical loading to emit is lower than the corresponding Rice predictions. Further, the elastic interaction between nucleating dislocation and the shear and opening cores present on other slip planes, is expected to affect the critical loading to emit. The relative ease of dislocation emission observed here will have to be analyzed in this light.

It should also be kept in mind that the results obtained here are static equilibrium configurations for the crack tip. In dynamic fracture, the conditions for dynamical emission of dislocations from the crack tip may be different.

REFERENCES

1. K. Cheung and S. Yip, Modelling and Simulation in Materials Science and Engineering **2**, 865 (1994).

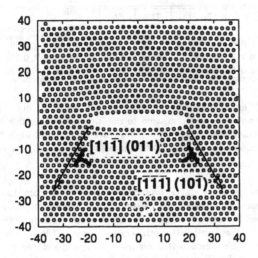

Figure 3: The relaxed atomic configuration associated with with $[1\bar{1}1](110)$ crack, loaded by an external stress, $\sigma^{ext} = 0.06ev/\text{Å}^3$. Dislocation emission occurs from both crack tips on $\langle 110 \rangle$ slip planes located on the lower side ($y < 0$) of the crack plane.

2. B. deCelis, A. Argon, and S. Yip, Journal of Applied Physics **54**, 4864 (1983).

3. R. Harrisson, F. Spaepen, A. Voter, and A. Chen, in *Innovations in Ultrahigh–Strength Steel Technology*, edited by G. Olson, M. Azrin, and E. Wright, page 651, Plenum, 1990.

4. R. A. Johnson, Implication of the EAM format, in *Many Atom Interactions in Solids*, volume 48, page 85, Springer Proc. in Phys., 1990.

5. S. Kohlhoff, P. Gumbsch, and H. F. Fischmeister, Philosophical Magazine A **64**, 851 (1991).

6. M. Finnis and J. Sinclair, Philosophical Magazine A **50**, 45 (1984).

7. G. Simonelli, R. Pasianot, and E. Savino, in *Materials Research Society Symposium Proceedings*, volume 291, page 567, 1993.

8. J. Hirth and J. Lothe, *Theory of Dislocations, 2nd edition*, Wiley & Sons, 1982.

9. F. Erdogan and G. D. Gupta, Q. Appl. Math. **29**, 525 (1972).

10. J. Rice, Journal of the Mechanics and Physics of Solids **40**, 239 (1992).

11. G. C. Sih and H. Liebowitz, Mathematical theories of brittle fracture, in *Fracture – An Advanced Treatise*, edited by H. Liebowitz, volume II, pages 69–189, Academic Press, New York, 1968.

12. S. Zhou, A. Carlsson, and R. Thomson, Physical Review Letters **72**, 852 (1994).

13. G. Xu, A. Argon, and M. Ortiz, Philosphical Magazine A **72**, 415 (1995).

Crack Growth and Propagation in Metallic Alloys

W. C. Morrey and L. T. Wille
Department of Physics, Florida Atlantic University, Boca Raton, FL 33431

ABSTRACT

Using large-scale molecular dynamics simulation on a massively parallel computer, we have studied the initiation of cracking in a Monel-like alloy of Cu-Ni. In a low temperature 2D sample, fracture from a notch starts at a little beyond 2.5% critical strain when the propagation direction is perpendicular to a cleavage plane. We discuss a method of characterizing crack tip position using a measure of area around the crack tip.

INTRODUCTION

Fracture in metals has been a field of study for many years [1]. Interest in this area has been sustained by the pervasive economic impact of even the smallest advances. Understanding metallic fracture is at the heart of improvements in safety, reliability, and cost of a vast range of human endeavor. Tremendous benefits accrue to manufacturing, transportation, and construction, among others, by improving a metal's resistance to fracture; understanding fracture mechanisms is key to devising that improvement.

Computer simulation may be used to study crack growth in order to isolate variables and save the time and expense of exhaustive destructive testing. Continuum models [1] are used to predict crack growth and are quite successful. The input to these models, however, depends on numbers developed by sample preparation and testing. Obtaining statistically reliable data is time consuming and costly. Many times the spread of results indicates hidden process and material variables. Often, data are reviewed and it is evident that what was perceived to be a consistent method of sample preparation is, in fact, insufficiently controlled to produce a narrow range of results. Of course, *insufficiently controlled* is an economic call, and may vary with the times. A 10% variation may be perfectly acceptable if the cost of conservative design is tolerable, while a 1% variation in another application may be prohibitively costly. Another shortcoming of the continuum approach is that a number of unwarranted assumptions about the nature of the cracks are made [2].

The path to illumination of these underlying mechanisms is microscopic simulations. Molecular Dynamics (MD) simulations using realistic potentials are now performed on a scale that is approaching the actual dimensions of the microscopic features of interest [3]. This breakthrough has only become possible in recent years because of the availability of scalable parallel computers.

In the present paper we investigate aspects of crack growth in Cu-Ni using MD on a MasPar MP-1 massively parallel computer. This is a follow-on of previous work [4] which was inspired by a study for pure elements by Wagner *et al.*,[5]. We briefly describe the implementation of the algorithm, present crack propagation snapshots and discuss methods of characterizing propagation speed.

METHODOLOGY

Technical details on the MasPar MP-1 are discussed by Trew and Wilson [6] and Hwang [7] and in papers quoted therein. In brief, the MasPar consists of a front-end running the Ultrix operating system and a data parallel unit (DPU). The latter is made up of a two-dimensional 64 x 64 grid of Processor Elements (PEs). Each PE receives the same instruction at every step. Activated PEs execute the instruction on variables that reside in local memory. Various strategies for parallelizing MD simulations have been reviewed by Nelson *et al.* [8] and Plimpton [9]. Of particular interest for the present paper are methods applicable to short-ranged potentials. When properly mapped to the parallel computer, these will require only short-range communications for data exchange.

Wagner *et al.* [5] used a combination of an analytic embedded-atom potential [10] and a Lennard-Jones (LJ) potential with a spline cutoff to determine the interatomic forces. More recently, Holian and Ravelo [11] have shown that the LJ potential alone can be used in fracture studies (see also Ref. [2]). The differences between brittle and ductile materials can be accommodated by a change in the shape of the cutoff tail. In this work we restrict ourselves to potentials of this LJ form.

Parallel computers using Single Instruction - Multiple Data (SIMD) structures are ideally suited to MD simulations. Each atom's position, potential, and force components are kept and Newton's equations of motion are integrated numerically. These calculations can be performed on each PE, P atoms at a time (where P is the number of processors). In the present study we use a geometric decomposition of the problem. Each PE element in the MasPar's 64 x 64 structure is divided up into C x C boxes of dimension equal to the cutoff range of the interaction potential. Searching a particle's box and its eight neighbors assures that all neighboring particles are counted. This eliminates the need for a linked neighbor list, although a linked list is used to keep track of the pointers to active particles in each PE.

All calculations discussed here were performed on an MP-1 with 4,096 PEs and 16 Kbytes of local memory per PE, or on an MP-2 running the same code 50% faster in a subset of the machine of the same configuration. Since the largest MP-1 and MP-2 contain 16,386 PEs and 64Kbytes of memory, we have the ability to increase the size of the simulation considerably.

RESULTS

A 2D arrangement of 224,727 atoms in a triangular lattice is used as the initial sample. All the atoms are originally spaced at an intermediate value obtained from pure fcc structured Cu and Ni. The atomic identification (type) is randomly assigned to represent Monel: disordered 28% Cu and 72% Ni. An atomically sharp notch is placed at the bottom surface in the middle of the piece. Distance is measured in r_0 units, the distance between Ni atoms in a 2D lattice, mass m is in Ni mass units, and the potential well depth, ε, is in electron volts. This leaves the characteristic time, t_0, in units of $r_0\sqrt{m/\varepsilon}$.

The sample is oriented such that the dislocation slip planes are +/- 30° and 90° to the direction of propagation. The system is quenched following substantially the same procedure as Ref. [11] by taking large time steps and resetting all velocities to zero after the system potential energy reaches a minimum and kinetic energy reaches a maximum. This procedure is repeated until the energy of the system reaches a low level (under 100 Kelvin).

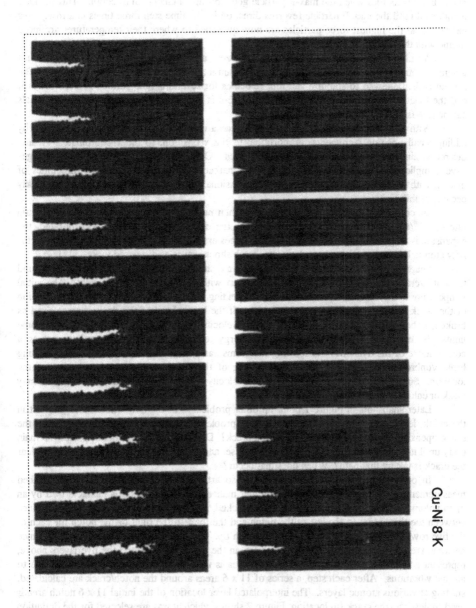

Cu-Ni 8 K

Figure 1: Cu-Ni at 8 K. First snapshot at t = 1500, time between snapshots = 250

At this point, we switch to tracking the particles with the highest force component, and reset all the velocities when the max-F particle goes through a change in direction. This tracking is followed until the max-F particle reverses direction in one time step three times in a row. This indicates that the highest force particles are now quenched to where their temperatures are more in line with the rest of the system.

We now revert to resetting all velocities to zero after the system potential energy reaches a minimum and kinetic energy reaches a maximum, until the average force is acceptably low. The system is then allowed to stabilize and mix velocities for several t_0 time periods at steps of .015, and the kinetic temperature is calculated. The system is saved at this point so that various trials can be run using the identical initial conditions.

With free boundary conditions, we introduce a constant strain rate in the x-direction by adding small position increments, proportional to the x position, to the x coordinate. At any desired strain, this expansion is merely discontinued. Other schemes involving velocity changes have complications at these boundaries and the mathematical and computational difficulties of dealing with clamping the edges and with compressional waves from these free surfaces quickly became prohibitive.

Accordingly, the base sample system is then numerically expanded at a constant strain rate of $10^{-4}/t_0$ until after crack initiation. After the desired excess strain is incorporated, the expansion is stopped and the left, right, and top edges are clamped. The velocities are ramped for fifty atomic layers from those boundaries to absorb shock waves as discussed in Ref. [11].

Snapshots every 250 time steps ($\sim 3.7\ t_0$) are shown in Figure 1. Only the atoms around the notch/crack are shown, and the snapshots start with t=1500 in the upper right. The initial temperature in this simulation was 8 Kelvin. In deciding how to describe the microscopic location of the crack tip, several factors come into play. If the layer ahead of the tip is characterized as broken, either through observing a high atomic velocity or an atomic separation beyond some limits, the layer can re-form due to loss of energy to the surrounding material or just as a consequence of the dynamics transporting the atoms back into proximity. In any case, assigning broken/unbroken to each layer results in binning of the crack tip due to quantization of the location. Subsequent measures of location and velocity will reflect this quantization, which may mask or enhance underlying phenomena.

Later snapshots in Figure 1 point up other problems. In the left column, bridges exist in the crack, leaving assignment of the tip location problematic. When a dislocation ahead of the crack opens up, is it considered the tip of the crack? Does the tip remain defined with the main body until it is clear that the dislocation won't close, and then jump ahead when it becomes clear the crack is taking that path? When does that occur?

In order to avoid these decisions and also artificial quantization we have approached measurement of the crack tip location in another manner. If the crack tip is characterized by an open volume (area in the 2D case) this can be tracked through the sample. We took the distance between the third atom to the left of the notch and the third to the right of the notch for each of the five rows from the row of the embedded notch tip. These five distances added to the distance between the six atoms centered at the notch tip, in the notch tip layer and the five layers above, represent a typical notch tip area. An 11 x 6 area is used to smooth out the variations due to atomic vibrations. After each step, a series of 11 x 6 areas around the notch/crack are calculated, starting at various atomic layers. The interpolated layer location of the initial 11 x 6 notch area is used to describe the crack tip location. Figure 2 shows which atoms are selected for the definition of crack tip area.

Figure 3 shows the plot of crack tip area location versus time. The vertical axis is the atom row number from the bottom of the sample, and the horizontal axis is time steps (.015 t_0).

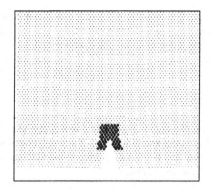

Figure 2: Area of Crack Tip

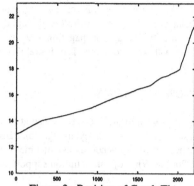

Figure 3: Position of Crack Tip

The defined nominal notch area starts at row 13, and gradually moves through the part merely as a function of the expansion. There is a slight rise in the slope just before step 1700, which corresponds to the time of the first bond breaking (see Fig. 4). This occurs after an expansion of about 2.5%. This value is very close to that found in the simulations for the elemental solids [2], [5]. The crack grows at a slow rate, then appears to stop proceeding forward. The slope returns to that associated with expansion. Then a dramatic change in slope at step 2020 indicates crack growth has reached another velocity regime, after about 3% expansion. Figure 4 shows a plot of the maximum valued x component of velocity near the notch tip. The first bond breaks and the atom rebounds to the right around step 1680.

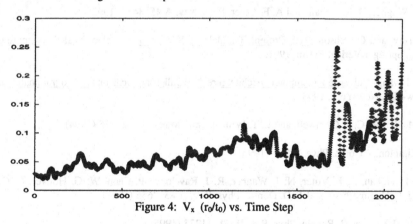

Figure 4: V_x (r_0/t_0) vs. Time Step

CONCLUSIONS

The visual determination of crack tip location is ambiguous because of the crack's leap-frogging ahead to dislocations. Assigning the tip position to a broken atomic layer leads to quantization of velocity. Looking at the movement of an area around the crack tip shows smooth crack propagation in two regimes: an initial slow crack growth, followed by an abrupt increase in velocity.

The next steps in this series of experiments are to investigate the transition between these two velocity regimes and to explore the effects of temperature and angle of propagation vs. slip planes on initiation and propagation velocity. Further refinements of objective location of the crack tip will be undertaken. The desirability of using a local density instead of an area will be evaluated.

ACKNOWLEDGMENTS

The authors wish to thank Dr. B. Holian at Los Alamos National Laboratory for extremely helpful discussions and Dr. C. Halloy and the Joint Institute for Computational Science at the University of Tennessee for the generous use of time on their MasPar MP-2. W. C. Morrey would like to thank Pratt & Whitney for its tuition support and schedule accommodations.

REFERENCES

1. L. B. Freund, <u>Dynamical Fracture Mechanics</u>, Cambridge University Press, Cambridge, 1990.

2. F. F. Abraham, D. Brodbeck, R. A. Rafey, and W. E. Rudge, Phys. Rev. Lett. **73**, 272 (1994).

3. N. Gronbech-Jensen, T. Germann, P. S. Lomdahl, and D. M. Beazly, IEEE Comp. Sci. Eng. **2**, (2), 4 (1995).

4. W. C. Morrey and L. T. Wille, Materials Science & Engineering B(in press).

5. N. J. Wagner, B. L. Holian, and A. F. Voter, Phys. Rev. A **45**, 8457 (1992).

6. A. Trew and G. Wilson, <u>Past, Present, Parallel - A Survey of Available Parallel Computing Systems</u>, Springer Verlag, Berlin, 1991.

7. K. Hwang, <u>Advanced Computer Architecture - Parallelism, Scalability, Programmability</u>, McGraw-Hill, New York, 1993.

8. K. M. Nelson, C. F. Cornwell, and L. T. Wille, Comp. Mater. Sci. **2**, 525 (1994).

9. S. Plimpton, J. Comp. Phys. **117**, 1 (1995).

10. B. L. Holian, A. F. Voter, N. J. Wagner, R. J. Ravelo, S. P. Chen, W. G. Hoover, C. G. Hover, J. E. Hammerberg, and T. D. Dontje, Phys. Rev. A **43**, 2655 (1991).

11. B. L. Holian and R. Ravelo, Phys. Rev. B, **51**, 11275 (1995).

SURFACE STRESS EFFECTS ON FRACTURE

R.C. CAMMARATA* and K. SIERADZKI**

*Department of Materials Science and Engineering, The Johns Hopkins University, Baltimore, MD 21218, USA
**Department of Mechanical and Aerospace Engineering, Arizona State University, Tempe, AZ 85287-6106, USA

ABSTRACT

The effects of surface stress on fracture are reviewed. Calculations by Thomson *et al.* seem to suggest that surface stresses do not affect cleavage fracture when the crack propagates in a self similar manner even though the surface stress can lead to large stresses near the crack tip. However, surface stresses may have an important effect on dislocation emission near the crack tip. A simple analysis is offered to show how these effects may be incorporated into a modified Rice ductile-brittle fracture criterion by adding a term to the unstable stacking energy that takes into account the energy to form an incipient ledge.

INTRODUCTION

In the thermodynamics of surfaces, there is a quantity γ that represents the excess free energy per unit area owing to the existence of a surface. Equivalently, it can be considered the reversible work per unit area needed to create a new surface by a process such as cleavage. The amount of reversible work dw performed to create new area dA of surface is

$$dw = \gamma \, dA. \tag{1}$$

The total work needed to create a planar surface of area A is equal to γA.

Gibbs [1] was the first to point out that for solids, there is another type of surface work, different from γ, that is associated with the reversible work per unit area needed to elastically stretch a surface. The relationship between this quantity and the surface free energy γ can be derived in the following manner [2]. Let the elastic deformation of a solid surface be expressed in terms of a surface elastic strain tensor ε_{ij}, where $i,j = 1,2$. Consider a reversible process that causes a small variation in the area through an infinitesimal elastic strain $d\varepsilon_{ij}$. The surface stress tensor f_{ij} relates the work associated with the variation in γA, the total excess free energy of the surface, due to the strain $d\varepsilon_{ij}$ (summing over each repeated index):

$$d(\gamma A) = A\, f_{ij}\, d\varepsilon_{ij}. \tag{2}$$

Since $d(\gamma A) = \gamma\, dA + A\, d\gamma$, and $dA = A\, \delta_{ij}\, d\varepsilon_{ij}$ (where δ_{ij} is the Kronecker delta), the surface stress tensor can be expressed as

$$f_{ij} = \gamma\, \delta_{ij} + \partial\gamma/\partial\varepsilon_{ij}. \tag{3}$$

Surface stress and surface free energy are completely different quantities. The surface stress is a second rank tensor while the surface free energy is a scalar. Since it is a tensor, the surface stress can be anisotropic. However, for a surface possessing a three-fold or higher rotation axis symmetry perpendicular to the surface, the surface stress is isotropic and can be taken as a scalar $f = \gamma + \partial\gamma/\partial\varepsilon$. It should be noted that the surface stress f can be positive or negative while the surface free energy γ always must be positive for a stable clean surface. The magnitude of the surface stresses can be relatively large. For example, first principles calculations for the surfaces stresses of Pt(111) and Au(111) gave values of 5.60 N/m and 2.77 N/m, respectively, which are more than twice as large as the calculated surface free energies [3].

Physically, the surface stress can be understood to result from the fact that the nature of the bonding (e.g., the number of bonds) is different for atoms at the surface compared to atoms in the interior. Because of this, the surface atoms would have an equilibrium interatomic distance different from that of the interior atoms if the surface atoms were not constrained to remain in atomic registry with the underlying lattice. For example, surface atoms in metals have an electron density smaller than the equilibrium bulk value. Therefore, the surface atoms would tend to contract in order to increase the electron density. The surface stress can be considered the force per unit length the underlying bulk must exert on the surface to constrain the surface atoms to have the bulk equilibrium interatomic distance. In fact, the surface stress (more correctly, the difference between the surface stress and surface free energy) may be so large that it is thermodynamically more favorable for the top monolayer of a solid to expand or contract and lose structural coherence with the underlying lattice. Such a surface reconstruction has been seen in (111)-oriented surfaces of Au and Pt and has been attributed to the influence of the surface stress [2].

Effects of the surface stress can become important when the size of the solid becomes small. Consider the case of a small spherical solid or radius r that has an isotropic surface stress f. The surface stress exerts a net hydrostatic pressure on the sphere ΔP. In equilibrium, the virtual work to expand or contract the solid by an amount dV will equal to the work to elastically strain the surface fdA. Therefore,

$$\Delta P = 2f/r. \tag{4}$$

Since f can be positive or negative, ΔP, called the Laplace pressure, can be positive or negative. (The Laplace pressure for solids is often incorrectly written as $\Delta P = 2\gamma/r$, an expression that would be correct for the case of a small fluid droplet.) Substituting Hooke's Law $\Delta P = -3K\varepsilon$, where K is the bulk modulus and ε is the radial strain, into Equation 4 gives

$$\varepsilon = -2f/3Kr. \tag{5}$$

Taking f = 2 N/m and K = 100 GPa, lattice strains greater than 0.1% in magnitude can be generated in solid particles when the diameter is reduced below 10 nm.

SURFACE STRESS AND FRACTURE

The fact that the surface stress can induce fairly large bulk stresses and strains in small solids has led some to suggest it may have an important influence near the tip of a sharp crack and therefore on the fracture behavior. Oriani [4] discussed the possibility of an anisotropic surface stress leading to both enhanced bond breaking and enhanced slip at the crack tip. This anisotropy of the surface stress may be intrinsic to the surface (if it is of low symmetry) or from effects of adsorption that breaks the symmetry of the bare surface. Oriani suggested that this latter possibility may be responsible for certain examples of environmentally induced failure. Based on this suggestion, Thomson et al. [5] performed a detailed analysis on the elastic fracture behavior incorporating the effects of surface stress and calculated forces on cracks and nearby dislocations using approximate stress function distributions. Their results led them to conclude that in general, surface stress is not expected to affect cleavage fracture but could significantly alter dislocation emission. Recently, a simple analysis [6] has been offered in the context of the Rice unstable stacking energy model [7] that illustrated the possible influence of surface stress on the dislocation emission criterion.

CLEAVAGE FRACTURE

It was noted previously that a substantial Laplace pressure can result from the surface stress when the size of the solid becomes small. In the case of a rounded crack tip with radius of curvature ρ, it might be expected that a stress field of magnitude f/ρ would be induced near the crack tip. Oriani [4] suggested that right at the surface, the stress resulting from f would be of magnitude f/d, where d is the interplanar spacing, which can be of order $E/10$, where E is Young's modulus. Thus, it might be expected that the surface stress could have a profound influence on cleavage fracture behavior. However, the analysis by Thomson *et al.* [6] indicated that the crack equilibrium depends only on the surface free energy γ and not on the surface stress as long as the crack configuration moves in a self similar configuration. They argued that as the crack moves, the sole result is the creation of new surface, and the energy required to do this is the relaxed surface energy for cleavage γ. Thus, the standard Griffith-Irwin criterion is retained even though surface stresses can affect the elastic strain fields at the crack tip. However, this argument is not completely convincing. While it is true that a surface stress does not change the work of fracture, it would nevertheless seem that it could contribute to the creation of this surface work, thereby increasing or decreasing (depending on the sign of f) the applied stress level necessary to cause fracture.

Thomson *et al.* pointed out that the surface stress could influence crack propagation if the crack moves at such a speed that the surface stress distribution cannot move rigidly with the tip. They suggested one way this may occur is if adsorbed chemical species cannot keep up with a fast moving tip, a scenario consistent with Oriani's original suggestion of environmental effects on surface stress affecting cleavage fracture. However, as Thomson *et al.* stated, since environmental effects will also significantly affect the surface free energy, it would be extremely difficult experimentally to separate out the contributions from these two effects.

DISLOCATION EMISSION

Thomson *et al.* [6] calculated how the forces on dislocations near a crack tip are affected by the surface stress and determined how the standard criterion for dislocation emission is altered. In simplified form, their analysis concerned the net contribution of the surface stress that affects slip, which can be denoted as an effective surface stress f_0. This effective surface stress results in a force for slip on a dislocation near the crack tip that can approximately be expressed as $(\mu b^2/4\pi r_0)(pf_0/\mu b)$, where μ is the shear modulus, b is the magnitude of the Burgers vector, r_0 is the dislocation core radius, and p is the number of atomic spacings from the crack tip the stress acting on the dislocation due to the surface stress reaches its maximum value. The criterion for dislocation emission will be affected by this term when it is comparable in magnitude to the image force acting on the dislocation that opposes slip. Since the image force is of order $\mu b^2/4\pi r_0$, the criterion for dislocation emission is modified by surface stress effects when $pf_0/\mu b$ is of order unity. Since $\mu b/\gamma$ is of order 5-20 in many crystals, and p is expected to be in the approximate range 2 to 10, surface stress effects will be important if f_0 is of order γ. Given that the surface stress and surface free energy are almost always expected to be of the same order of magnitude, the conclusion is that the surface stress will significantly affect dislocation emission in many if not most cases.

In several classic papers, different analyses for a ductile-brittle criterion have been offered [8-10] that all lead to the parameter $\mu b/\gamma$. One way to interpret this ratio is as a term proportional to the dislocation formation energy divided by the energy to form a new surface by cleavage. When this ratio exceeds a certain critical value ≈ 10, brittle fracture is predicted, and when this ratio is less than this critical value, dislocation emission is preferred. Rice has recently offered an alternative criterion [7]. He has defined a quantity called the unstable stacking energy γ_{us} that represents the interface energy per unit area of an incipient edge dislocation that is being nucleated by localized sliding on a slip plane. γ_{us} can be considered the activation energy barrier to nucleating a dislocation. The new ductile-brittle criterion involves the ratio γ_{us}/γ. When this ratio exceeds a certain critical value determined by the type of loading and geometrical parameters associated with the slip plane and direction, cleavage fracture is preferred; when this ratio is less than the critical value, dislocation emission is expected. Recently, Zhou *et al.* [11] conducted computer simulations that indicated that at least in certain cases dislocation emission during Mode I loading depends on a term that ostensibly could be added to γ_{us} that was related to the energy of formation for a ledge that blunts the crack at its tip. However, since γ_{us} refers to the state of an incipient dislocation while a ledge energy corresponds to an already formed dislocation, it may be more appropriate to include a ledge energy term into the numerator of the Rice-Thomson $\mu b/\gamma$ criterion.

More in the spirit of the Rice analysis is the consideration of the energy of the incipient ledge that is formed when the incipient dislocation is in the activated state. One can define a ledge stress f_L that presumably has a value close to that of the surface stress f and a surface strain ε that varies from 0 to a value of ε_{us} equal to the local strain when the incipient dislocation is in the activated unstable stacking configuration. Thus, the work to stretch the surface bonds is $\int f_L d\varepsilon$ which up to first order in strain equals $f_L \varepsilon_{us}$. (Since ε_{us} can presumably be relatively large, of order 0.1 or more, it may be necessary to include higher order terms in ε_{us}.) The total activation energy to form an incipient dislocation would therefore be $\gamma_{us} + f_L \varepsilon_{us}$ and the modified criterion would involve the ratio $(\gamma_{us} + f_L \varepsilon_{us})/\gamma$. It should be noted the ledge stress term can either assist or impede dislocation emission depending on the sign of f_L. If $f_L > 0$, then the work to form the incipient ledge is positive and therefore adds to the activation barrier, while for $f_L < 0$, the work to form the incipient ledge is negative and therefore reduces the activation energy. If the magnitude of f_L is assumed to be of order 2 N/m and the strain ε_{us} is of order 0.2, $f_L \varepsilon_{us}$ will be of order 0.4 N/m which is the same order of magnitude for γ_{us}. In fact, it is conceivable that $f_L \varepsilon_{us}/\gamma_{us}$ could be less than -1, meaning that the surface stress alone, without any applied external stress, is enough to load the crack and cause emission. In this event a bare crack would not exist. It is noted that, in general, the surface stress leads to mixed Mode I/Mode II loading of the crack which is important for the dislocation emission process. (It should also be pointed out the above discussion would not be relevant to the simulations of Zhou et al. since the potentials they used lead to a surface stress of identically zero [12]).

As a final note, some experimental work on small particles of LiF indicated that the lattice parameter increased with decreasing particle size, behavior that would correspond to a negative surface stress [13,14]. In the case of Si and Ge, Meade and Vanderbilt [15,16] have performed first principles calculations showing that the unreconstructed (111) surfaces have large negative surface stresses. Negative ledge stresses may account for the observation that these materials seem to display a greater facility for dislocation emission at crack tips than would be expected from the conventional ductile-brittle criteria. This issue could presumably be addressed most directly by computer simulations.

REFERENCES

[1] J.W. Gibbs, *The Scientific Papers of J. Willard Gibbs*, Vol. 1 (Longmans-Green, London, 1906) p. 55.

[2] R.C. Cammarata, Prog. Surf. Sci. **46**, 1 (1994).

[3] R.J. Needs and M. Mansfield, J. Phys. (Cond. Matter) **1**, 7555 (1989).

[4] R.A. Oriani, Scripta Metall. **18**, 165 (1984).

[5] R. Thomson, T.J. Chuang, and I.H. Lin, Acta Metall. **34**, 1133 (1986).

[6] K. Sieradzki and R.C. Cammarata, Phys, Rev, Lett, **73**, 1049 (1994).

[7] J.R. Rice, J. Mech. Phys. Solids **40**, 239 (1992).

[8] R.W. Armstrong, Mater. Sci. Eng. **1**, 251 (1966).

[9] A. Kelly, W.R. Tyson, and A.H. Cottrell., Phil. Mag. **15**, 567 (1967).

[10] J.R. Rice and R. Thomson, Phil. Mag. **29**, 73 (1974).

[11] S.J. Zhou, A.E. Carlsson, and Robb Thomson, Phys. Rev. Lett. **72**, 852 (1994).

[12] S.J. Zhou, A.E. Carlsson, and Robb Thomson, Phys. Rev. Lett. **73**, 1050 (1994).

[13] J.S. Halliday, T.B. Rymer, and K.H.R. Wright, Proc. Roy. Soc. **A255**, 548 (1954).

[14] T.B. Rymer, Nuovo Cimento (Suppl.) **6**, 294 (1957).

[15] R.D. Meade and D. Vanderbilt, Phys. Rev. B **40**, 3905 (1989).

[16] R.D. Meade and D. Vanderbilt, Phys. Rev. Lett. **63**, 1404 (1989).

REFERENCES

EFFECT OF CRACK BLUNTING ON SUBSEQUENT CRACK PROPAGATION

J. SCHIØTZ*, A. E. CARLSSON*, L. M. CANEL*, and ROBB THOMSON**
*Department of Physics, Washington University, St. Louis, MO 63130-4899
**National Institute of Standards and Technology, Gaithersburg, MD 20899

ABSTRACT

Theories of toughness of materials depend on an understanding of the characteristic instabilities of the crack tip, and their possible interactions. In this paper we examine the effect of dislocation emission on subsequent cleavage of a crack and on further dislocation emission. The work is an extension of the previously published Lattice Greens Function methodology[1, 2, 3]. We have developed a Cavity Greens Function describing a blunt crack and used it to study the effect of crack blunting under a range of different force laws. As the crack is blunted, we find a small but noticeable increase in the crack loading needed to propagate the crack. This effect may be of importance in materials where a dislocation source near the crack tip in a brittle material causes the crack to absorb anti-shielding dislocations, and thus cause a blunting of the crack. It is obviously also relevant to cracks in more ductile materials where the crack itself may emit dislocations.

INTRODUCTION

When a sharp crack in a material is loaded until it deforms plastically at the crack tip, two fundamentally different modes of deformation can occur. The crack may propagate (possibly leading to cleavage of the specimen), or it may emit a dislocation. In the first case the material is said to be intrinsically brittle, in the second it is intrinsically ductile. Even for single crystals, many other aspects of the microstructure of the material influence the behavior of the material, such as dislocation activity in the material surrounding the crack and shielding of the stress fields by the dislocations present in the vicinity of the crack. In this paper we will however concentrate on the intrinsic behavior of the crack.

The intrinsic behavior of the crack can be predicted by comparing the critical stress intensity factors necessary for cleavage and for dislocation emission. The cleavage is well described by the Griffith criterion[4], but several criteria have been proposed for the dislocation emission. Rice[5] has proposed an emission criterion given by a well-defined solid-state parameter, the unstable stacking fault energy (γ_{us}), characterizing the barrier to displacement along the slip plane. In this model the emission criterion becomes a "balance" between two energies: The surface energy (γ_s) and the unstable stacking energy. Zhou *et al.*[2, 6] extended this model to include the energy of the small ledge at the end of the crack, and showed that for most physically realistic force laws this term dominates the energetics, leading to an emission criterion containing both γ_s and γ_{us} and a ductility criterion that is independent of γ_s.

These ductility criteria all assume that the crack is sharp at the atomic level. However, emission of a dislocation will change the local geometry of the crack, possibly changing both the emission and cleavage criteria. Furthermore, a crack may *absorb* a dislocation, coming from sources in the neighborhood of the crack tip. This may also lead to blunting of the crack. In the following we will discuss the effects such blunting may have on the dislocation emission and cleavage of the crack.

METHODOLOGY

Ordinary molecular dynamics (MD) simulations of cracks are very difficult, due to the extremely slow decay of the stress fields ($\sigma \propto r^{-1/2}$). This makes it necessary to use very large systems to minimize the effects of boundary conditions; millions of atoms may be required even for two-dimensional simulations. On the other hand all of the interesting phenomena will be concentrated near the crack tip, and most of the atoms are just propagating the elastic field. This leads to a very inefficient use of computer resources.

Our solution is to model the elastic response of the surrounding media by a Green's function. The atoms are divided in two classes. The atoms near the crack tip interact with each other through a non-linear force law (the *non-linear zone*), whereas the atoms far from the crack tip interact only through linear forces. This *linear zone* can then be fully described by a lattice Green's function $G_{ij}(\mathbf{r}, \mathbf{r}')$, describing the response of the atom at $\mathbf{r}$ to a force acting on the atom at $\mathbf{r}'$. One needs only the Green's function elements involving the nonlinear zone and a defect zone. The Green's function can be calculated in a computationally efficient way. The procedure for calculating the Green's function, and for introducing a defect (a crack in this case) has been discussed by Zhou *et al.*[1]. The total energy of the system can then be described as the sum of the energy in the elastic far field (calculated from the Green's function) and in the non-linear interactions. In this way the total energy of the system is described as a function of a much reduced number of degrees of freedom (the positions of the 10^2–10^3 atoms in the non-linear zone, compared to the 10^6–10^7 atoms in the full problem). The problem can then be treated with conventional minimization techniques[3].

We use this method to study the deformation modes of blunt cracks with up to seven atomic layers of blunting, for a range of force laws. Figure 1a show a typical initial configuration. We have a region in front of the crack where the atoms are allowed to

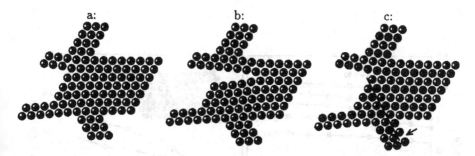

Figure 1: The two deformation modes. (a) shows the configuration before plastic deformation. The crack, three atomic layers thick, extends out of the left of the picture. Only atoms in the non-linear zone are shown, see text. (b) shows the crack propagating by cleavage. (c) shows the crack emitting a dislocation. The dislocation has traveled along the black line. Since it cannot leave the non-linear zone, it has been pinned at the bottom (indicated by the arrow). To make the path of the dislocation visible, the atoms were given two different colors. Prior to emission the atoms were lined up in rows of the same color. Where the dislocation has traveled lines of atoms of different color meet.

move freely, and two spurs are added, alog which dislocations generated at the crack tip can move away from the crack. Since bonds cannot be broken and reformed in the linear zones, dislocations will be unable to leave the non-linear zone.

The inter-atomic interactions in the non-linear zone are described by the UBER pair potential:

$$F(r) = -k(r - r_0) \exp\left(\frac{r - r_0}{\beta}\right) \tag{1}$$

where r is the inter-atomic separation, r_0 is the separation in equilibrium. and β is a range parameter. We cut off the interactions so that only nearest-neighbor interactions are included, and shift the potential slightly to avoid a step in the force law. Further, a small scaling of the force law is used to preserve the elastic constants, thus enabling us to use the same Green's function for all the force laws. The force law thus becomes

$$F(r) = C\left[-k(r - r_0) \exp\left(\frac{r - r_0}{\beta}\right) - F_0\right] \tag{2}$$

where $F_0 = -k(r_{\text{cutoff}} - r_0) \exp\left((r_{\text{cutoff}} - r_0)/\beta\right)$ assures that the force is zero at the cutoff distance r_{cutoff} (in this work $1.7r_0$). This is only a slight perturbation of the force law as the scaling factor C only differs from unity by a few percent. Since this work does not study specific materials, no attempt is made to use realistic many-body potentials.

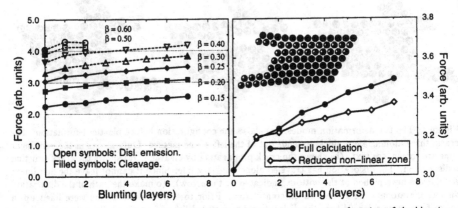

Figure 2: Left: The force required for cleavage or dislocation emission as a function of the blunting, for seven different force laws. Right: The effect of reducing the non-linear zone at the crack tip. When the atoms shown in black are removed from the non-linear zone (i.e. they are restricted to interact through linear forces), the effect of the blunting is reduced. The range parameter in the force law is $\beta = 0.25$.

RESULTS

When we load the cracks until plastic deformation occur, two different deformation modes are observed: cleavage and dislocation emission, see figure 1. The dislocation emission always occurs in the downwards direction, as shown in the figure, except for the case of a single layer of blunting, where the crack geometry is symmetric. Furthermore, the emission occurs in such a way that the asymmetric crack geometry is preserved. This means that further dislocations can be expected to be emitted in the same direction, provided that the dislocation can move so far away that its effect on the local stress field at the crack tip is small. When a sufficient number of dislocations have been emitted, their combined screening may prevent further emission, and cause the crack to propagate by cleavage or begin emitting in the opposite direction. This may have been observed experimentally[7, 8].

Figure 2 (left) shows the force required for cleavage or dislocation emission as a function of the blunting, for seven different force laws. All force laws result in cleavage in the case of a sharp crack, but for a range of force law parameters dislocation emission becomes favored as soon as the crack is blunted. In all cases the required force for cleavage increases slightly with increasing blunting. When dislocation emission occur the force also increases slightly with the blunting as soon as any blunting occurs, but for these force laws the force required to emit a dislocation from the sharp crack is *larger* that for the blunt crack. The tendency for the sharp crack to propagate is so large that it is difficult to measure the dislocation emission criterion accurately.

Using a conformal mapping technique, we have been able to calculate the stress field around a blunt crack (to be published elsewhere), but only in anti-plane strain (mode III loading). The result is as could be expected: far from the crack the stress field is unperturbed by the blunting, and decays as for the sharp crack ($\sigma \propto r^{-1/2}$), but near the crack tip the stress singularity is similar to what is found near a wedge-shaped crack[9]:

$$\sigma \propto r^{\alpha}, \qquad \alpha = \frac{\pi}{2\pi - \theta} - 1 \qquad (3)$$

i.e. $\alpha = -2/5$ for the given configuration, a $60°$ wedge, as opposed to $-1/2$ for a sharp crack ($\theta = 0$). This leads to a reduction of the stresses near the crack tip, and thus to an increase in the crack loading needed to break the bonds. However, this leads us to expect an increase in the load of approximately 25–50% over the range of blunting investigated here, versus the observed increase of only $\sim$ 10%. This is caused by the difference between the mode I and mode III configurations. Williams[10] has solved the mode I wedge configuration and the stress singularity is significantly different. He finds $\alpha = -0.478$ for this geometry, leading to a stress singularity that is indistinguishable from the sharp crack. This would lead us to expect that there should be no effect from the blunting. So we have to seek the explanation of the observed effect elsewhere than in linear elasticity.

Figure 2 (right) shows the result of reducing the number of atoms interacting through non-linear forces. This strongly suppresses the effect of the blunting, indicating that it is mostly a non-linear effect. When the crack is loaded the bonds at the blunt end of the crack are stretched into their non-linear regime. They are thus stretched more than they would have been if the bonds had been purely linear, and the bonds at the corner where the crack appear do not have to stretch as much, and therefore do not break as early. As the blunting increases, more non-linear bonds become available for relieving the stress concentration at the corner.

CONCLUSIONS

We have shown that blunt cracks have a stronger tendency to emit dislocations than do sharp cracks. This may be important in intrinsically brittle materials, where the natural tendency for a sharp crack is to propagate by cleavage rather than to emit dislocations. In these materials a dislocation may be blunted by *absorbing* a dislocation from a nearby source, and thereby be turned into an emitting crack, preventing further cleavage.

We also observe, that dislocation emission from an asymmetric blunt crack preferentially occurs to one side, and that the emission preserves the shape of the crack. This leads to emission of multiple dislocations to the same side, even in situations where emission on two symmetrical slip planes should be equally favorable, and where a simple shielding argument would result in emission of every second dislocation on alternate planes.

ACKNOWLEDGMENTS

The authors would like to thank Rob Phillips for many useful discussions. This work was in part supported by the National Institute of Standards and Technology under award 60NANB4D1587, and by the Office of Naval Research under Grant Nos. N00014-92-J-4049 and N00014-92-F-0098.

REFERENCES

[1] R. Thomson, S. J. Zhou, A. E. Carlsson, and V. K. Tewary, Phys. Rev. B **46**, 10613 (1992).

[2] S. J. Zhou, A. E. Carlsson, and R. Thomson, Phys. Rev. Lett. **72**, 852 (1994).

[3] L. M. Canel, A. E. Carlsson, and R. Thomson, Phys. Rev. B **52**, 158 (1995).

[4] A. A. Griffith, Philos. Trans. R. Soc. London Ser. A **221**, 163 (1920).

[5] J. R. Rice, J. Mech. Phys. Solids **40**, 239 (1992).

[6] S. J. Zhou, A. E. Carlsson, and R. Thomson, Phys. Rev. B **47**, 7710 (1993).

[7] H. Vehoff and P. Neumann, Acta Metall. **28**, 265 (1980).

[8] S. M. Ohr, Mater. Sci. Eng. **72**, 1 (1985).

[9] R. Thomson, Solid State Physics **39**, 2 (1986).

[10] M. L. Williams, J. Appl. Mechanics **19**, 526 (1952).

ISOTROPIC MD SIMULATIONS OF DYNAMIC BRITTLE FRACTURE

PEP ESPAÑOL, MIGUEL A. RUBIO, AND IGNACIO ZÚÑIGA
Departamento de Física Fundamental, Universidad Nacional de Educación a Distancia,
Madrid, Spain

ABSTRACT

We present results obtained by molecular dynamics simulations on the propagation of fast
cracks in triangular 2D lattices. Our aim is to simulate Mode I fracture of brittle isotropic
materials. We propose a force law that respects the isotropy of the material. The code
yields the correct imposed sound $c_{\parallel}$, shear $c_{\perp}$ and surface V_R wave speeds. Different notch
lengths are systematically studied. We observe that initially the cracks are linear and always
branch at a particular critical velocity $c^* \approx 0.8 V_R$ and that this occurs when the crack tip
reaches the position of a front emitted from the initial crack tip and propagating at a speed
$c = 0.68 V_R$.

INTRODUCTION

Pioneering [1] and recent [2] molecular dynamics computer simulations of fracture have
focused on Lennard-Jones particles in solid phase forming a triangular 2D lattice. These
simulations are useful for understanding crystal fracture dynamics, which is essentially
anisotropic.

Our interest is in understanding fracture dynamics of isotropic amorphous materials like
PMMA used in recent experiments of dynamic brittle fracture [3],[4]. The question is: which
law of force one has to use for simulating isotropic solids with molecular dynamics? We will
answer this question by resorting to a discretization of the isotropic continuum equations of
linear elasticity.

The linear equations of elasticity that govern the displacement field $\mathbf{u}(\mathbf{r}, t)$ in an homogeneous material subject to small deformations are

$$\ddot{\mathbf{u}}(\mathbf{r}, t) = c_{\perp}^2 \nabla^2 \mathbf{u}(\mathbf{r}, t) + (c_{\parallel}^2 - c_{\perp}^2) \nabla(\nabla \cdot \mathbf{u}(\mathbf{r}, t)) \tag{1}$$

where the transverse $c_{\perp}$ and longitudinal $c_{\parallel}$ sound speeds are material properties related
to Young's modulus E and Poisson's coefficient σ. Surface waves propagating on the free
boundaries of the material travel at the Rayleigh wave speed

$$V_R \approx c_{\perp} (0.874 + 0.162\sigma) \tag{2}$$

DISCRETIZATION OF CONTINUUM EQUATION

We have devised a method for discretizing partial differential equations in an arbitrary
mesh (including random lattices) [5]. For the particular case of a triangular mesh, the
continuum equations (1) become in a particular node i

$$\ddot{\mathbf{u}}_i(t) = \left[\frac{c_{\perp}^2 - c_{\parallel}^2/3}{a^2} \right] \sum_{j=1}^{6} (\mathbf{u}_j - \mathbf{u}_i) + \frac{4(c_{\parallel}^2 - c_{\perp}^2)}{3a^2} \sum_{j=1}^{6} (\mathbf{u}_j - \mathbf{u}_i) \cdot \hat{\mathbf{r}}_{ji}^0 \hat{\mathbf{r}}_{ji}^0 \tag{3}$$

101

Mat. Res. Soc. Symp. Proc. Vol. 409 ⊕ 1996 Materials Research Society where $\hat{\mathbf{r}}_{ji}^0$ are the lattice unit vectors joining particle i and its nearest neighbours j and a is the lattice spacing. The continuum equations become equations for point particles interacting with a linear law of force. This law of force will respect the isotropy of the initial continuum equations. In general, real springs or Lennard-Jones interactions between particles (two popular choices in computer simulations) are not isotropic and are not well suited to study isotropic elasticity of amorphous materials.

The fracture criteria we select is that if two particles separate more than a certain critical distance, the force between the particles is set to zero. This corresponds in the continuum picture to a criterium involving the eigenvalues of the deformation tensor $\gamma = \frac{1}{2}[\nabla \mathbf{u} + \nabla \mathbf{u}^T]$.

MD SIMULATIONS

We perform ordinary molecular dynamic simulations of 2D triangular lattices of point particles interacting with (3). A conventional velocity Verlet algorithm is used. The units of space and velocity are a and V_R, respectively. We have carefully checked that the propagation of sound and shear waves is according to $c_\parallel$ and $c_\perp$: a localized pulse in the center of an unstreched 2D plate propagates as an ellipse with semiaxis equal to $c_\parallel t$ and $c_\perp t$.

We will consider finite rectangular plates of dimensions $L_x \times L_y$ that contain an initial sharp notch of zero width, and length L_n located in the middle of the upper boundary. The plate is subject to a displacement ΔL_x in the x direction of one of the (vertical) boundaries, thus producing a Mode I load with nominal deformation $\epsilon = \Delta L_x / L_x$. This loading is assumed to be quasistatic in the sense that prior to the initiation of fracture, the loading is so slow that the stress field assumes its equilibrium value form.

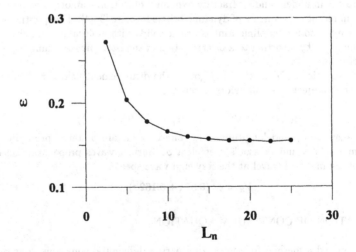

Fig. 1: The breaking deformation ϵ for different notch lengths. Shorter notches require higher deformations.

The most streched bond is at the tip of the notch. We select a critical deformation of $\gamma_c = 0.01$ and compute at which plate deformation ϵ the most streched bond reaches the deformation threshold γ_c. In Fig. 1 ϵ is plotted for different notch lengths. The lattice is allowed to relax to the equilibrium position without breaking and then the simulation starts allowing this bond to break.

RESULTS

We consider two lattices with the same aspect ratio $L_y/L_x = 1.16$ and composed of 2500 and 40000 particles, respectively. A typical realization is shown in Fig. 2 for the small lattice.

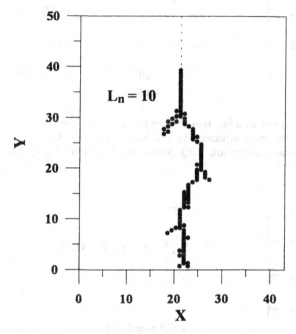

Fig. 2: A typical crack showing a first linear regime that eventually branches.

The first thing we observe is that the cracks start moving in a straight line before branching occurs. The time of branching is always smaller than the time needed by sound waves to reflect from the boundaries. The branches do not follow the lattice directions which is a consequence of the isotropy of the simulated lattice.

We focus on the initial destabilization of the linear crack. Linear cracks move always at a velocity smaller than V_R. Actually, the linear cracks accelerate; this can be seen in Fig. 3, and the acceleration depends on notch length. They branch when they reach a velocity around $c^* = 0.77 V_R$, independent of notch length as shown in Fig. 4.

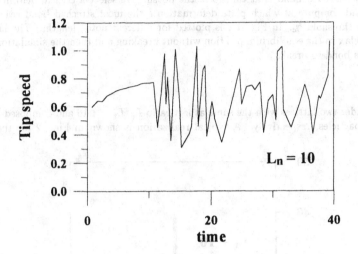

Fig. 3: Crack tip speed as a function of time for $L_n = 10$. The tip speed is calculated from the position of the most advanced broken bond. The smooth curve up to $t \sim 10$ corresponds to the linear propagation; the following oscillations are due to crack branching.

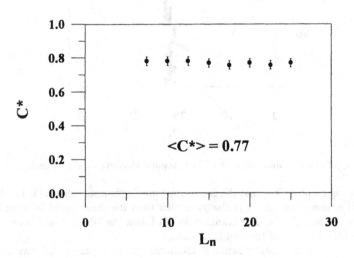

Fig. 4: Crack tip speed at which branching occurs for different notch lengths.

In Fig. 5a we plot the distance from the initial notch at which branching occurs versus the time at wich it occurs, for different notch lengths. The points fall onto a straight line of slope $c = 0.686$. Plotting $y_{tip} - ct$ as a function of t as in Fig. 5b it appears that branching occurs when the crack tip reaches a front emitted at the initial notch and propagating at $c = 0.686$. When this happens, the velocity of the crack tip is always $c^* = 0.77$.

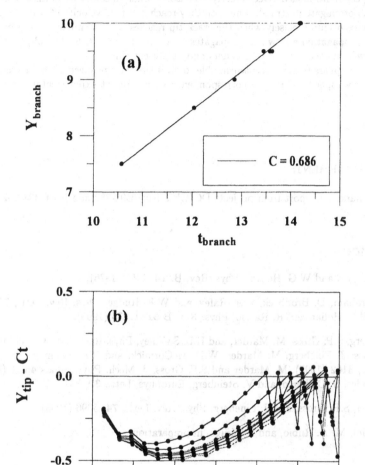

Fig. 5: (a) Distance from the initial notch at which branching occurs versus the time at wich it occurs, for different notch lengths. The solid line is a linear fit with slope $c = 0.686$. (b) The difference between the position of the crack tip and the front Ct as a function of time, for different notch lengths. Short notches produce faster cracks.

CONCLUSIONS

We have deviced a molecular dynamics simulation for studying the dynamic fracture of isotropic amorphous brittle materials in Mode I loading. The force law between the particles representing the material is derived directly from the continuum equations of linear elasticity.

All cracks propagate in straight line until they reach a critical velocity $c^* = 0.77V_R$. This critical velocity occurs precisely when the crack tip reaches a front emitted at the begining of the crack propagation and which propagates at a velocity $c = 0.686$. The physical origin of this new velocity scale is not well understood at present time.

Branching appears to be a mechanism able to limit the average speed of the cracks below the surface wave speed. This is in correspondence with a recent experimental observation [4].

ACKNOWLEDGEMENTS

Partial financial support from projects DGICYT No PB94-382 and No PB93-292 is acknowledged.

References

[1] W.T. Ashurst and W.G. Hoover, Phys. Rev. B. **14**, 1465 (1976).

[2] F.F. Abraham, D. Brodbeck, R.A. Rafey, and W.E. Rudge, Phys. Rev. Lett., **73**, 272 (1994), B.L. Holian and R. Ravelo, Phys. Rev. B **51**, 11275 (1995).

[3] J. Fineberg, S.P. Gross, M. Marder, and H.L. Swinney, Phys. Rev. Lett., **67**, 457 (1991). S.P. Gross, J. Fineberg, M. Marder, W.D. McCormick, and H.L. Swinney, Phys. Rev. Lett., **71**, 3162 (1993). M. Marder and S.P. Gross, J. Mech. Phys. Solids **43**, 1 (1995). J.F. Boudet, S. Ciliberto, and V. Steinberg, Europhys. Lett., **30**, 337 (1995).

[4] E. Sharon, S.P. Gross, and J. Fineberg, Phys. Rev. Lett., **74**, 5096 (1995).

[5] P. Español, M. A. Rubio, and I. Zúñiga, in preparation.

MECHANISM OF THERMALLY ASSISTED CREEP CRACK GROWTH

LEONARDO GOLUBOVIĆ and DOREL MOLDOVAN
Department of Physics, West Virginia University, Morgantown, WV 26506

ABSTRACT

We use atomistic Monte-Carlo simulations to investigate the dynamics of cracks which sizes are *smaller* than the Griffith length. We demonstrate that such cracks can *irreversibly* grow proviso their size is larger than a certain critical length which is smaller than the Griffith length, as recently suggested [L. Golubović and A. Peredera, Phys. Rev. E51, 2799 (1995)]. We show here that this thermally assisted creep crack growth is dominated by irreversible changes in the region of the crack tip, primarily in the form of dislocation emissions and nucleation of microcavities and voids. These processes act together during the crack growth: the crack tip region acts as a source for emissions of dislocations which subsequently serve as seeds for creation of vacancy clusters in a region away but still close to the crack tip. Eventually, passages between these vacancy clusters and the mother crack are formed and the crack thus increases in size. As this process repeats, the crack grows.

INTRODUCTION

Stressed solids can be treated as a metastable state of matter analogous to, say, supercooled liquids [1]-[6]. The failure treshold corresponds to a metastability limit, or spinodal point, at which the external stress σ as a function of the strain reaches its maximum, σ_{max} [2]-[4]. If the external tensile stress is smaller than σ_{max}, a stressed sample will still break, however, with a time-delayed fracture. The sample lifetime depends on the temperature and the applied stress [7]. This phenomenon is believed to be directly related to the processes of *microcrack nucleation* and growth [2]-[6]. Microcrack nucleation is phenomenologically similar to that of the stable phase droplets in a metastable state [8]. The seminal work of Griffith on fracture mechanics [9], already contains all the elements to construct a phenomenological theory of microcrack nucleation. The critical, Griffith crack behaves like a critical droplet: Cracks larger than the Griffith crack grow irreversibly in a rapid fashion [10]. On the other hand, the growth of cracks which are smaller than the Griffth crack size is energetically disfavored. It is believed, however, that their size can still change in time, however, in a slow fashion via an activational dynamics [4]-[6]. Nature of this thermally assisted creep crack growth remains unclear, in spite of its practical significance [7].

Here we use atomistic Monte-Carlo simulations to investigate the dynamics of cracks which sizes are *smaller* than the Griffith length. We demonstrate that such cracks can *irreversibly* grow proviso their size is larger than a certain critical length which is smaller than the Griffith length, as recently suggested [5][6]. We show here that this thermally assisted creep crack growth is dominated by irreversible changes in the region of the crack tip, primarily in the form of dislocation emissions and nucleation of microcavities. These two tipes of processes act together to produce the crack growth: the crack tip region acts as a source for emissions of dislocations which subsequently serve as seeds for creation of vacancy clusters (microcavities) in a region away but still close to the crack tip. Eventually, passages between these mirocavities and the mother crack are formed and the crack thus increases in size. As this process repeats, the crack grows.

THEORIES OF FRACTURE NUCLEATION

We begin by reviewing first the conventional theory of fracture nucleation inspired by the pioneering work of Griffith.[9] In this picture microcracks play a role analogous to that of the stable phase droplets in a metastable state. Griffith established a criterion for crack growth by estimating the energy cost of creating a brittle crack of length L in a solid under a uniaxial stress σ perpendicular to the crack. Creation of the crack, for example, in a two-dimensional solid, costs an energy of the order

$$E(L) \; = \; gL \; - \; \frac{\sigma^2 L^2}{2Y}. \tag{1}$$

The first term in (1) is the energy cost of creating crack's edges by breaking atomic bonds. Thus ga, with a the atomic size, is of the order of a bond energy. After crack creation, its edges will separate, with maximal opening displacement of the order $d = \sigma L/Y$, where Y is the Young modulus. The crack opening relaxes the stress in a domain of size L^2 and lowers the elastic energy of the stressed solid by an amount of the order $L^2\sigma^2/2Y$. This yields the second term in (1), which, in contrast to the first one, energetically favors crack growth. Crack energy (1) reaches its maximum at L equal to the critical Griffith length $L_g = gY/\sigma^2$, corresponding to the energy (1) of the order $E_g = E(L_g) = g^2Y/\sigma^2$. The crack state with $L = L_g$ is unstable: For $L > L_g$, the crack growth decreases $E(L)$. This leads to the well-known irreversible, very rapid crack growth [10]. For $L < L_g$, an increase of L costs a positive amount of energy (1). This hinders the crack growth for $L < L_g$. This picture resembles that of standard nucleation phenomena [8], with E_g playing the role of the nucleation energy barrier E_b. The nucleation rate R_N - or the time t_N needed for the fracture to be nucleated by thermal fluctuations - can be, in general, estimated by the Arrhenius law $R_N = 1/t_N \sim exp(-E_b/k_BT)$.[8] Within the conventional fracture nucleation picture E_b is identified with E_g. Thus $E_b = E_g = g^2Y/\sigma^2$.

This conventional theory was criticized by Golubović and Feng (GF) [5]. They consider *surface processes* such as surface diffusion [11], which restructure crack edges and may *inhibit* healing of microcracks shorter than the Griffith length $L = L_g$. These processes become active as soon as the crack opening displacement d becomes larger than the atomic size a. This happens for $L > L_{min}$ with $L_{min} = aY/\sigma$, or, as $g \approx aY$, $L_{min} = g/\sigma$. By considering surface processes inhibiting crack healing, GF argued that the effective energy barrier for the fracture nucleation is of the order $E(L_{min})$. Thus, $E_b = E(L_{min}) = gL_{min} = g^2/\sigma$, in the phenomenological theory of fracture nucleation proposed in GF. As $E_g/E(L_{min}) \approx L_g/L_{min} \approx Y/\sigma \approx \sigma_{max}/\sigma$, the nucleation rate predicted by GF is, for weak stresses, $\sigma << \sigma_{max}$, enormously larger than that of the conventional theory.

More recent Monte-Carlo dynamics simulations of Golubović and Peredera provided, however, a different prospect on the same problem [6]. They indicate that *microcavities* (i.e., clusters of vacant sites) rather than microcracks (i.e., lines or surfaces of "broken bonds") are major nucleated defects in the delayed fracture regime. Thus, the major kinetic process in the time-delayed fracture is the nucleation of vacancy clusters or microcavities (possibly having a non-zero Burgers vector), *not* of critical Griffith-type microcracks. Energetics of a vacancy cluster nucleation is rather different from that of the Griffith crack: Consider, for example, a vacancy cluster in a 2d solid. If R is its linear size, the cluster involves$\approx (R/a)^2$ vacant sites. Its size-dependent energy is of the form

$$E_{vac}(R) = gR - \sigma R^2. \tag{2}$$

The first term in (2) is, as in Eq. (1), the surface energy contribution($\sim$perimeter$\sim R$ in 2d). The second term in (2) is a stress-induced volume contribution($\sim$area$\sim R^2$ in 2d) [6][12]. This volume term can be rationalized as follows: Consider a solid under a tensile stress σ with no vacancy cluster initially present. As the number of atoms is conserved, the creation of the cluster induces an increase of the sample's linear size preferentially along the direction of the applied stress [6][12]. This lowers the energy by $\delta E =$stress$\times$cluster volume$\sim \sigma R^2$ yielding the second term of the equation (2). By Eq. (2), the size of the critical vacancy cluster, maximizing $E_{vac}(R)$, is of the order $R_c = g/\sigma$, whereas the energy barrier for the vacancy cluster nucleation is of the order $E_b = E_{vac} = g^2/\sigma$. Interestingly, this energy-barrier scale for fracture nucleation coincides with that proposed by Golubović and Feng [5], however, from seemingly different arguments. Thus, the microcavity nucleation yields fracture nucleation barriers and corresponding rates identical to those proposed before, in a different manner, by Golubović and Feng [5]. This agreement is probably *not* accidental. These authors invoke in their discussion processes which start to restructure microcrack edges as soon as the microcrack size reaches the lengthscale L_{min}. This lengthscale coincides with the size of the critical vacancy cluster, R_c. As $R_c = L_{min}$, one may argue that the processes restructuring microcrack edges would transform a microcrack of the size L_{min} into a microcavity of the size R_c which then continues to grow irreversibly. In this way one can rationalize the suggestion of GF that a crack with $L = L_{min}$ never heals and continues to grow irreversibly. Thus, L_{min} is the true critical size of defects in solids under weak tensile stresses.

GROWTH OF CRACKS SHORTER THAN THE GRIFFITH SIZE

Whereas the previous works indicate that a crack of the length L in the range $L_g > L > L_{min}$ is likely to slowly grow irreversibly due to thermal fluctuations (seemingly violating Griffith's original ideas), these works say little about the actual nature of this growth. To clarify the nature of this thermally activated (creep) crack growth we performed atomistic (off lattice) Monte-Carlo simulations of a 2d Lennard-Jones solid under a tensile stress, similar to recent simulations in Refs. [4] and [6]. Here we simulate behavior of cracks shorter than the Griffith length, $L < L_g$. In Figures 1 and 2 we give the time evolution of the solid initially containing such a crack. The sample is under an external tensile force along the vertical direction. The time unit used in these figures is one Monte-Carlo Cycle (MCC) [$1MCC$ involves updates of *all* particles' positions via the standard Metropolis algorithm].

Figure 1 is an example from our simulations illustrating creation of microcavities (vacancy clusters) nucleated from dislocations emitted by the crack tip. These emissions quickly blunt initially sharp crack tips. More importantly, these emissions frequently produce defects having character of *dislocation-vacancy pairs* - see the first atomic configuration in Fig. 1. This pair dissociates and the free dislocation continues to glide [see $t = 1333$]. Eventually, this same dislocation acts as a seed for creation of another vacancy [see $t = 5333$], which evolves into a microcavity [see the left bottom corner at $t = 10000$]. Meanwhile, the crack tip gets further blunted by the emission of another dislocation [see $t = 10000$] which temporarily gets trapped by the nearby vacancy forming a dislocation-vacancy pair close to the crack tip [see t between 10000 and 12000].

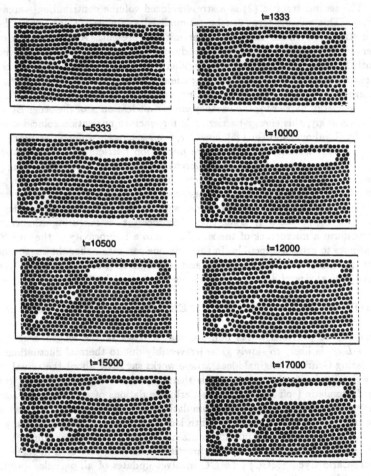

Figure 1: Creation of cavities nucleated from dislocations emitted by the crack tip.

Eventually, this pair dissociates, and the free dislocation approaches the left bottom corner where the microcavity is waiting for it. Once they had come close to each other, a new microcavity was formed sitting next to the old one [see $t = 15000$ and $t = 17000$]. Figure 2 depicts the further evolution of this sample [between 27000 and 34500]. We see that the two microcavities from $t = 17000$ joined into a single one [see $t = 27000$]. Finally, a *passage* between this cavity and the mother microcrack is nucleated [see $t = 34500$].

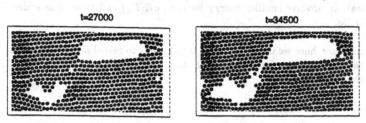

Figure 2 : Nucleation of a passage between two microcavities.

These simulations show that the thermally assisted creep crack growth is a complex combination of three types of processes going on in the region of the crack tip: (i) emissions of dislocations from the crack tip, (ii) these dislocations subsequently serve as seeds for creation of vacancy clusters (microcavities) in a region close to the crack tip, and (iii) nucleation of passages between these mirocavities and the mother crack. These subsequent processes form a cycle by which the crack grows. This picture suggests a simple kinetic model for the thermally assisted creep growth of the crack length L as a sequence of steps like to that in Figures 1 and 2: During each step the crack tip produces one or few cavities of the size $R \approx g/\sigma_{loc}$ and the energy $E(R) \approx gR$. Here, σ_{loc} is local stress in the vicinity of the blunt tip with radius$\approx R$. Thus, $\sigma_{loc} \approx \sigma(L/R)^{1/2}$, where σ is the remote (external) stress. By combining these relations, one finds $R(L) \approx R_c^2/L$ and $\sigma_{loc}/\sigma \approx L/R_c$, where $R_c = g/\sigma$ is the critical cavity size (in the absence of the crack) discussed in the previous section. Note a strong dependence of the cavity size $R(L)$ on the crack length L: For $L \approx R_c = L_{min}$ (the minimal possible size of a crack [5][6]): $R \approx R_c$ and $\sigma_{loc} \approx \sigma$. For $L \approx L_g$: $R \approx g/Y \approx a$ and $\sigma_{loc} \approx Y \approx \sigma_{max}$. Thus, as the crack size L grows, the size of the cavities it nucleates decreases, and becomes the atomic size a when L reaches the Griffith length L_g. This growth is step-like. During each step L increases by $\Delta L \approx R(L)$ during the cavity nucleation time interval $\Delta t \sim exp(gR/k_BT)$. Thus, the average speed of the crack $v(L) = \Delta L/\Delta t$ behaves as $v(L) \sim exp(-gR(L)/k_BT) = exp(-gR_c^2/k_BTL)$. By using this, the time scale needed for a crack with the initial size L_o to reach the Griffith size behaves as

$$t(L_o) \sim exp\left(\frac{\sigma_{max}}{\sigma}\frac{T_m}{T}\frac{L_{min}(\sigma)}{L_o}\right), \tag{3}$$

with $T_m = ga/k_B \approx$bond breaking temperature, and $L_{min}(\sigma) = R_c = g/\sigma \approx (\sigma_{max}/\sigma)a$. For the smallest possible cracks with $L_o = L_{min}$, the time scale (3) behaves as the microcavity nucleation time [6], whereas for $L_o = L_g \approx (\sigma_{max}/\sigma)L_{min}$, the time scale (3) becomes microscopic, $t(L_g) \sim exp(T_m/T)$ ~thermal bond breaking time.

For $\sigma \ll \sigma_{max}$, the growth of L is a sequence of many steps (as the one in Figures 1 and 2) with *decreasing* sizes $L_{n+1} - L_n = R(L_n) = R_c^2/L_n$, for the n-th step. Thus $L_n^2 - L_o^2 = nR_c^2$. This implies that during the growth of L from the inital L_o to the Griffith size L_g there is a large number of about $N = (L_g^2 - L_o^2)/R_c^2 = (\sigma_{max}/\sigma)^2[1 - (L_o/L_g)^2]$ steps. The time scale (3) is actually dominated by the time it takes to grow from $L_{n=o}$ to $L_{n=1}$, i.e., by the time it takes for the *first* nucleation event to occur. All subsequent cavity nucleations involve smaller energy barriers $gR(L_n)$ and thus time scales smaller than that in Eq. (1).

To summarize, here we used numerical simulations to reveal a complex nature of the thermally assisted growth of cracks shorter than the Griffith size and suggest a simple kinetic model for this growth.

ACKNOWLEDGMENTS

This work is suported by the NSF/WV EPSCoR program.

REFERENCES

1. K. Nishioka and J. K. Lee, Philos. Mag. A **44**, 779 (1981).
2. R. L. B. Selinger, Z.-G. Wang, W. M. Gelbart, and A. Ben-Shaul, Phys. Rev. A **43**, 4396 (1991)
3. Z.-G. Wang, U. Landman, R. L. Blumberg Selinger, and W. M. Gelbart, Phys. Rev. B **44**, 378 (1991).
4. R. L. B. Selinger, Z.-G. Wang, and W. M. Gelbart, J. Chem. Phys. **95**, 9128 (1991).
5. L. Golubović and S. Feng, Phys. Rev. A **43**, 5223 (1991).
6. L. Golubović and A. Peredera, Phys. Rev. E **51**, 2799 (1995).
7. S. S. Brenner, in Fiber Composite Materials (American Society for Metals, Metals Park, OH, 1965), p. 11.
8. E. M. Lifshits and L. P. Pitaevski, Physical Kinetics (Pergamon, Oxford, 1981), p. 427-431.
9. A. A. Griffith, Philos. Trans. R. Soc. London Ser. A **227**, 163 (1920); see also, L. D. Landau and E. M. Lifshits, Theory of Elasticity, 2nd ed. (Pergamon, Oxford, 1970), pp. 144-149.
10. N. F. Mott, Engineering **165**, 16 (1948).
11. C. Herring, J. Appl. Phys. **21**, 301 (1950).
12. C. Herring, J. Appl. Phys. **21**, 437 (1950); F. R. N. Nabarro, Report of a Conference on the Strength of Solids (Phys. Soc., London, 1948), p. 75.

Part II
Dislocations—Theory/Simulation Approaches

DYNAMIC SIMULATION OF CRACK PROPAGATION
WITH DISLOCATION EMISSION AND MIGRATION

N. ZACHAROPOULOS*, D.J. SROLOVITZ*, R.A. LeSAR**
*Department of Materials Science and Engineering, University of Michigan, Ann Arbor, MI
48109-2136, nikzach@umich.edu, srol@umich.edu
**Los Alamos National Laboratory, Los Alamos, NM 87545, ral@lanl.gov

ABSTRACT

We present a simulation procedure for fracture that self-consistently accounts for dislocation emission, dislocation migration and crack growth. We find that the dislocation microstructure in front of the crack tip is highly organized and shows a complex temporal-spatial evolution. The final dislocation microstructure and the number of emitted dislocations immediately proceeding fracture varies rapidly with the loading rate. For high loading rates, fracture occurs at smaller loads with increasing loading rate. However, the load at fracture shows a maximum with respect to loading rates.

INTRODUCTION

As is well known, crack propagation and dislocation plasticity are intimately linked in metallic alloys. Despite years of study, however, our understanding of crack growth with dislocation generation and propagation is still in its infancy. Rice and Thomson [1] proposed that dislocation emission from the crack tip determines how ductile (or brittle) fracture will be. Other studies [2-8] have shown that the dislocations shield the crack from the applied stress. These theoretical studies have either considered the direct elastic interaction between a crack and dislocations [2-7] or have addressed crack tip shielding through more continuum plasticity based approaches [8]. Simulations of crack growth with dislocation emission have been used to describe ductile-to-brittle transitions in multilayers [9]. Nonetheless, the synergism between dislocation emission and crack tip shielding has not self-consistently accounted for the true dislocation microstructure.

Dislocation emission from the crack tip has been observed using *in situ* deformation in a transmission electron microscope [10-12]. Unfortunately, since such studies were, by necessity, performed on extremely thin samples, it is unclear how these observations relate to fracture in bulk materials. Although post crack propagation observations of materials have provided hints as to the nature of the dislocation microstructure associated with fracture, it is difficult to extrapolate back to the actual dislocation microstructure development and crack shielding that actually took place immediately before and following fracture. Therefore, both models and experiments that examine the emission of dislocations from the crack tip and their subsequent migration will be central.

Recently, simulations have been performed which account for the motion of large numbers of dislocations and the formation of dislocation microstructures [13-17]. In the present study, we apply such a simulation method to the situation of a crack subject to mode III loading to account for dislocation emission from the crack tip, dislocation microstructure formation and evolution and crack propagation in an attempt to bridge the gap between discrete crack tip shielding models, dislocation emission models and continuum plasticity models.

SIMULATION METHOD

In this study, we employ the fast multipole method (FMM) developed by Greengard and Rokhlin [18] in order to describe dislocation interactions in two dimensions (we consider only parallel, straight dislocations). The basic concept of the multipole method is that the pairwise interactions between dislocations in two different, *well-separated* regions of the material can be replaced by the multipole expansion of the field from all dislocations in one region evaluated at the center of the other region. Regions, or sets, of dislocations are considered well-separated when

115

the centers of the sets are separated by a distance greater than 3R, where R is the radius of each, individual region.

This approach is based upon a tree hierarchy. The highest level of the tree (level 0) is the simulation cell. This is divided into quadrants, thus creating level 1 cells. Each of the level 1 cells is again divided into quadrants which represent level 2 cells. This procedure is carried out to a level in which the cells (leaves) contain an optimal number of dislocations (see [16] for details). A level i cell is the parent of the level i+1 cells that it is divided into; the latter are the former's children.

Starting at the lowest level, we calculate the multipole expansion at the center of each leaf. We then shift the multipole of each cell to the center of its parent, summing the contributions from all four children to yield the multipole moment of the parent. By repeating the procedure, the multipoles for all of the cells at each level are attained. Thus, if we are interested in the stress on a dislocation we only need to consider the direct interactions with the dislocations within the same leaf and the first nearest neighbor leaves. The well-separated dislocations are accounted for via the multipole moments. The advantage of this algorithm is that the computational operations are O(N), where N is the number of dislocation in the entire simulation cell.

MODE III CRACK

In this study, we consider the relatively simple case of a semi-infinite crack under anti-plane shear loading. We describe the simulation cell in terms of complex coordinates $z=x+iy$ and make use of a simple coordinate transformation. The negative x semi-axis of the simulation cell is mapped into the $\xi>0$ half-space in the complex $\zeta = \xi + i\eta = \sqrt{z}$ transform plane. Hence, the condition of traction-free crack faces is satisfied by introducing image dislocations. The tractions on the free surface created by the stress field of a dislocation located at $\zeta_i = \xi_i + i\eta_i$ are negated by the stress field of an image dislocation of opposite Burgers vector situated at $\zeta_i' = -\xi_i + i\eta_i$. The dislocation-dislocation interactions are calculated in the transform plane by way of the FMM using the multipole expansion of the screw dislocation potential, Ω

$$\Omega(\zeta) = b\ln(\zeta) \qquad , \qquad (1)$$

where b is the Burgers vector. We write the dislocation stress field in the z plane as a combination of the real and imaginary parts of the derivative of $\Omega(\zeta)$

$$\sigma^d(z) = \frac{\mu b}{2\pi}\Omega'(\zeta)\frac{1}{2\sqrt{z}} \qquad , \qquad (2)$$

where $\sigma^d(z) = \sigma^d_{yz} + i\sigma^d_{xz}$ and μ is the shear modulus. The total stress field is then the linear superposition of $\sigma^d(z)$ and the stress field of the loaded crack, $\sigma^c(z)$

$$\sigma^c(z) = \frac{K_{III}}{\sqrt{2\pi z}} \qquad , \qquad (3)$$

where K_{III} is the applied mode III stress intensity factor.

The force on a dislocation is given by the Peach-Köhler relation

$$\vec{F} = (\vec{b} \cdot \vec{\sigma}) \times \vec{\xi} \qquad , \qquad (4)$$

where $\vec{\sigma}$ is the total stress tensor and $\vec{\xi}$ the dislocation line direction. The presence of dislocations modifies the stress intensity factor at the crack tip from the applied value K_{III} to

$$K_{tip} = K_{III} - \sum_i \frac{\mu b_i}{2\sqrt{2\pi}} \left(\frac{1}{\sqrt{z_i}} + \frac{1}{\sqrt{\bar{z}_i}} \right) \tag{5}$$

where b_i, z_i refer to the burgers vector and position of the ith dislocation, respectively, and the bar denotes the complex conjugate. The second term in Eq. (5) is the shielding term.

The criterion for crack growth is the satisfaction of the classical Griffith condition

$$K_{tip} > 2\sqrt{\mu\gamma} \tag{6}$$

where γ is the surface energy. In these simulations, the crack is always brought/kept at equilibrium before any emission or motion of dislocations occurs, since we assume that crack propagation is much faster than dislocation motion. Screw dislocations of positive Burgers vector are emitted in pairs at slip planes inclined at $\pm 30°$ to the crack plane (similar to the approach used in [9]) at a distance of $4b$ away from the tip. Emission takes place when the force on the incipient dislocation pair exceeds the Peierls force (lattice friction) and there are no dislocations present in a radius of $2b$ from the point of emission. Once emitted, the dislocations move with a velocity, $v = MF$, where F is the Peierls force, and the dislocation motion is not confined to any particular plane. The time step in these simulations is determined dynamically and is controlled by the slowest process (crack growth or emission). Dislocations whose trajectory crosses the crack faces are annihilated.

RESULTS AND DISCUSSION

The evolution of the dislocation microstructure and the fracture behavior were investigated for different values of the applied loading rate, $\dot{K}_{III}$, presented here in units of $MPa^2\sqrt{m} \cdot M$, where the dislocation mobility is in units of $MPa^{-1} \cdot s^{-1}$. Figure 1 shows the evolution of the dislocation microstructure for $\dot{K}_{III} = 0.5$. The physical parameters (b, μ, γ) are chosen to match those of copper. The first micrograph is shown for time t=t_0 ($t_0 \approx .72$ $MPa^{-1} \cdot M^{-1}$), where the dislocation microstructure is restricted to a small region in front of the crack tip. As time progresses, the size of the dislocated region grows and the microstructure undergoes a transition from a relatively scattered configuration of dislocations ($3.0t_0$) to one with well-defined branches radiating from the crack tip ($3.5t_0$). Farther from the crack tip, the branches bifurcate into a less well defined structure ($5.0t_0$). These branches become less well defined as the dislocation density increases ($5.5t_0$), although they persist to just before K_{tip} reaches the critical value for crack growth ($8.5t_0$).

The final microstructures, just before the crack propagates catastrophically are presented in Fig. 2 for different loading rates. For large $\dot{K}_{III}$, the dislocation microstructures are limited to the immediate neighborhood of the crack tip and are significantly less developed than the dense, extended ellipse-shaped microstructures that are observed at low loading rates. These microstructures are nearly identical with those seen during the microstructural evolution of a low $\dot{K}_{III}$ simulation, where time increases clockwise starting from the micrograph in the lower right. In all cases, we note the existence of a dislocation free zone directly ahead of the crack. This zone is a result of the fact that any dislocation ahead of the crack at y≠0, experiences a force that drives it further away from the y=0 line.

Figure 3 shows the number of emitted dislocations as a function of the normalized, applied stress intensity K_{III}/K_c, where $K_{III}>K_c$ is the condition for crack growth in the absence of any dislocations. The four curves in this figure correspond to different values of $\dot{K}_{III}$ and terminate at the value of the applied loading at which cleavage occurs. We observe that the increase in the number of emitted dislocations per increase in loading is greater for small values of $\dot{K}_{III}$ than it is for large values. This is because the same change in the applied stress intensity factor occurs over a longer time period at low loading rates, hence the emitted dislocations have more time to move away from the crack tip such that more dislocations can be nucleated. At large loading rates, the value of K_{III}/K_c at which the crack propagates through the system increases as the loading rate is decreased because more dislocations are emitted, increasing the crack tip shielding. However, the high density of dislocations produced at low loading rates eventually limits the nucleation of additional dislocations. This occurs because the back stress these dislocations create at the

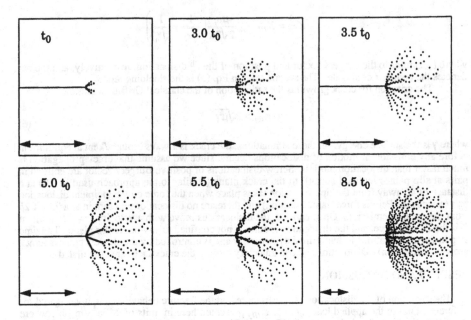

Figure 1. Evolution of the dislocation microstructure in front of the crack tip loaded at $\dot{K}_{III}=0.5$. The double-arrowed bars represent the relative scales.

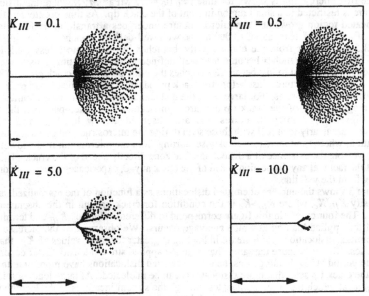

Figure 2. Final micrographs for four different values of $\dot{K}_{III}$. The double-arrowed bars represent the relative scales.

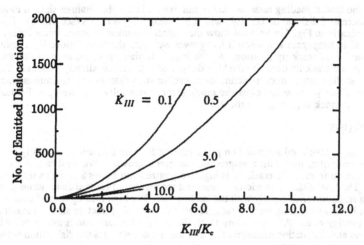

Figure 3. Number of emitted dislocations as a function of the applied stress intensity factor normalized by $2\sqrt{\mu\gamma}$ for four different loading rates.

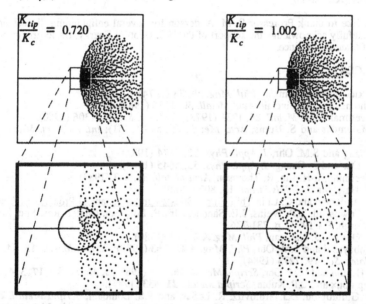

Figure 4. Two micrographs that show the fracture behavior for $\dot{K}_{III}$=0.1. The region in the vicinity of the crack tip is magnified for clarity.

emission sites near the crack tip sufficiently balances the stress field due to the crack at those sites. As a result, a crack loaded at a low rate will produce a high density of dislocations but the dislocation generation is eventually shut-off. Therefore, the material loaded at the lowest rate fractures at relatively low K_{III}/K_c.

At the lowest loading rates, the crack can extend in a stable manner since it moves into a region of increased dislocation density, which is absent at higher rates (see Fig. 2). This may be seen more clearly in Fig. 4, where we show the dislocation microstructure immediately after the crack begins to propagate and when it has grown well into the high dislocation density region ahead of the initial crack tip position. As the load is further increased no new dislocations are emitted from the crack tip owing to the high density of dislocations already present in that region. However, the dislocation microstructure does evolve such that those dislocations ahead of the crack move slightly above or below the crack plane, essentially opening up a dislocation free channel for the crack to propagate through.

CONCLUSIONS

We have developed a simulation procedure for fracture that self-consistently accounts for dislocation emission, dislocation migration and crack growth. We find that the dislocation microstructure in front of the crack tip is highly organized and shows a complex temporal-spatial evolution. The final dislocation microstructure and the number of emitted dislocations immediately proceeding fracture varies rapidly with the loading rate. For high loading rates, fracture occurs at smaller loads with increasing loading rate. However, the load at fracture shows a maximum with respect to loading rates. At low loading rates, the smaller fracture load is associated with stable crack growth and dislocation microstructure evolution that decreases the dislocation emission rate and forms an easy crack path.

ACKNOWLEDGMENTS

We would like to thank Professor P. M. Anderson for several enlightening discussions. The authors gratefully acknowledge the support of the U.S. Dept. of Energy, Basic Energy Sciences, Division of Materials Science.

REFERENCES

1. J.R. Rice and R. Thomson, *Phil. Mag.* **29**, 73 (1974).
2. R. Thomson and J. Sinclair, *Acta Metall.* **30**, 1325 (1982).
3. J. Weertman, *Acta Metall.* **26**, 1731 (1978); *J. Mater. Sci.* **15**, 1306 (1980).
4. B. Majumdar and S. Burns, *Acta Metall.* **29**, 579 (1981); *Int. J. Fract. Mach.* **21**, 229 (1983).
5. S. Chang and S.M. Ohr, *J. Appl. Phys.* **52**, 7174 (1981).
6. S.M. Ohr and S. Chang, *J. Appl. Phys.* **53**, 5645 (1982).
7. J. Weertman, I.-H. Lin, R. Thomson, *Acta Metall.* **31**, 473 (1983).
8. E.W. Hart, *Int. J. Solids Struct.* **16**, 807 (1980).
9. P.M. Anderson and C. Li in Thin Films: Stresses and Mechanical Properties IV, edited by P.H. Townsend, T.P. Weihs, J.E. Sanchez, Jr., P. Børgesen (Mater. Res. Soc. Proc. 308, Pittsburgh, PA 1993), p. 731-736.
10. S.M. Ohr and J. Narayan, *Phil. Mag. A* **41**, 81 (1980).
11. S. Kobayashi and S.M. Ohr, *Phil. Mag. A* **42**, 763 (1980); *Scripta Metall.* **15**, 343 (1981); *J. Mater. Sci.* **19**, 2273 (1984).
12. J.A. Horton and S.M. Ohr, *Scripta Metall.* **16**, 621 (1982); *J. Mater. Sci.* **17**, 3140 (1982).
13. J. Lepinoux and L.P. Kubin, *Scripta Metall.* **21**, 833 (1987).
14. A.N. Gulluoglou, D.J. Srolovitz, R. LeSar, and P.S. Lomdahl, *Scripta Metall.* **23**, 1347 (1989); in Simulation and Theory of Evolving Microstructures, edited by M.P. Anderson and A.D. Rollet (The Minerals, Metals & Materials Society 1990), p. 239-247.
15. R.J. Amodeo and N.M. Ghoniem, *Phys. Rev. B* **41**, 6958 (1990); in Modeling the Deformation of Crystalline Solids, edited by T.C. Lowe, A.D. Rollet, P.S. Follansbee and G.S. Daehn (The Minerals, Metals & Materials Society 1991), p. 125-143.
16. H.Y. Wang and R. LeSar, *Phil. Mag. A* **71**, 149 (1995).
17. D.B. Barts and A.E. Carlsson, *Phys. Rev. B* **52**, 2195 (1995).
18. L. Greengard and V. Rokhlin, *J. Comp. Phys.* **73**, 325 (1987).

SIMULATION OF DISLOCATIONS IN ORDERED Ni₃Al BY ATOMIC STIFFNESS MATRIX METHOD

Y. E. HSU, T.K. CHAKI
State University of New York, Department of Mechanical and
Aerospace Engineering, Buffalo, NY 14260

ABSTRACT

A simulation of structure and motion of edge dislocations in ordered Ni₃Al was performed by atomic stiffness matrix method. In this method the equilibrium positions of the atoms were obtained by solving a set of linear equations formed by a stiffness matrix, whose terms consisted of derivatives of the interaction potential of EAM (embedded atom method) type. The superpartial dislocations, separated by an antiphase boundary (APB) on (111), dissociated into Shockley partials with complex stacking faults (CSF) on (111) plane. The core structure, represented by the Burgers vector density distribution and iso-strain contours, changed under applied stresses as well as upon addition of boron. The separation between the superpartials changed with the addition of B and antisite Ni. As one Shockley partial moved out to the surface, a Shockley partial in the interior moved a large distance to join the lone one near the surface, leaving behind a long CSF strip. The decrease in the width of the APB upon addition of B and antisite Ni has been explained by a reduction of the strength of directional bonding between Ni and Al as well as by the dragging of B atmosphere by the superpartials.

INTRODUCTION

Nickel aluminide. Ni₃Al, is an intermetallic compound with L1₂-type ordered crystal structure. The single crystal of Ni₃Al is ductile at room temperature and its yield strength increases with temperature, reaching peak values at temperature in the range 700-1000 K. The generally accepted mechanism [1-4] for this flow stress anomaly involves immobilization of screw superpartial dislocations by cross-slip from {111} to {010} planes, forming Kear-Wilsdorf [5] locks. In order to verify the theoretical models [1-4] and study the dislocation core structure, a few atomistic computer simulations of screw superdislocations have been performed [6-9]. The simulations (except those by Yoo et al. [8]) showed non-planar core structures of screw superpartials.

It has recently been recognized that the plastic strain in Ni₃Al at relatively low temperatures is controlled by free movement of edge dislocations on {111} planes [4,10,11]. The transmission electron microscopic studies [10,11] of Ni₃Al specimens deformed at room temperature have shown the presence of numerous edge dipoles. However, no atomistic simulation of edge dislocations in Ni₃Al has been reported. Furthermore, even though the beneficial effect of B doping on the ductility in polycrystalline Ni₃Al (slightly Ni rich) at room temperature has been documented [12,13] quite a few years ago, no report of the effects of B and excess Ni on the dislocation core structure is known to us. Here we present the results of a computer simulation (performed by a new algorithm) of the core structure and mobility of edge dislocations in Ni₃Al. The simulations are also performed in Ni₃Al containing B and antisite Ni.

Mat. Res. Soc. Symp. Proc. Vol. 409 ©1996 Materials Research Society

COMPUTATIONAL METHOD

First, we constructed a parallelopiped block of Ni_3Al crystal. The coordinate axes were chosen parallel to the sides of the block with the x-axis along [10$\bar{1}$], y-axis along [111] and z-axis along [1$\bar{2}$1]. The dimensions of the block along x, y and z axes were 18.048 nm (143 layers), 3.507 nm (18 layers) and 0.657 nm (8 layers), respectively. A periodic boundary condition was applied along the z-axis while the other two surfaces were free. An edge superdislocation was introduced along the z-direction by removal of two (10$\bar{1}$) half-planes of atoms near the middle of the block. The block with the superdislocation contained 5112 atoms. Then, one empty half-plane was displaced with respect to the other along [10$\bar{1}$] so that two superpartials with a (111) APB in between them was created. To study the effect of B doping on the dislocations, we constructed a B-doped Ni_3Al block by introducing a B atom in the octahedral interstitial site of every 27th lattice cell. We preferentially put B atoms on the cells containing the glide plane. We also constructed a Ni-rich, B-doped Ni_3Al block by introducing B and antisite Ni atoms [14] in every 64th cell in such a way that the antisite Ni and intersititial B did not occupy adjacent cells. The concentrations of B and antisite Ni are close to the optimum numbers [15] which provide the maximum ductility.

The model block was relaxed by the atomic stiffness matrix method. In this method the positions of the atoms are determined by modified Newton's method [16]. Let N interacting atoms be in equilibrium under an external force f, which has 3N components corresponding to N atoms. The potential energy (U) of the system is a function of the position coordinates (denoted by vector x in 3N dimensional space). Now, let f be changed to f+Δf. Due to this increment in force, let the displacements of the atoms be u, which is a vector with 3N components. u is determined iteratively by the following equation [16]:

$$u^{s+1} = -\alpha^s [K^s]^{-1} (\Delta U^s - f - \Delta f) \tag{1}$$

where the superscript s denotes the iteration number, α is a scalar multiplier and [K] is named as the atomic stiffness matrix, in analogy with the finite element method [17] of the continuum mechanics. [K] is given by

$$K_{pq} = K_{qp} = \frac{\partial^2 U}{\partial x_p \partial x_q}, \qquad p, q = 1, 3N \tag{2}$$

The interaction potential used here is an embedded atom method (EAM) potential developed by Voter and Chen [18]. The iterations are continued until the force components on all unconstrained atoms are less than 2.7×10^{-3} eV nm^{-1}. In order to reduce the number of iterations, α is chosen at every iteration (starting with 1.0) so that U at a particular iteration becomes the smallest. Instead of taking the inverse of [K] directly, we solved for u in eqn. (1) by the Cholesky factorization method [19]. With the availability of large memory in the present-day computers, the storing of elements of [K] matrix does not pose any problem. and consequently the present technique offers an alternative (and also efficient) method for static simulations.

For visualization of the dislocation, the positions of the atoms of certain layers were plotted by projecting them onto to the slip plane. The location of the dislocation was quantitatively determined by the Burgers vector density distribution

ρ, which is the spatial derivative of the disregistry in the positions of the atoms above and below the slip plane [20]:

$$\rho_x(x) = \frac{\Delta}{\Delta X}[\delta u_x(x)] \tag{3}$$

where δu_x is the disregistry in the x-component of the displacement. The core structure of the dislocation was represented by iso-strain contours. The strain components at an atomic location were calculated from the distortion of a cell formed by 12 nearest neighbor atoms around a reference atom. The strain was calculated by taking the finite difference of the following strain-displacement relation:

$$\epsilon_{ij} = \frac{1}{2}\left(\frac{\partial u_i}{\partial x_j} + \frac{\partial u_j}{\partial x_i}\right) \tag{4}$$

RESULTS AND DISCUSSION

During relaxation the superpartials glided on (111) plane, changing the width of APB. They also split into Shockley partials with complex stacking fault (CSF) on (111) in between a pair. Fig. 1(a), (b) and (c) show the atomic configurations around the glide plane in pure Ni_3Al, B-doped Ni_3Al and B-doped $Ni_{76}Al_{24}$ blocks, respectively, with arrows indicating the locations of the Shockley partials. Fig. 2(a), (b) and (c) show the iso-strain (ϵ_{12}) contours on the glide plane around the superpartials in pure Ni_3Al. B-doped Ni_3Al and B-doped $Ni_{76}Al_{24}$ blocks, respectively. In all three cases the dislocation cores were essentially planar, spreading on the slip plane and extending towards the slip direction. Fig. 3 shows the iso-strain contour plots representing the core structures of the superpartials in pure Ni_3Al under an external shear stress of 507 MPa. It is seen that under external stresses the cores remained planar, but changed somewhat in shape.

Fig. 1 shows that the equilibrium positions of the Shockley partials are different in three blocks and they are not symmetric in B-doped blocks. The equilibrium separations between the superpartials. measured (from Burgers vector density distribution) as the distance between the centers of the two CSF regions. were 7.92, 5.42 and 5.90 nm in pure Ni_3Al, B-doped Ni_3Al and B-doped $Ni_{76}Al_{24}$, respectively. The initial separation between the superpartials. created by removal of half-planes. was 4.03 nm. It should be noted that the equilibrium separations between the superpartials in B-doped materials were smaller than that in pure Ni_3Al. In earlier papers [15] we explained B-enhanced ductility in Ni_3Al by proposing that strong Ni-B bonding causes weakening in Ni-Al directional bonding. In fact, the weakening of Ni-Al bonding by B decreased the shear modulus (G) which controls the equilibrium separation between the partials. The values of G in dislocation-free blocks of pure Ni_3Al. B-doped Ni_3Al and B-doped $Ni_{76}Al_{24}$ were simulated to be 75, 68 and 70 GPa, respectively.

Another important effect of strong Ni-B interactions is that the core structure of the superpartials became distorted in the presence of B. as seen in the iso-strain contour plots of Fig. 2. Away from the dislocations. local distortion of the lattice around B atoms is also seen (Fig. 2). Strong Ni-B interactions can produce Cottrell atmosphere [21] around the superpartial core. The difficulty in dragging such atmospheres will cause a reduction in the equilibrium separation between

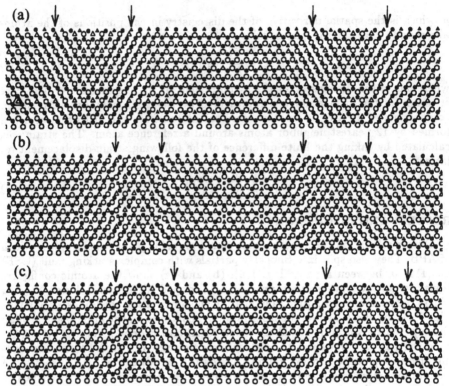

Fig. 1. Projections of two (111) planes on the glide plane. Open and closed symbols denote atoms above and below the glide plane, respectively. (a) Ni_3Al: (b) B-doped Ni_3Al: (c) B-doped $Ni_{76}Al_{24}$. Circle: Ni: Triangle: Al: Square: B.

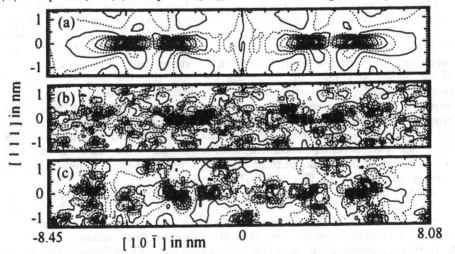

Fig. 2. Iso-strain (ϵ_{12}) contours. (a)Ni_3Al:(b)B-doped Ni_3Al:(c)B-doped $Ni_{76}Al_{24}$.

the superpartials in B-doped Ni₃Al. The asymmetry in the positions of the Shockley partials in the unstressed blocks of B-doped materials can be traced to arbitrary locations of B atoms on the glide plane (fig. 1).

Under increasing external stresses, four Shockley partials initially glided together, maintaining constant inter-separations. However, as the outermost right Shockley partial was about to move out to the free surface, there was a dramatic rearrangement of the remaining partials. The movement of the Shockley partials was illustrated in a plot (fig. 4) of the positions of the partials in pure Ni₃Al against the applied shear strain. As the outermost right partial came close (within about 2 nm) to the free surface, it felt attractive image force and moved rapidly to the free surface (curve a in fig. 4). At this stage the inner right partial kept on moving in a usual way, but the inner left partial moved rapidly (curve c in fig. 4) to reduce the APB width. The inner right partial eventually went out to the free surface (curve b in fig. 4). At this stage the outermost left partial moved rapidly (curve d in fig. 4) and caught up the inner left partial. This pair of the Shockely partials then moved together until they moved out of the block.

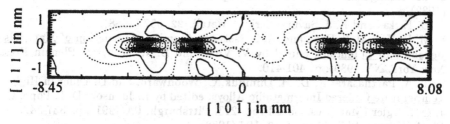

Fig. 3. Iso-strain (ϵ_{12}) contour plots representing cores of stressed (507 MPa) superpartials in pure Ni₃Al.

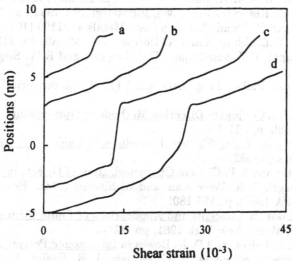

Fig. 4. Positions of Shockley partials against the applied shear strain. Curves a and b correspond to the right superpartial, and c and d to the left superpartial.

125

CONCLUSIONS

The core structure, dissociation into partials and glide of an edge superdislocation in Ni_3Al have been simulated by the atomic stiffness matrix method. The core structure of the superpartials changed in shape under external stresses as well as due to the presence of B atoms, but remained planar on (111). Under external stresses large movement and rearrangement of the Shockley partials took place.

REFERENCES

1. P. H. Thornton, R. G. Davies, and T. L. Johnston, Metall. Trans. **1A**, 207 (1970).
2. S. Takeuchi and E. Kuramoto, Acta Metall. **21**, 415 (1973).
3. V. Paidar, D. P. Pope, and V. Vitek, Acta Metall. **32**, 435, 1984.
4. P. B. Hirsch, Phil. Mag. A **65**, 569 (1992).
5. B. H. Kear and H. G. F. Wilsdorf, Trans. Metall. Soc. **224**, 382 (1962).
6. M. Yamaguchi, V. Paidar, D. P. Pope, and V. Vitek, Phil. Mag. A **45**, 867 (1982).
7. D. Farkas and E. J. Savino, Scripta Metall. **22**, 557 (1988).
8. M. H. Yoo, M. S. Daw, and M. I. Baskes in Atomistic Simulation of Materials - Beyond Pair Potentials, edited by V. Vitek and D. J. Slorovitz (Plenum Press, New York, NY 1989), p. 401-410.
9. T. A. Parthasarathy, D. M. Dimiduk, C. Woodward, and D. Diller in High Temperature Ordered Intermetallic Alloys, edited by L. Johnson, D. P. Pope, and J. O. Stiegler (Mat. Res. Soc. Proc. 213, Pittsburgh, PA 1991), p. 337-342.
10. A. Korner, Phil. Mag. A **58**, 507 (1988).
11. X. Shi, G. Saada, and P. Veyssiere in High Temperature Ordered Intermetallic Alloys VI, edited by J. A. Horton, I. Baker, S. Hanada, R. D. Noebe, and D. S. Schwartz (Mat. Res. Soc. Proc. 364, Pittsburgh, PA 1995), p. 701-706.
12. K. Aoki and O. Izumi, J. Japan Inst. Metals **43**, 1190 (1979).
13. C.T. Liu, C.L. White, and J.A. Horton, Acta Metall. **33**, 213 (1985).
14. A. Dasgupta, L.C. Smedskjaer, D.G. Legnini, and R.W. Siegel, Mater. Lett. **3**, 457 (1985).
15. T.K. Chaki, Philos. Mag. Lett. **61**, 5 (1990); Mater. Sci. Eng. A. **190**, 109 (1995).
16. M. Hestenes, Conjugate Direction Methods in Optimization, Springer-Verlag, New York, 1980, pp. 17-18.
17. K.-J. Bathe, Finite Element Procedures in Engineering Analysis, Prentice Hall, New Jersey, 1982.
18. A. F. Voter and S. P. Chen in Characterization of Defects in Materials, edited by R. W. Siegel, J. R. Weertman, and R. Sinclair (Mat. Res. Soc. Proc. 82, Pittsburgh, PA 1987), p. 175-180.
19. R. D. Cook in Concepts and Applications of Finite Element Analysis, 2nd edition, John Wiley, New York 1981, pp. 41-45.
20. V. Vitek, L. Lejcek, and D. K. Bowen in Interatomic Potential and Simulation of Lattice Defects, edited by P. C. Gehlen, J. R. Beeler, Jr., and R. I. Jaffe (Plenum Press: New York 1972), p. 493.
21. J. P. Hirth and J. Lothe, Theory of Dislocations, 2nd ed., Wiley, New York.

Critical Evaluation of Atomistic Simulations of 3D Dislocation Configurations

Vijay B. Shenoy, Rob Phillips

Division of Engineering, Brown University, Providence, RI 02912

Abstract

Though atomistic simulation of 3D dislocation configurations is an important objective for the analysis of problems ranging from point defect condensation to the operation of Frank-Read sources, such calculations pose new challenges. In particular, use of finite sized simulation cells produce additional stresses due to the presence of fixed boundaries in the far field which can contaminate the interpretation of these simulations. This paper discusses an approximate scheme for accounting for such boundary stresses, and is illustrated via consideration of the lattice resistance encountered by straight dislocations and simulations of 3D bow out of pinned dislocation segments. These results allow for a reevaluation of the concepts of the Peierls stress and the line tension from the atomistic perspective.

Introduction

Three dimensional dislocation configurations such as dislocation loops and Frank-Read sources play an important role in determining the plastic behavior of crystalline solids. Atomistic simulations of such defect configurations are essential for a) validation of the continuum theories that are widely used to model these defects and b) computation of fundamental parameters (e.g. line tension) that are required in the continuum analysis. Unfortunately, atomistic simulations are restricted by the size of the simulation cell, necessitating approximate treatments of the boundary conditions. One approach used with much success is to adopt the linear elastic solution for the defect of interest as the initial trial solution for the atomistic energy minimisation with the atoms on the boundary of the simulation cell fixed at the positions dictated by the elastic solution. This may not be a serious drawback if the goal of the simulation is the determination of the static dislocation core structure. However, in response to an external stress the defects move within the simulation cell, and this motion is strongly influenced by the boundary. Thus if the boundary forces are not explicitly accounted for, the results of the simulation will not be simulation cell-size independent leading to difficulty in assigning meaning to quantities such as the Peierls stress or the line tension.

In this paper, we develop a method that explicitly accounts for the boundary stress, and is illustrated through the example of a screw dislocation. It is found that the boundary force scales as the inverse square of the simulation cell size. This formulation is applied to the calculation of the lattice resistance curve for a screw dislocation in aluminum. In addition, a simple 3D configuration of bowing out of a pinned screw dislocation under an applied stress is also considered where it is found that continuum theory provides a quantitative description of the mechanics of such bow out.

Boundary Stress

In this section we describe the methodology for explicitly accounting for the boundary force. We restrict attention to the case of a straight screw dislocation for the sake of clarity. The simulation cell is chosen to be a cylinder with axis along the x_2-direction as shown in fig. 1. The atoms are displaced using the linear elastic solution which is taken as the trial solution for the atomistic energy minimisation. During the minimisation step the atoms in a region F (as shown in fig. 1) are held at the positions dictated by the linear elastic solution that has been imposed; only the atoms in the region M of radius R (*dynamic region*) enclosed by F (cf. fig. 1) are allowed to move. The relaxed atomistic core of the dislocation is thus

127

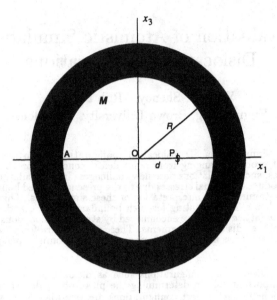

Figure 1: Side view of geometry of simulation cell. Shaded area denotes region where atoms are fixed according to linear elastic solution

obtained. To simulate the interaction of the dislocation with a homogeneous applied shear stress, we apply a homogeneous strain corresponding to the given stress on the *relaxed* configuration of the dislocation, by displacing the atoms both in F and M in the *relaxed* configuration by the displacements corresponding to the homogeneous strain. Thus, far away from the core the displacements are in agreement with linear elasticity in that they are the sum of the displacements due to the dislocation and those due to the applied homogeneous stress. Again, during the energy minimisation process, the atoms in F are held fixed and the atoms in M are allowed to move. The dislocation moves through the crystal to a new position P (cf. fig. 1), which is determined by the equilibrium condition

$$\tau_{app} + \tau_L(d) + \tau_b(d) = 0. \tag{1}$$

In eq. (1), $d = OP$ as shown in fig. 1, τ_{app} is the resolved shear stress which causes a Peach-Koehler force $\tau_{app}b$ in the x_1-direction (b is the Burgers vector), τ_L is the *lattice resistance function* (cf. [1]) and τ_b is the *boundary stress*. The last two terms are functions of the position d of the dislocation. While the lattice resistance is periodic, the boundary term is monotonically increasing.

What is the origin and significance of the boundary stress term? On the application of a homogeneous strain, the net displacement of the atoms before relaxation in region F is equal to the sum of elastic displacements due to the dislocation at O ($\mathbf{u}_O$) and the displacements due to the homogeneous strain. After relaxation the dislocation moves to the point P, therefore if the dislocation were present in an infinite crystal, the displacements of the atoms in F would have been the sum of the elastic displacements due to the dislocation at P ($\mathbf{u}_P$) and the displacement due to the homogeneous strain. Thus the boundary conditions (displacements in region F), *after relaxation*, are not consistent with the fields induced by the presence of the dislocation at P. The net effect of this inconsistency is to spuriously increase the total energy stored in the system. This additional energy cost gives rise to an additional energetic stress on the dislocation which we denote the *boundary stress* and which tends to repel the dislocation from the boundary. This stress is important in the simulation of dislocations in crystals with low lattice resistance (FCC, HCP). The significance of the boundary stress is that it is size dependent and decreases with increasing R , leading to size dependence of quantities deduced from the simulation such as the Peierls stress or the line tension. Indeed, Basinski et al. [2] observed such effects in their calculation of the Peierls stress in a model sodium lattice. Thus it is crucial that this stress is accounted for explicitly in cases with low lattice resistance. In the case of dislocations in materials with high lattice resistance (e.g. BCC metals) this term may be neglected since it is dominated by the lattice resistance term which is a few orders of magnitude larger.

An outline of the method used in the computation of the boundary stress τ_b is given below. Details may be found in [3]. Let ∂M_e denote the external boundary of the moving region M while ∂M_s denotes the slip surface AO and $\partial M_{s'}$ denotes the slip surface corresponding to the dislocation at P (AP) (cf. fig. 1). Let $()_O$ represent the fields associated with the dislocation at O and $()_P$ those associated with the *unconstrained* (as though in an infinite crystal) dislocation at P. We now move the dislocation from O to P while keeping the displacements on ∂M_e fixed at $\mathbf{u}_O$, i.e., a *constrained* movement of the dislocation. If $\Delta W(d)$ is the elastic energy cost associated with this operation, then the boundary force f_b (force due to the boundary being fixed) is given approximately by

$$f_b = -\frac{\partial \Delta W}{\partial d}$$

$$\tau_b = f_b/b. \tag{2}$$

To obtain $\Delta W(d)$, we consider two configurations of the dislocation. First, the energy is computed when the dislocation is at O. The total elastic energy (W_1) of this configuration is given by

$$W_1 = \frac{1}{2}\int_{\partial M_s} \mathbf{t}_O \cdot [\mathbf{u}_O]dS + \frac{1}{2}\int_{\partial M_e} \mathbf{t}_O \cdot \mathbf{u}_O dS, \tag{3}$$

where $\mathbf{t}_O$ is the traction vector on the boundary due to the stress field of the dislocation at O and $[\ ]$ represents the jump in the displacement across the slip surface (i.e. the Burgers vector). Secondly, we consider the energy when the dislocation is at P, with the displacements on ∂M_e fixed at $\mathbf{u}_O$. This configuration may be taken as the superposition of the two fields $\mathbf{u}_P$ and $\Delta\mathbf{u}(=\mathbf{u}_O - \mathbf{u}_P)$ and yields an elastic energy W_2,

$$W_2 = \frac{1}{2}\int_{\partial M_{s'}} \mathbf{t}_P \cdot [\mathbf{u}_P]dS + \frac{1}{2}\int_{\partial M_e} \mathbf{t}_P \cdot \mathbf{u}_P dS + \int_{\partial M_e} \mathbf{t}_P \cdot \Delta\mathbf{u}dS + \frac{1}{2}\int_{\partial M_e} \Delta\mathbf{t} \cdot \Delta\mathbf{u}dS \tag{4}$$

Here $\mathbf{t}_P$ is the traction on the boundary due to the stress field of the dislocation at P and $\Delta\mathbf{t}$ corresponds to the traction of the stresses associated with the fields $\Delta\mathbf{u}$. The additional energy stored in the system due to the inconsistent boundary is now given by $\Delta W = W_2 - W_1$ and the boundary stress is obtained using eq. (2). A plot of the normalised boundary stress (magnitude) as a function of d/R for the case of aluminum is given in fig. 2, where it is seen that for small values of d/R the force is linear and diverges as d/R approaches unity. In most applications the exact result may be approximated by a linear function of the form

$$\frac{\tau_b(d)}{\mu} = -\frac{Ab}{4\pi^2 R^2}d \tag{5}$$

where μ is the shear modulus and A is the boundary force coefficient which is equal to 6.3 for screw dislocations in isotropic solids. It is clear from eq. (5) that for a given d, τ_b depends on the inverse square of the size of the simulation region M.

Applications

Our preliminary applications of these results to atomistic simulations are performed using the embedded atom potentials for aluminum developed by Ercolessi and Adams[5]; the

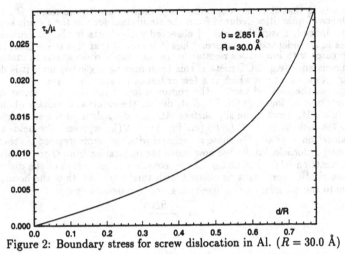

Figure 2: Boundary stress for screw dislocation in Al. ($R = 30.0$ Å)

dislocation considered is of screw type with Burgers vector $\frac{a_0}{2}\langle 110 \rangle$. Two applications are discussed, the first being the calculation of the lattice resistance function for screw dislocations and the second the bow out of a pinned dislocation. In all of the calculations below the position d of the dislocation is obtained by computing the value of x_1 at which the slip ($[u_1]$) equals $b/2$. For the dislocations considered here, this measure accurately reflects the dislocation position.

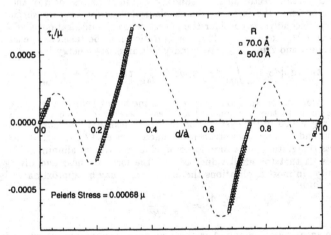

Figure 3: Lattice Resistance Function for screw dislocation in Al. The dashed line is a guide to the eye.

Lattice Resistance Function and Peierls Stress

The lattice resistance function is computed using the equilibrium condition of eq. (1). The initial point O (cf. fig 1.) is chosen to be in a Peierls valley (point of equilibrium under absence of external stress). The screw dislocation undergoes a splitting reaction into two Shockley partials. Thus the lattice resistance curves presented here are for the extended dislocation. Fig. 3. shows the lattice resistance curves computed with various cell-sizes. It is seen that the lattice resistance curve is size-independent. The Peierls stress associated with these curves is 0.00068μ. The resistance curve is multi-welled unlike that supposed by the Peierls-Nabarro model.

Bow out of a pinned dislocation segment under applied stress

The atomistic simulation of a pinned dislocation is an important step towards the simulation of a Frank-Read source. Previous work in this direction has been restricted to continuum formulations. For example, Foreman [6] used Brown's self stress[7] approach in his simulation of dislocation bow out. A simpler continuum theory adopts a line tension model (Nabarro [8]), where it is supposed that a dislocation behaves like a flexible string characterized by a line tension T. Our goals in these simulations are a) to examine the validity of the concept of line tension and to compute its value and b) to compare the shape of the bow out predicted by the continuum theory with that obtained from atomistics.

The simulation of bow out of a pinned segment under an applied stress is achieved as follows. A straight screw dislocation is placed along the axis of a cylindrical cell of atoms of length L (the axis of the cylinder is the x_2-axis in fig. 1 and goes into the plane of the paper), with the radius of the dynamic region M set to R. A strain corresponding to the applied stress τ_{app} is now applied to this configuration and the atomistic energy is minimised. In the energy minimisation step, in addition to the atoms in the region F, the atoms along the planes $x_2 = 0$ and $x_2 = L$ are held fixed emulating a pinning effect on the dislocation.

The equilibrium shape of the bow out as predicted by continuum theory may be obtained as follows. We consider a dislocation that is on the x_2-axis pinned at the points $x_2 = 0$ and $x_2 = L$. The homogeneous applied stress exerts a Peach-Koehler force $\tau_{app}b$ on the dislocation in the x_1 direction. Let the equilibrium shape of the segment be given by $x_1 = f(x_2)$. Assuming a constant line tension T, the line shape is given by that function that minimises the functional

$$I(f) = \int_0^L \left(T\sqrt{1 + f'^2} - \tau_{app}bf + \frac{\lambda}{2}f^2 \right) dx_2, \qquad (6)$$

where f' is the derivative of f with respect to x_2 and $\lambda = A\mu b^2/(4\pi^2 R^2)$, where A is the boundary force coefficient. The quadratic term in f accounts for the additional energy due to the rigid boundary. The solution f is given by the catenary (for details see ref. [3])

$$f(x_2) = \frac{\tau_{app}b}{\lambda} \left(1 - \frac{\sinh\sqrt{\frac{\lambda}{T}}(L - x_2) + \sinh\sqrt{\frac{\lambda}{T}}x_2}{\sinh\sqrt{\frac{\lambda}{T}}L} \right) \qquad (7)$$

It is evident that the maximum bow out d occurs at $x_2 = L/2$.

We adopt the following strategy in computing line tension from atomistic simulations. We "measure" the maximum bow out d from the results of the simulation. On knowing d the line tension may be calculated as

$$T = \frac{\lambda L^2}{4\left(\cosh^{-1}\left[\left(1 - \frac{\lambda d}{\tau_{app}b}\right)^{-1} \right] \right)^2} \qquad (8)$$

using eq. (7) (see ref. [3]). If the effect of the boundary is not explicitly accounted for ($A = 0$) one may obtain an *uncorrected* estimate of the line tension as $T_u = \tau_{app}bL^2/8d$.

Table I shows the computed values of line tension for various lengths of the dislocation. It is seen that the value of T computed using eq. (8) are size independent, while those computed using the uncorrected estimate are strongly size dependent. Further, fig. 4 shows a comparison of the catenary solution which is derived from continuum arguments with the atomistic solution where it is evident that the agreement is excellent.

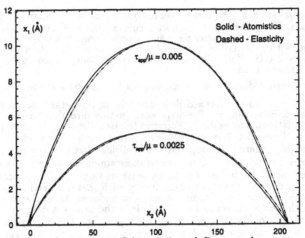

Figure 4: Comparison of Atomistic and Catenary bow outs.

Table I: Line Tension for Screw Dislocation in Al (R=40.0 Å)

L (Å)	$T_u/\mu b^2$ (without boundary effects)	$T/\mu b^2$ (with boundary effects)
53.94	0.43	0.42
108.52	0.55	0.43
211.20	0.95	0.46

Conclusion

The significance of additional stresses that act on a dislocation due to fixed boundaries is discussed. These boundary stresses can be estimated on the basis of a linear elastic treatment of the interaction of the fields due to the perturbed dislocation and the frozen boundaries. This scheme has been applied to computing the lattice resistance curve of a screw dislocation in Al. The bow out of a pinned screw dislocation in Al is studied using atomistics and it is found that the concept of line tension is meaningful. Even in the simplest of 3D configurations such as the bow out of a dislocation, it is seen that the boundary force cannot be neglected. Thus the explicit accounting of the boundary stresses is important for obtaining meaningful results from atomistic simulations of 3D dislocation configurations.

Acknowledgements

The authors thank V. Bulatov, A. Carlsson and M. Khantha for their valuable comments. Support for this work by NSF under Grant No. CMS-9414648 is gratefully acknowledged.

References

[1] U. F. Kocks, A. S. Argon, M. F. Ashby, *Progress in Material Science*, **19**, (1975).
[2] Z. S. Basinski, M. S. Duesbery, R. Taylor, *Can. J. Phys.*, **49**, p. 2160, (1971).
[3] V. B. Shenoy, R. Phillips, Manuscript under preparation.
[4] M. S. Daw, M. I. Baskes, *Phys. Rev. Lett.*, **50**, p. 1285, (1983).
[5] F. Ercolessi, J. Adams, *Europhys. Lett.*, **26**, p. 583, (1993).
[6] A. J. E. Foreman, *Phil. Mag.*, **15**, p. 1011, (1967).
[7] L. M. Brown, *Phil. Mag.*, **10**, p. 441, (1964).
[8] F. R. N. Nabarro, *Theory of Crystal Dislocations*, Oxford University Press, (1967).

REPRESENTATION OF FINITE CRACKS BY DISLOCATION PILEUPS : AN APPLICATION TO ATOMIC SIMULATION OF FRACTURE

Vijay Shastry and Diana Farkas
Department of Materials Science and Engineering,
Virginia Tech., Blacksburg VA 24061.

ABSTRACT

The elastic displacement field solution of a semi-infinite crack in an anisotropic body, calculated using a complex variable approach due to Sih and Liebowitz, is usually used by atomistic simulations of fracture. The corresponding expression for the displacement field of a finite crack is numerically cumbersome since it involves multiple square roots of complex numbers. In this study, displacement field of the crack is calculated by superposing the displacements of dislocations in an equivalent double pileup, equilibrated under mode I conditions. An advantage of this method is its extensibility to atomistic studies of more complex systems containing multiple cracks or interfaces. The pileup representation of the finite crack is demonstrated as being equivalent to its corresponding continuum description using the example of a double ended crack in α-Fe, loaded in mode I. In these examples, the interatomic interaction in α-Fe is described by an empirical embedded atom (EAM) potential.

Introduction

Molecular statics, atomistic simulations of fracture require the initial displacements of atoms in the simulation cell before relaxing the atomic configuration in accordance with interatomic force laws. This elastic field is also used to obtain atomic positions in the outer boundary region, far from the crack tip. Traditionally, these simulations [1, 2, 3] have used a specific crack geometry in which a single ended crack is embedded in an anisotropic material. The initial displacements of atoms are approximated by that prescribed by the elastic displacement field of a *semi-infinite* crack in the anisotropic body. Equivalently, one can also use a finite double ended crack geometry to study the fracture processes at the two tips provided the appropriate elastic solution for the displacement field is used. The displacement field of a mode I double ended crack can be obtained using a complex variable approach by Sih and Liebowitz [4]. The numerical implementation of this solution is cumbersome, because of the multiplicity of roots involved in the expressions for displacements.

The elastic displacement field is obtained more conveniently, in certain special cases, by determining the displacement field due to a double pileup of opening Volterra dislocations equilibrated by external loading. The pileup displacement field is obtained by the simple superposition of displacements of these dislocations. The algebraic expressions for the displacements of dislocations located in cubic materials are simple, if the line and Burgers vectors lie along directions of even-fold symmetry [5]. This method can also be used to construct the displacement field of cracks with bonds linking the faces. Further, the method can be extended to study systems containing multiple cracks and cracks in inhomogeneous materials.

The next section discusses how the dislocation distribution in the equivalent pileup, loaded under mode I and pinned at two ends, is obtained. A simple comparison between the displacement fields predicted by the Sih formalism and the pileup method is shown. Next, the effect of bonding between the crack faces on the dislocation distribution in the pileup is discussed for cases where the external loading is at least as high as the Griffith value for a crack with traction free faces. This study of bonding effects is useful in verifying the accuracy of traction free boundary conditions when the crack is found to be stable above the Griffith loading. The atomic arrangement, which uses the pileup displacement field, is shown for a crack in α-Fe modeled using the Simonelli EAM potential [6].

The elastic displacement field of the double pileup

The complex variable formulae suggested by Sih and Liebowitz [4] can be used to calculate the displacement field of the double ended crack and they are

$$u_x = \Re \left[\frac{\sigma^{ext}}{s_2 - s_1} \left\{ s_2 p_1 (\sqrt{z_1^2 - (\ell/2)^2} - z_1) - s_1 p_2 (\sqrt{z_2^2 - (\ell/2)^2} - z_2) \right\} \right] \qquad (1)$$

$$u_y = \Re \left[\frac{\sigma^{ext}}{s_2 - s_1} \left\{ s_2 q_1 (\sqrt{z_1^2 - (\ell/2)^2} - z_1) - s_1 q_2 (\sqrt{z_2^2 - (\ell/2)^2} - z_2) \right\} \right] . \qquad (2)$$

Here $z_1 = x + s_1 y$ and $z_2 = x + s_2 y$. Here, s_1, s_2 p_1, p_2, q_1, and q_2 are constants which depend upon the elastic compliances and are not reproduced here for brevity. $\ell/2$ is the half length of the double ended crack and σ^{ext} is the externally applied stress.

In this work, the 2D displacement field of a finite mode I crack is calculated by linearly superposing the displacement fields of model opening Volterra dislocations in an equivalent double pileup pinned at the two tips. The displacement fields of individual Volterra dislocations are obtained from the sextic theory of Stroh described by Hirth and Lothe [5]. The crack plane is located midway between two adjacent atomic planes which are then displaced symmetrically in opposite directions, by amounts prescribed by the opening profile of pileup. This method of constructing the crack preserves the bonds that link atoms located on either side of the crack plane. Therefore, bonding tractions are present on the crack faces along the entire length of the crack.

The crack is represented as a region stretching from $-\ell/2 \leq x \leq \ell/2$, which obeys a non-linear constitutive law, and is bounded on both sides by elastic half spaces. The discrete dislocation representation of the loaded crack implies that the continuity of normal tractions is evaluated at selected points along the crack. The condition for continuity of normal tensile tractions is

$$\sigma(x) = \sigma^{ext} + \frac{\mu^{eff}}{2\pi(1-\nu)} \int_{-\ell/2}^{\ell/2} \frac{\phi(t)}{(x-t)} dt - \sigma^*(x), \qquad (3)$$

$$= 0 \ (-\ell/2 \leq x \leq \ell/2). \qquad (4)$$

The shear components of tractions are zero along the plane of a mode I crack. $\phi(x) = d[\Delta u_y(x)]/dx$ is the dislocation distribution function, where $\Delta u_y(x)$ is the crack opening at location x. μ^{eff} is the effective shear modulus for the cleavage plane. When the coordinate axes fixed on a crack embedded in a cubic material, lie parallel to directions with even-fold symmetry, μ^{eff} is phrased in terms of elastic constants (C_{ij}),

$$\mu^{eff} = \frac{\overline{C}_{11} - C_{12}}{(C_{22}/C_{11})^{1/4}\overline{C}_{11}} \left[C_{22} + C_{12} \left(\frac{C_{22}}{C_{11}} \right)^{1/2} \right] \left[\frac{\overline{C}_{11}C_{66}}{\overline{C}_{11}^2 - C_{12}^2 + 2C_{66}(\overline{C}_{11} - C_{12})} \right]^{1/2} . \qquad (5)$$

Here, C_{ij} are transformed to the coordinate system fixed on the crack. The coordinate system has basis vector $\hat{y}$ oriented along the normal to the crack plane, whereas $\hat{z}$ contains the crack front and

$\hat{x} = \hat{y} \times \hat{z}$. $\overline{C}_{11} = [C_{11}C_{12}]^{1/2}$. In the example considered here, the untransformed elastic constants corresponding to the Simonelli potential [6] for α-Fe are $C_{11} = 1.51ev/\mathring{A}^3$, $C_{12} = 0.91ev/\mathring{A}^3$ and $C_{44} = 0.7ev/\mathring{A}^3$.

$\sigma^*(x)$, the resistive stress due to bonding, depends upon the total opening at x. One functional form of $\sigma^*(x)$, suggested by Beltz and Rice [7] is

$$\sigma^{ext}(x) = 2\frac{\gamma_s}{L}\frac{\Delta u_y(x)}{L}e^{-\Delta u_y(x)/L}, \tag{6}$$

where γ_s is the surface energy of the cleavage plane, L, the Fermi-Rose decohesion length. For the Simonelli potential based on α-Fe, the {011} plane has constitutive properties $\gamma_s = 0.09ev/\mathring{A}^2$ and $L = 0.33\mathring{A}$.

Eqn. (4) is solved using a method developed by Erdogan and Gupta [8] at discrete points, $x_i = \ell/2\cos(\pi i/N)$, for $\phi(x)$. Here, $\phi(x)$ is assumed to be singular at the pinning points $x = \pm\ell/2$. The total number of discrete points, $N = 160$, here. The Burgers vectors of dislocations located at x_i, are obtained by integrating $\phi(x)$ over a length interval assigned to x_i.

Figure 1 compares the displacements due to a traction free crack, predicted by the Sih and Liebowitz and the pileup methods, for atoms located on the atomic plane $y = 2.072\mathring{A}$ which forms the upper crack face. The crack lies on the $(01\bar{1})$ plane at $y = 1.014\mathring{A}$ and the crack front is parallel to [100]. The two approaches predict identical values of displacement due to the crack tip for each of the atoms shown. The crack has half length $\ell/2 = 22.3\mathring{A}$ and is embedded in an iron crystal, which is described by the Simonelli potential. The external loading on the crack is $\sigma^{ext} = 0.07ev/\mathring{A}^3$. The Griffith value of external stress for this case is $0.058ev/\mathring{A}^3$. When the atomic ensemble was allowed to relax using molecular statics scheme [9], the critical loading for cleavage was observed to lie between $0.06 - 0.07ev/\mathring{A}^3$.

Effect of bonding on the dislocation distribution

Figure 2 shows the relative opening between crack faces for different values of σ^{ext} for two types of cracks in iron. In one case, bonding exists between crack faces whereas the other crack is traction free. The crack front lies along [100] and the crack plane is $(01\bar{1})$. $\phi(t)$ for a "traction free" crack is obtained by setting σ^* to zero in Eqn. (4). Here, $\mu^{eff} = 0.65ev/\mathring{A}^3$ and $\ell/2 = 22.3\mathring{A}$. The opening profiles are drawn only for one half of the crack and are symmetric about its center. When $\sigma^{ext} \geq 0.07ev/\mathring{A}^3$, the opening profiles for the cracks with and without bonding are very similar. At high levels of remote loading, the region immediately behind the tip, where bonding tractions are largest, is small. Here, bonding tractions are large since the opening is comparable to L (decohesion length). The opening increases with further distance away from the tip and the effect of bonding tractions diminishes rapidly. However, when σ^{ext} is lowered, the opening increases less rapidly with distance from away the tip and the effect of bonds on the opening profile is more pronounced. The effect of bonding on the opening between crack faces and therefore, the dislocation distribution in the pileup is found to be small.

Figure 3 illustrates the variation of bonding tractions with distance from the pinning point, $x = +\ell/2$. The maximum in bonding tractions occurs when $\Delta u_y = L$ and its location is taken to be the position of the real crack tip. As σ^{ext} is lowered, the crack tip retreats under the influence of bonding. When this retreat is large, i.e. greater than $0.6\mathring{A}$ here, the singularities at the pinning points are lost. Then, solutions for $\phi(x)$, singular at the two pinning points, cannot be found.

Finally, Fig. 4 shows the initial atomic arrangement around a crack in α-Fe at $\sigma^{ext} = 0.07ev/\mathring{A}^3$.

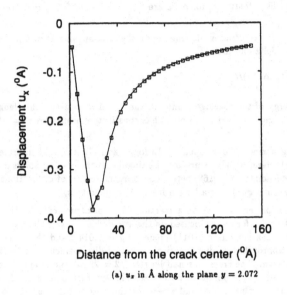

(a) u_x in Å along the plane $y = 2.072$

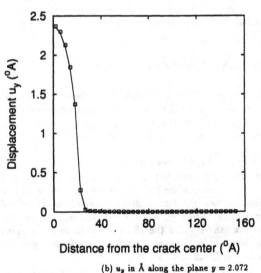

(b) u_y in Å along the plane $y = 2.072$

Figure 1: Displacements u_x and u_y along the atomic plane $y = 2.072$Å. The solid lines are the displacement values predicted by the Sih and Liebowitz formula whereas the □ symbols denote the displacements of individual atoms predicted by the pileup method. The crack is located along the atomic plane $y = 1.014$Å.

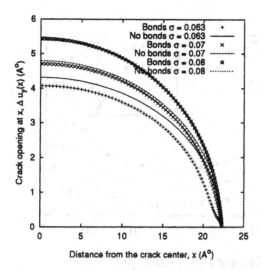

Figure 2: Symbols +, × and * represent relative opening between crack faces when bonding tractions are present, for three values of $\sigma^{ext} = 0.063$, 0.07, $0.08 ev/\mathring{A}^3$. The dashed lines represent the corresponding profiles for the traction free crack.

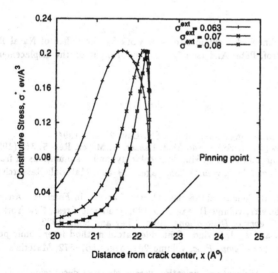

Figure 3: Variation of constitutive stress, σ^*, with distance from the positive crack tip when bonding is present between the crack faces. For brevity, the variation is shown only for the region immediately behind the tip. Stress σ in $ev/\mathring{A}^3$.

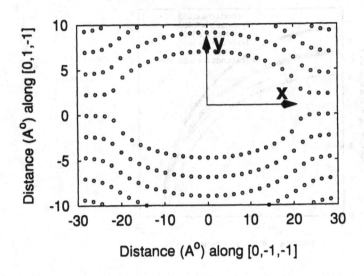

Figure 4: Initial (unrelaxed) Atomic arrangement around a [100](01$\bar{1}$) crack in α-Fe.

Acknowledgments

The authors gratefully acknowledge support for this work by the Office of Naval Research (ONR). We are grateful to Prof. Peter Anderson for his helpful input on the displacement fields of finite cracks.

REFERENCES

1. K. S. Cheung and S. Yip, Modelling Simul. Mater. Sci. Eng. **2**, 865 (1994).
2. R. G. Hoagland, M. S. Daw, S. M. Foiles, and M. I. Baskes, J. Mater. Res. **5**, 313 (1990).
3. C. S. Becquart, P. C. Clapp, and J. A. Rifkin, Molecular dynamics simulations of fracture in RuAl, in *Mat. Res. Soc. Symp. Proc*, volume 288, pages 519–524, Materials Research Society, 1993.
4. G. C. Sih and H. Liebowitz, Mathematical theories of brittle fracture, in *Fracture – An Advanced Treatise*, edited by H. Liebowitz, volume II, pages 69–189, Academic Press, New York, 1968.
5. J. P. Hirth and J. Lothe, *Theory of Dislocations*, John Wiley and Sons, 1982.
6. G. Simonelli, R. Pasianot, and E. J. Savino, Embedded-atom-method interatomic potentials for bcc-iron, in *Mat. Res. Soc. Symp. Proc*, volume 291, pages 567–572, Materials Research Society, 1993.
7. G. E. Beltz and J. R. Rice, Dislocation nucleation versus cleavage decohesion at crack tips, in *Modeling the deformation of Crystalline Solids*, edited by T. C. Lowe, A. D. Rollet, P. S. Follansbee, and G. S. Daehn, pages 457–480, Warrendale, PA, 1991.
8. F. Erdogan and G. D. Gupta, Q. Appl. Math. **29**, 525 (1972).
9. V. Shastry and D. Farkas, Manuscript in preparation .

Part III

Fracture at Interfaces—Role of Impurities

Embrittlement of Cracks at Interfaces.

Robb Thomson, Emeritus
Materials Science and Engineering Laboratory
National Institute of Standards and Technology
Gaithersburg, MD, USA 20899
and
A. E. Carlsson
Department of Physics
Washington University
St. Louis, MO 63130

Abstract

This paper presents a synopsis of already published work for cracks on interfaces and new results on the effect of changing the chemistry of the atoms at the interface. Once the appropriate cut-off for the oscillatory singularity in the interface analysis is determined, the physics of the crack is determined in terms of an appropriate driving force for a particular event of interest, and a lattice resistance to the event. Analytic approximations are available for the driving forces, and the lattice resistance is determined on the basis of a block construction for the lattice. We show that for the case where the chemical bonding at the interface is different from that in the matrix, the same ideas are applicable, provided the different chemistry is incorporated correctly in the lattice resistance construction.

1. Introduction

The proposition on which this paper is based is that the crack, whether on an interface or not, is governed by a balance between the elastic driving forces determined by the far field continuum elastic fields, and the lattice resistance exerted in the core of the crack by the nonlinear bonds of the lattice. The template for this idea is the classic Griffith relation which says that the crack is in equilibrium with respect to cleavage when the elastic crack extension force is balanced by the surface tension of the cleavage surface being created at the crack tip by the nonlinear bond breaking. Our thesis is that this idea can be generalized to the case of lattice breakdown by shear at the crack tip when a dislocation is formed. Rice[1] was the first to demonstrate that such a balance was realized in Mode II dislocation emission in which the crack is loaded in pure Mode II, and the dislocation is emitted on the crack plane ahead of the crack. But the principle can also be applied to the case when the dislocation is emitted in a blunting configuration, at an angle to the original crack plane, and the crack is loaded in mixed Modes I and II, and also when the crack is on an interface. Rice and coworkers have discussed the blunting case from the point of view

Mat. Res. Soc. Symp. Proc. Vol. 409 © 1996 Materials Research Society

of the "tension shear coupling" which is expected o operate when the crack is loaded in Mode I[2], but Zhou, etal[3] showed that the important effect in blunting mission is the ledge formation at the tip of the blunted crack.

2. Elastic Driving Forces.

In three recently published papers[4,5,6], we have shown that the principal difficulty associated with the phase singularity which mixes the modes at the tip of the crack in a logarithmically divergent fashion, can be alleviated by expressing the elastic driving forces for both cleavage and emission in terms of stress intensity factors evaluated at the crack core. That is, an atomic cut-off is introduced for the phase singularity factor in the analysis. We found that the appropriate cut-off distance was simply the range parameter of the force law binding the atoms together. Thus, the core elastic driving force, $\mathcal{G}$ appropriate to either cleavage or emission is given, respectively, by

$$\mathcal{G}_c = \frac{k_I^2 + k_{II}^2}{2\mu'}$$

$$\mathcal{G}_e = \frac{k_{II}^2}{2\mu'},$$

$\qquad(1)$

where lower case k refers to the core stress intensity factor described earlier, and are related to the load stress intensity factors, K by the relation

$$k = Ke^{i\eta},$$

$\qquad(2)$

where both k and K are complex stress intensity factors, $k = k_I + ik_{II}$, etc. The phase factor, η is given by

$$e^{i\eta} = \left(\frac{2a}{r}\right)^{i\epsilon}$$

$$\epsilon = \frac{1}{2\pi}\ln\frac{\kappa_1\mu_2 + \mu_1}{\kappa_2\mu_1 + \mu_2} = \frac{1}{2\pi}\ln\frac{11(\mu_1/\mu_2) + 5}{11 + 5(\mu_1/\mu_2)},$$

$\qquad(3)$

where r is the radial distance from the crack tip along the cracking plane (which might be a branching plane) for cleavage, or the radial distance along the slip plane for dislocation emission. a is the half length of the parent crack. The elastic constant, μ' is given by

$$\mu' = \frac{2(1-\nu)\mu_1\mu_2}{\mu_1 + \mu_2}.$$

$\qquad(4)$

Subscripts on elastic coefficients refer to a particular sublattice, and μ, κ, and ν are the standard elastic coefficients for isotropic elasticity. We note that the cleavage driving force is an invariant in the sense that it does not depend on the phase angle, but the emission driving force most certainly does depend on the core phase angle.

The relations for transforming from the parent load stress intensity factors, K to the effective stress intensity factors, k' for an inclined plane are given by the proceedure first proposed by Cotterell and Rice[7], and given for the interface by Rice, Suo and Wang[8], which we repeat below.

$$k'_I = \sigma_{\theta\theta}\sqrt{2\pi r} = \mathcal{K}_I \Sigma_{\theta\theta}^I + \mathcal{K}_{II} \Sigma_{\theta\theta}^{II}$$

$$k'_{II} = \sigma_{r\theta}\sqrt{2\pi r} = \mathcal{K}_I \Sigma_{r\theta}^I + \mathcal{K}_{II} \Sigma_{r\theta}^{II}$$

$$\mathcal{K} = \mathcal{K}_I + i\mathcal{K}_{II} = |\mathcal{K}|e^{i(\psi-\eta)}$$

$$\Sigma_{\theta\theta}^I = \frac{1}{\cosh \pi\epsilon}\left[\sinh((\pi-\theta)\epsilon)\cos\frac{3\theta}{2} + e^{-(\pi-\theta)\epsilon}\cos\frac{\theta}{2}(\cos^2\frac{\theta}{2} - \epsilon\sin\theta)\right]$$

$$\Sigma_{\theta\theta}^{II} = -\frac{1}{\cosh \pi\epsilon}\left[\cosh((\pi-\theta)\epsilon)\sin\frac{3\theta}{2} + e^{-(\pi-\theta)\epsilon}\sin\frac{\theta}{2}(\sin^2\frac{\theta}{2} + \epsilon\sin\theta)\right] \qquad (5)$$

$$\Sigma_{r\theta}^I = \frac{1}{\cosh \pi\epsilon}\left[\sinh((\pi-\theta)\epsilon)\sin\frac{3\theta}{2} + e^{-(\pi-\theta)\epsilon}\sin\frac{\theta}{2}(\cos^2\frac{\theta}{2} - \epsilon\sin\theta)\right]$$

$$\Sigma_{r\theta}^{II} = \frac{1}{\cosh \pi\epsilon}\left[\cosh((\pi-\theta)\epsilon)\cos\frac{3\theta}{2} + e^{-(\pi-\theta)\epsilon}\cos\frac{\theta}{2}(\sin^2\frac{\theta}{2} + \epsilon\sin\theta)\right].$$

With the equations (5), the elastic driving force can be determined for any of the events, 1) cleavage on the parent plane, 2) nonblunting emission on the parent plane (called Mode II emission above), 3) cleavage on a plane inclined to the parent crack plane, or 4) blunting emission of a dislocation on an inclined plane. The crack may lie on an interface separating two dissimilar elastic media, or it may be in homogeneous material.

3. Lattice Resistance.

The lattice resistance half of the problem is simply

$$\mathcal{R}_c = 2\gamma_s \qquad (6)$$

for cleavage (including branching cleavage), according to the Griffith relation. Rice(1) proposed that for emission, the lattice resistance is similarly given by the "unstable stacking fault", which is simply the theoretical shear strength of the lattice expressed as an energy,

$$\mathcal{R}_e = \gamma_{us}. \qquad (7)$$

As noted, this shear energy must be augmented when the (blunting) dislocation slip plane is inclined to the parent crack plane.

The construction for this augmented lattice resistance is shown in Fig. 1. In the figure, the bonds are cut on the cleavage plane to the crack tip, and a block of atoms is defined by the cleavage plane and the inclined slip plane. The lattice resistance is computed by shearing the block along the slip plane. For atoms far from the crack tip, the energy changes during shearing are just those computed for the γ_{us} by Rice[1]. However, at the crack tip, as the block is sheared, one atom bond is broken as the crack becomes blunted[3]. In the case of the actual crack, as shear takes place on the slip plane, a continuum distribution of dislocations, β, moves out of the tip. The lattice resistance is the misfit stress exerted on these dislocations. Thus,

$$\mathcal{R}_e = \int_0^\infty \sigma\beta(x)\,dx$$
$$= \sum_0^\infty f_i b_i,$$

(8)

where in the second equation, a lattice sum is taken over the bonds crossing the slip plane multiplied by the local dislocation density, b_i. The force, f_i, is the bond force component parallel to the slip plane. If the dislocation density is highly localized near the crack tip, then nothing else can be done. However, if the dislocation density is rather evenly spread over a large number of atoms at the tip, then the sum can be broken into a series of terms corresponding to the uniform shear displacement, and an end correction,

$$\mathcal{R}_e \approx f_{max}(0) + \int_1^\infty \sigma\beta(x)\,dx$$
$$= f_{max}(0) + \gamma_{us}.$$

(9)

$f_{max}(0)$ is the maximum bond force exerted in the first bond at the crack tip.

This form of the lattice resistance must be taken instead of the simple γ_{us} whenever the dislocation emsision blunts the crack. If the bonds on the cleavage plane are different from those in the matrix, then $f_{max}(0)$ corresponds to the segregated chemical bond, while the γ_{us} refers to the bonds in the matrix. This construction is made for nearest neighbor pair forces, and when the force law is more general, then Eqn. (8) or (9) must be extended in an obvious manner. It is also obvious how to extend the construction for more physical crystal lattices than the simple hexagonal lattice shown. But the simple generic case illustrated will be used to test the ideas with actual simulations in the 2D hexagonal lattice.

3. Lattice Simulations.

Our simulations were accomplished with a lattice Green's function technique described in an earlier paper[9], and were made for a homogeneous lattice, for simplicity. That is, the spring constants of the lattice are everywhere equal. Fig. 2 shows the results of a series of simulations in which the matrix bonding was held fixed, and the bonding on the interface was varied. In the figure, the total $\mathcal{G}$ in the parent crack system is plotted on the ordinate, and f_{max} on the absissa. The dashed line represents the critical values for emission. The second line is a plot of twice the bond strength of the cleavage plane bonds as a function of the same f_{max}, and is proportional to the Griffith $\mathcal{G}_c$. The emission line lies lower than the Griffith line to the right (emission is stable relative to cleavage), and the two curves lie on top of one another to the left of their intersection. The points to the right correspond to loading the crack in pure Mode I. When the two lines intersect, the crack can still be made to emit rather than cleave by adding Mode II loading, up to that value of Mode II where the lattice is unstable in shear on the cleavage plane, and a dislocation is emitted on the cleavage plane instead of the inclined slip plane. The emission induced by Mode II

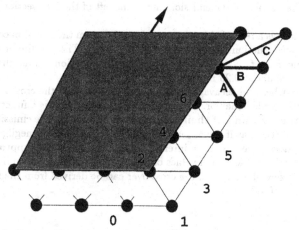

Figure 1. Block diagram for computation of lattice resistance. The bonds across the cleavage plane are cut up to the crack tip, and the block is rigidly sheared in the direction of the arrow along the slip plane. As the shear progresses, the energy changes in the bonds between the numbered atoms is given by Eqn. (8). After some manipulation, this energy can be written approximately as Eqn. (9). The energy changes in the bonds labelled A,B,C correspond to the standard γ_{us}.

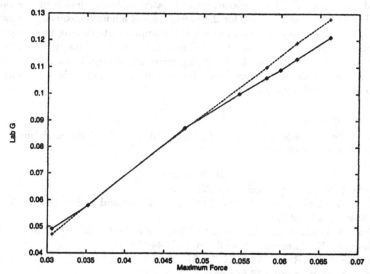

Figure 2. Simulation results for the hexagonal 2D lattice for fixed force laws in the matrix, and variable force law on the cleavage plane. The absissa is the slip plane component of the maximum force, f_{max}, developed in the bond between atoms 1 and 2, and the ordinate is the driving force, $\mathcal{G}$ as measured in the "lab" or parent crack coordinates. The full line corresponds to emission results, and the dashed line is a plot of twice the bond strength of the 12 bond plotted as a function of f_{max}.

is shown by the portion of the emission line to the left of the intersection with the cleavage line.

Figure 2 shows a remarkable correspondence between the emission condition predicted by the block construction, and the simulation results. That is, the emission criterion is accurately linear in f_{max}. From the slope of the emission curve, the value of b_0 can determined to be about 0.25 in atomic units.

The point where the cleavage and emission lines meet is the crossover of the material from brittle to ductile behavior. In our simulations, this point is a function of the relative slopes of the two lines, **and of their intercepts** in the plots. The emission line is found to to have zero intercept, as if the γ_{us} contribution in Eqn. (9) were negligible. The cleavage line, however, does have a nonzero intercept, because f_{max} is only approximately linear in the bond strength, and for small values of f_{max} departs strongly from linearity, going into the origin with zero slope. Thus, the cross over can be derived from the force law, and the computed value of b_0.

4. Conclusion.

This simple model throws a strong light on the physical mechanisms underlying chemical embrittlement. Although its specific predictions must be followed up further before general criteria for chemical embrittlement can be made; nevertheless, it is clear that the simple intuitive idea that embrittlement is caused by a weakening of the crack plane by chemical segregation is far too simple. A crack plane can certainly be forced by lowering the bonding at the plane, but whether a crack on that plane will cleave or emit a dislocation will be determined by much subtler considerations associated with the details of the (altered) bonding. The results described here for the special case of a homogeneous lattice have also been shown by us to apply to the interface, if the appropriate driving force expressions and core stress intensity factors are factored into the analysis. Finally, it appears that simple ideas associated with the elastic driving force, and a rigid block lattice resistance construction may be useful in determining the ductility in real materials, without actual calculation of the crack structure.

5. Acknowledgement.

A.E.C. gratefully acknowledges support from the Office of Naval Research under Grants Number N00014-92-J-4049.

References
1. J. R. Rice, J. Mech. Phys. Solids **40**, 239 (1992).
2. J. R. Rice, G. E. Beltz, and Y. Sun, Topics in Fracture and Fatigue, ed. A. S. Argon, Springer (New York) (1993).
3. S. J. Zhou, A. Carlsson and R. Thomson, Phys. Rev. Lett., **72**, 852 (1994).
4. S. J. Zhou and R. Thomson, Phys. RevB, **49B** 44 (1994).
5. R. Thomson, Phys. RevB., **52** 7124 (1995).
6. R. Thomson, Phys. RevB., **52** in press.
7. B. Cotterell and J. R. Rice, Int. J. Fract., **16** 155 (1980).
8. J. R. Rice, Z. Suo and J. -S Wang in Metal-Ceramics Interfaces, edited by M. Ruhle, A. G. Evans, M. F. Ashby, and J. Hirth, Acat Scripta Met. Proc. Series 4 (Pergamon, New Yourk, 1990), p. 269.
9. R. Thomson, S. J. Zhou, A. E. Carlsson, and V. K. Tewary, Phys. Rev. B, **46B**, 10613 (1992).

MECHANISM OF DUCTILE RUPTURE IN THE AL/SAPPHIRE SYSTEM ELUCIDATED USING X-RAY TOMOGRAPHIC MICROSCOPY

WAYNE E. KING*, GEOFFREY H. CAMPBELL*, DAVID L. HAUPT**,
JOHN H. KINNEY*, ROBERT A. RIDDLE**, AND WALTER L. WIEN*
*Chemistry and Materials Science Department
*Mechanical Engineering Department
University of California, Lawrence Livermore National Laboratory, Livermore, CA 94551-9900

ABSTRACT

The fracture of a thin metal foil constrained between alumina or sapphire blocks has been studied by a number of investigators. The systems that have been investigated include Al [1,2], Au [3], Nb [4], and Cu [5]. Except for Al/ Al_2O_3 interfaces, these systems exhibit a common fracture mechanism: pores form at the metal/ceramic interface several foil thicknesses ahead of the crack which, under increasing load, grow and link with the initial crack. This mechanism leaves metal on one side of the fracture surface and clean ceramic on the other. This has not been the observation in Al/ Al_2O_3 bonds where at appropriate thicknesses of Al, the fracture appears to proceed as a ductile rupture through the metal.

The failure of sandwich geometry samples has been considered in several published models, e.g., [6,7]. The predictions of these models depend on the micromechanic mechanism of crack extension. For example, Varias et al. proposed four possible fracture mechanisms: (i) near-tip void growth at second phase particles or interfacial pores and coalescence with the main crack, (ii) high-triaxiality cavitation, i.e., nucleation and rapid void growth at highly stressed sites at distances of several layer thicknesses from the crack tip, (iii) interfacial debonding at the site of highest normal interfacial traction, and (iv) cleavage fracture of the ceramic. Competition among the operative mechanisms determines which path will be favored.

This paper addresses the question of why the fracture of the Al/Al_2O_3 system appears to be different from other systems by probing the fracture mechanism using X-ray tomographic microscopy (XTM). We have experimentally duplicated the simplified geometry of the micromechanics models and subjected the specimens to a well defined stress state in bending. The bend tests were interrupted and XTM was performed to reveal the mechanism of crack extension.

EXPERIMENT

The system of pure aluminum bonded to sapphire exhibits low X-ray absorption, chemical compatibility, and a tendency to rupture in a ductile manner. Sapphire avoids the problems associated with grain pull-out and porosity present in polycrystalline alumina. The sapphire was purchased as cylinders of 16 mm diameter by 20 mm height with faces polished parallel to (0001) $\pm$ 1° and λ/10 ($\lambda \approx$ 550 nm) flatness. The aluminum foil was purchased as 99.999% pure and 50 μm thick. The foil was laser cut into disks of 19 mm diameter. A support ring assembly made of commercially pure aluminum was attached to the perimeter of the disk by laser welding. Since bonding of the aluminum foil between two cylinders was to be accomplished by ultrahigh vacuum diffusion bonding [8], the support ring facilitated the remote handling of the disks inside the diffusion bonding machine. Otherwise, the materials were used in the as-received state.

147

The surfaces of the sapphire to be bonded and both sides of the Al foil were sputter cleaned with 1 keV Xe⁺ impinging at 15° above the horizontal. Specimens were rotated during sputtering. The surface elemental composition was characterized using Auger electron spectroscopy. All surface contaminants thus measured, including the oxide layer on the metal, were removed to below the detection limit of the technique. Previous experience with Al shows that many hours of exposure are necessary to reform a contamination layer in the UHV environment (< 10^{-10} torr). The specimens were stacked and a load of 10 MPa applied. The specimens were heated to 600°C and held for 38 h. Relatively long bonding times were required due to variations in the thickness of the Al foil which requires creep to bring the surfaces into contact and eliminate all void space.

Bend beams were cut from the diffusion bonds using diamond cutting. The beams had nominal dimensions of $3 \times 3 \times 40$ mm. To ensure that failure initiated in the metal foil rather than at a flaw in the ceramic, a thin notch was placed within the metal. The notch root diameter was 40 μm and was 500 μm deep. The beams were loaded in four-point loading with inner span of 10 mm and outer span of 30 mm. Several beams were loaded to failure to characterize failure stresses.

Specimens examined with XTM were first characterized in the as-notched condition. The beam was then loaded until the first sign of non-linearity in the load versus cross-head displacement plot was observed. The mechanical test was interrupted, the specimen removed, and an XTM scan performed. The mechanical test was subsequently resumed to strain the beam further. This iterative sequence was continued for as long as was practical.

Ductile metals are opaque to the wavelengths of visible light. Hence, observations of the mechanisms of crack extension have not been possible in these materials. The advent of synchrotron-based XTM [9] has made it possible to determine the three-dimensional variations in density of specimens of several millimeters in dimension for low atomic number materials at resolutions approaching 1 μm. In performing the XTM scans, the X-rays from a synchrotron source (the 8-pole wiggler beamline 4-2 at the Stanford Synchrotron Radiation Laboratory) were passed through a Si(220) monochromator to select 25 keV photons. After passing through the specimen, the X-rays struck a CdWO₄ scintillator. The light emitted from the scintillator passed through magnifying optics and was collected by a CCD camera at a magnification sufficient to give the volume elements (voxels) in the reconstructed image the size 5.3 μm on a side. Radiographs were collected at 0.5° increments of sample rotation over a range of 180°. Reference images of the unobstructed X-ray beam were acquired every 3 degrees to correct the radiographs for beam inhomogenieties, beam instabilities, and pixel-to-pixel gain variations in the CCD array. A three-dimensional image, which represents the product of the pixel size and the absorption coefficient, was reconstructed as slices delineated by the rows of the CCD array using a filtered backprojection algorithm. Defects in the scintillator can give rise to bright or dark spots in radiographs which are not compensated for through normalization to the reference image. Such spots result in the appearance of bright or dark (usually dark) "rings" in the reconstructed slice. Most rings were removed using an interpolation algorithm on experimental sinograms. However some rings remained which can, in severe cases, give rise to contrast resembling voids, i.e., relatively low values of the absorption coefficient-thickness product. Such severe rings were found in the scans of the as-prepared sample and after the first deformation.

Locations of interfaces are delineated by pairs of bright and dark fringes in the XTM contrast. Although the origin of this fringe is currently not fully understood, the point of contrast reversal has been empirically shown to mark the interfaces. Voids, which have a relatively low values of the absorption coefficient-thickness product compared with the sapphire or aluminum are readily revealed by XTM. Voids exceeding two voxels in volume were identified and counted with an efficient, single-pass cluster labeling algorithim developed by Hoshen and Kopelman [10] and described in more detail elsewhere [11]. All void contrasts, including those from ring artifacts, are displayed in Figure 1. Although it is possible to follow the growth of isolated voids, this was not done in this initial study.

RESULTS

As-prepared sample

XTM revealed that the sample prepared in the 4-point bend geometry exhibited two dense interfaces between the aluminum and the sapphire. No resolvable voids were observed outside of three regions affected by ring defects. Figure 1a shows a rendering of the volume studied with XTM. The profile of the notch was extracted from this data and was used as input for finite element calculations to calculate the stress state of the sample.

First deformation

Figure 2a shows the load-displacement curve for the first in a series of three deformations. Loading was relaxed at the first indication of the onset of plasticity. XTM revealed four resolvable (size greater than 2 voxels) voids located at the interfaces between the metal and the sapphire. These voids were of volume between 3 and 7 voxels. Three of the voids were lenticular in nature. The fourth, and largest, had an aspect ratio (ratio of the radius of gyration normal to the interface to that in the plane of the interface) of about 1, i.e., the largest void was spherical. Figure 1b show the rendered volume with voids (including those resulting from ring artifacts and those one voxel in size). Finite element modeling results for the maximum in triaxiality, maximum in tensile traction at the interface, and their locations relative to the notch root are shown in Table I.

Table I. Finite element modeling results for the maximum in triaxiality, maximum in tensile stress at the interface, and their locations relative to the notch root relative to the foil thickness.

Loading	σ_m/σ_o	X_m/h	σ_{yy}/σ_o	X_{yy}/h
1	4.1	2.3	4.5	2.3
2	5.4	2.8	5.9	2.6

Yield stress used in this calculation was 12.2 MPa, foil thickness was 50 μm, and notch depth was 630 μm.

Second deformation

Figure 2b shows the load-displacement curve for the second deformation. XTM revealed 53 resolvable voids located at the interfaces between the metal and the sapphire. Figure 3 shows a slice through the volume revealing several voids at the interfaces. A common feature of is that they are all at the interfaces and voids are never seen to nucleate on opposite sides of the foil directly across from each other; they are usually staggered with respect to each other. Voids of size less than ~30 voxels are usually lenticular in shape. Between volumes of 30-100 voxels, the

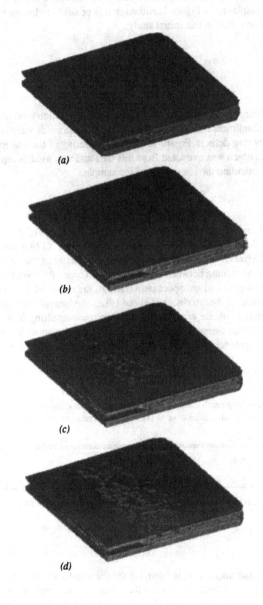

(a)

(b)

(c)

(d)

Fig. 1. Rendering of (a) undeformed volume, (b) after first loading, (c) after seco nd loading, (d) after third loading. Voids appear as lighter contrast features in the volume.

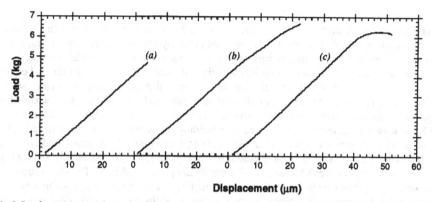

Fig. 2. Load as a function of cross-head displacement for (a) first loading, (b) second loading, (c) third loading.

aspect ratio grows to ~0.6. For volumes in excess of 100 voxels, the aspect ratio decreases to ~0.2. Figure 1c shows a rendering of the void distribution and illustrates the increase in the void population compared with the first deformation.

Third deformation

Figure 2c shows the load-displacement curve for the third deformation. XTM revealed 131 resolvable voids located at the interfaces between the metal and the sapphire. While some small voids do take on a spheroidal shape, the general trend for the smallest and the largest voids is to have aspect ratios of <0.2. Voids in the range of 200-900 voxels in volume tend to have aspect ratios approaching 0.4. The void population has dramatically increased and the voids have spread ahead of the notch (Figure 1d).

DISCUSSION

XTM has been used to elucidate the mechanism of ductile rupture of a 50 μm thick Al foil bonded between two sapphire cylinders subjected to Mode I loading. At the onset of plasticity, small lenticular voids form at each interface at a location consistent with the maxima in

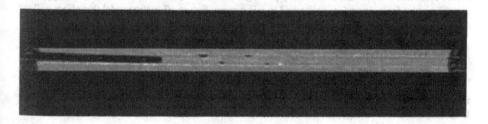

Fig. 3. Slice through volume perpendicular to notch after second deformation showing voids on both interfaces.

151

triaxiality and tensile traction at the interface. As such, this work reveals that debonding plays an important role in the failure at the earliest stages of plasticity. Previous observations of rupture in this system revealed the presence of voiding on the fracture surfaces, e.g. see Ref. 1 and 2. However, the authors were only able to link the origin of some voids with the interfaces, i.e., the generalization of this observation to all voids was not clear. In this investigation, all voids were initially associated with an interface. Further, using conventional fractography, the mechanism of void nucleation and growth could not be revealed. The lenticular nature of these voids as revealed by XTM suggests that they may be formed by debonding and propagation of an interface crack rather than by nucleation and growth via dislocation agglomeration. The existence of debonds at interfaces in Al is consistent with observations in other systems, e.g., Au [3], Nb [4], and Cu [5], which debond but do not exhibit the extensive ductile dimpling found in Al. The observations of this work suggest that the tendency of a system to rupture in a ductile or brittle fashion may be controlled by the interplay of the interface bond strength and the yield strength of the metal. The formation of a void at an interface locally relaxes the constraint thus eliminating the possibility that a void will nucleate directly across the foil on the opposing interface. This increases the constraint nearby giving rise to the tendency of the voids to be staggered with respect to each other.

As deformation proceeds, voids of intermediate size tend to become spherical under the influence of the triaxial stress state which serves to relieve the constraint locally. It is reasonable to conclude that the maximum in the constraint then shifts to locations farther in advance of the notch root where more small debonds form. At high strains, spherical voids continue to grow until they interpenetrate which eventually leads to the formation of the ductile ligaments observed in fractographs. There appears to be a second class of voids that remain lenticular throughout the deformation. These seem to be lenticular voids that have grown past some perhaps critical size above which the tendency to become spherical is diminished.

CONCLUSIONS

Ultrahigh vacuum diffusion bonding has been used to make model specimens in the Al/ sapphire system. The miocromechanics of failure for this system were observed by XTM. The primary findings are that (i) damage ahead of the notch initiates by interface debonding at a location coinciding with maxima in triaxiality and tensile traction at the interface, (ii) debonding occurs at the most early stages in the observation of plasticity, (iii) the debond expands for a limited distance, likely arresting due to crack tip blunting, (iv) this lenticular debond then becomes spherical with further strain and (v) intergrowth of the spherical voids leads to the typical ductile rupture fracture surfaces observed in this system. Future efforts to model this effect will need to include the debonds at the interface which alter the state of stress.

ACKNOWLEDGEMENTS

The authors gratefully acknowledge Professor R. H. Dauskardt for providing the mechanical testing equipment used in this investigation. This work performed under the auspices of U. S. Department of Energy and the Lawrence Livermore National Laboratory under contract No. W-7405-Eng-48.

REFERENCES

1. B. J. Dalgleish, K. P. Trumble, and A. G. Evans, Acta Metall. **37**, 1923 (1989).
2. W. E. King, G. H. Campbell, S. L. Stoner, and W. L. Wien, Cer. Eng. Sci. Proc., **15**, 769 (1994).
3. I.E. Reimanis, B.J. Dalgleish, and A.G. Evans, Acta Metall. Mater. **39**, 3133 (1991).
4. I.E. Reimanis, Scripta Metall. Mater. **27**, 1729 (1992).
5. T.S. Oh, J. Rödel, R.M. Cannon, and R.O. Ritchie, Acta Metall. **36**, 2083 (1988).
6. A. G. Varius, Z. Suo, and C. F. Shih, J. Mech. Phys. Solids **39**, 963 (1991).
7. V. Tvergaard and J. W. Hutchinson, Phil. Mag. A **70**, 641 (1994).
8. W. E. King, et al., Mat. Res. Soc. Symp. Proc. **314**, 61 (1993).
9. J. H. Kinney and M. C. Nichols, Ann. Rev. Mater. Sci. **22**, 121 (1992).
10. J. Hoshen and R. Kopelman, Phys. Rev. B,**15** 3438 (1976).
11. J. H. Kinney, N. E. Lane, and D. L. Haupt, J. Bone and Miner. Res. **10**, 254 (1995).

SIMULATION OF THERMAL STRESSES, VOIDS AND FRACTURE AT THE GaAs/CERAMIC INTERFACE

NICKOLAOS STRIFAS AND ARIS CHRISTOU
Materials & Nuclear Engineering Department, University of Maryland,
College Park, MD 20742   Tel (301) 405-5208   Fax (301) 314-2029

ABSTRACT

Stresses induced at the GaAs-Al_2O_3 interface by large ΔT excursions have been investigated by finite element simulation and have been correlated with experimental results. The effects of power and temperature cycling on crack propagation at the die attach are investigated. The FEA (finite element analysis) method is used to simulate the effect of die attach voids on the peak surface temperature and on the die stresses. These voids in the die attach are identified to be the major cause of die cracking. It was found that stresses developed on the die because of the environmental temperature changes and their dissipation as part of an effective thermal management is necessary to ensure reliable performance.

INTRODUCTION

Generally, in electronic devices, the thermal stresses due to mismatches of material thermal properties will be generated in the device, even if the device is free from external mechanical loading [1]. Moreover, when defects and imperfections are present at material interfaces, such as trapped voids and impurities contained in a solder layer, the thermal stress concentrates at the entrapped defect. It is also possible that the thermal stress becomes singular at the edges of the interfaces between a solder and an electronic component and between a solder and a substrate [2]. Consequently, it is necessary to evaluate the effects of defects on the thermal stress distributions in the material attach layers.

Computer-aided thermomechanical modeling was applied to die attach assemblies in order to assess the role of voids under thermal and mechanical loads. These techniques are very flexible and can handle highly complex models for which an analytical approach is not practical. Such simulation software is most efficient in modeling the thermal stresses induced in a given device due to temperature cycling, in assessing the impact of a defect or design change on the reliability of a given device.

ACOUSTIC MICROSCOPY OF DIE ATTACH VOIDS

Acoustic microscopy was used successfully in the present investigation to measure the volume of the die attach voids in GaAs microwave monolithic integrated circuits (MMICs) before and after temperature cycling. These MMICs consist of the GaAs die, attached with AuSn eutectic solder (25 μm thick) to an Al_2O_3 substrate. The temperature cycling tests were performed by cycling the packages from room temperature to 225 °C for 80 cycles, maintaining the temperature constant for one hour. The tests were carried out in an environmental test oven configured for testing electronic packages. The transition time from 25 °C to 225 °C was 30 minutes. Void density as a function of number of temperature cycles was measured and is shown in figure 1. Typical configurations of voids are shown in figures 2 and 3, indicating that voids extend to the edge of the die or are totally centrally enclosed.

155

Mat. Res. Soc. Symp. Proc. Vol. 409 ® 1996 Materials Research Society

Effect of Thermal Cycles on Void Density

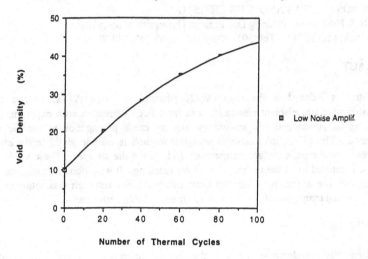

Figure 1. Void density as a function of temperature cycles. Void density is expressed as percent of total die-substrate interfacial area.

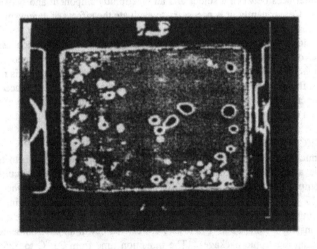

Figure 2. SAM images of a GaAs die bond sample with edge voids before 40 cycles of thermal shock.

The analytical results [3] which indicate that the highest shearing stresses are at the edge of the die were validated by the observation of die cracks which were initiated in the locations of edge voids. Figure 3 shows such an observation for a 40 cycle test. In this MMIC, edge voids are present prior to testing and subsequent cycling resulted in vertical cracks in the vicinity of the edge voids.

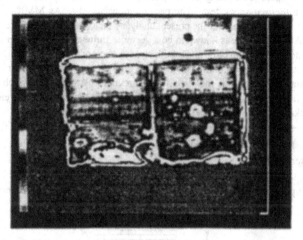

Figure 3. SAM images of a GaAs die bond sample with edge voids after 40 cycles of thermal shock, after which it suffered a crack at the corner of the die.

The majority of the die crack failures can be traced to imperfect die-attachment (voiding). Voids cause local stress concentration, whose magnitude is very much dependent on the location of the voids. Using finite element analysis, we have previously reported [3] that the presence of an edge void at the die-attach interface changes the local stress and creates a tensile longitudinal stress field. For a center void, the longitudinal stress at that location becomes less compressive than the average stress obtained without the void. This leads to the conclusion that for die-attachment without voids or with some center voids, die cracking will not occur, whereas specimens with voids near the edges of the dice are likely to have die cracks. Experimental observations of the actual production samples, as well as of samples with purposely formed voids confirmed this conclusion. It has also been further confirmed by other reported observations [4]. Correlations have also been found between voids in the bonding layer and the die cracking due to thermal shock tests. Therefore, the results of this investigation experimentally confirm the role of die attach voids in the mechanical failure of die and die attach assemblies. Temperature cycling initiates fracture at the edge-void location. In the following sections FEA models will be applied in order to further understand the experimental results.

MODELING OF VOIDS BY FEA

In the present study, thermal stress distributions in die attach regions of an electronic device (Gallium Arsenide die bonded on an Al_2O_3 substrate using AuSn as die attach material) are investigated analytically. Specifically, in order to examine the effect of void shape on the thermal stress distribution, die attaches with circular and elliptical voids are examined. The thermal stress distribution in the die attachment is analyzed using the two dimensional finite element method. The effects of the location, void shape in the die attachment and temperature changes on the stress distribution at the periphery of the defects and at the interface are examined. In addition, the finite element method results are compared to analytical results.

The finite element method was used to simulate the thermal stresses induced in the package due to temperature cycling of the package from 25 °C to 125 °C. Such a test is characteristic of a thermal shock test. The package consists of GaAs MMIC, which is epoxy mounted to an Al_2O_3 base plate. For the present analysis, the codes selected were ABAQUS, and PATRAN. The ABAQUS code is known to be a versatile finite element analyzer that can handle linear as well as non-linear problems, under both steady state and transient conditions. It also has some pre and post processing capabilities, but in this study it was used for obtaining the solution and post processing. PATRAN is mainly a strong pre and post processor which also has some analytical capability, herein used for mesh generation only. In the present finite element analysis (FEA) of the package there are a few basic assumptions that were made:

· The material properties were assumed temperature dependent (100-300 °C).

· The interfaces between different materials were assumed to have perfect adhesion at their interface.

· All the materials were considered to be time independent linearly elastic in nature.

The analysis was performed in 2-D for simplicity and to allow maximum mesh resolution in the

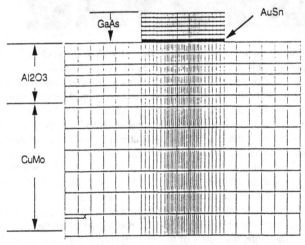

CuMo = 500 microns,   Al2O3 = 250 microns
AuSn = 25 microns,   GaAs = 100 microns

Figure 4. FEA grids prior to the simulated stresses.

vicinity of the void. Mesh divisions were held to a minimum close to the center of the die and were made progressively more dense towards the void edges. The package was discretized using axi-symmetric tetragonal elements. Each tetragonal element has four nodal points. Each nodal point has two degrees of freedom.

Total number of elements = 1250
Total number of nodes   = 1810

The displacement boundary conditions to enforce the symmetry of the package were defined by prescribing zero displacement in the Y-direction for all nodes along the X-direction at Y=0. The temperature loads to the package were applied by assigning the same temperature at every node of the package, and were varied from 25 °C to 125 °C. The FEA analysis was further used to simulate the effect of die attach voids on the Von Mises stress in each of the device layers. The simulated voids were placed at the edge of the chip and at the center directly underneath the heat generating transistor.

Two types of voids were simulated in order to determine which void may initiate cracks in the die. The structures with voids were also compared to control simulations of unvoided die attach assemblies. Figure 4 shows the FEA grid prior to the simulated stresses. The applied temperature loads of 25 °C to 125 °C resulted in high shearing stresses in the case of edge voids, higher than those present in the case of central voids, as shown in figures 5 and 6. The shear stress typically initiated horizontal cracks at the die-attach interface, while the normal stress initiated vertical cracks. The location of the voids were typical of die attach voids found in the GaAs structures analyzed by scanning acoustic microscopy. Voids beneath the location of the power transistors were found to significantly change (10% increase) the surface temperature by disrupting the thermal dissipation paths, resulting in an increase in thermal resistance and die temperature. The increase in surface temperature can significantly reduce the die reliability.

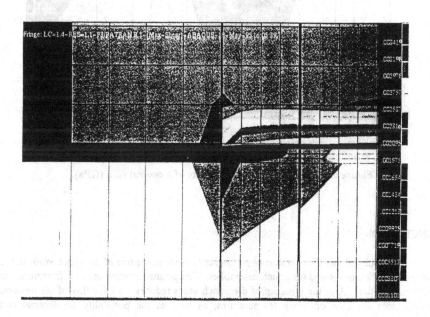

Figure 5. Shear stresses in the vicinity of an edge void (Gpa).

The central void is shown to increase the Von Mises stresses from a peak of 129 MPa to 150 MPa. The effect of an edge void was also found to increase the Von Mises stresses. Voids result in stress concentration which may increase as a result of temperature cycling and can result in die fracture.

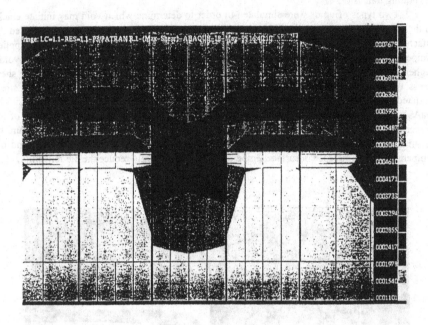

Figure 6. Shear stresses in the vicinity of a central void (GPa).

CONCLUSION

The results of this investigation experimentally confirm the role of die attach voids in the mechanical failure of die and die attach assemblies. Temperature cycling initiates fracture at the edge-void location. Also the presence of die attach voids reduces the reliability of the devices, and increases the chip operating temperatures, as well as the possibility for thermal and mechanical stress failures.

REFERENCES

[1] J. H. Lau, "Thermal Stress and Strain in Microelectronics Packaging", Van Nostrand New York, pp. 385-565, 1993.

[2] N. Strifas and Aris Christou, "Die Attach Adhesion and Void Formation at the GaAs Substrate Interface", MRS, Fall 1994.

[3] N. Strifas and Aris Christou, "Failure Mechanisms, Thermal Analysis and Finite Element Modeling of Medium Band Low Noise Amplifiers", ASME, Winter Annual Meeting. 1994.

[4] N. Strifas, "Die Attach Voids, Measurements and Simulation", Master's Thesis, University of Maryland, 1995.

MIXED MODE INTERFACE TOUGHNESS OF METAL / CERAMIC JOINTS

YUEGUANG WEI, JOHN W. HUTCHINSON

Division of Applied Sciences, Harvard University, Cambridge, MA 02138

ABSTRACT

A mechanics study of the interface toughness of joints comprised of ceramic substrates joined by a thin ductile metal layer is carried out for arbitrary combinations of mode I and mode II loading. The crack lies on one of the metal/ceramic interfaces, and the mechanism of separation at the crack tip is assumed to be atomic decohesion. The SSV model proposed by Suo, Shih and Varias is invoked. This model employs a very narrow elastic strip imposed between the substrate and the ductile layer to model the expected higher hardness of material subject to high strain gradients and possible dislocation-free zone in the immediate vicinity of the crack tip. The criterion for crack advance is the requirement that energy release rate at the crack tip in this narrow elastic strip be the atomistic work of fracture. The contribution of plastic dissipation in the metal layer to the total work of fracture is computed as a function of the thickness and yield strength of the layer and of the relative amount of mode II to mode I. Ductile joints display exceptionally strong thickness and mixed mode dependencies.

INTRODUCTION

Plastic dissipation at the tip of a crack constitutes a major fraction of the total work of fracture for many ductile structural metals and tough metal/ceramic interfaces. A quantitative understanding of interface toughness is just beginning to emerge [1,2]. Recent experiments on the toughness of metal ceramic interfaces [3-5] have helped to elucidate the relationship between the total work of fracture and the work of the fracture process. For interfaces which fail by an atomic decohesion mechanism, the total work of fracture can be hundreds of times the atomistic work of separation. Nevertheless, the atomistic work of separation sets the magnitude of the contribution from plastic dissipation, and small changes in the former can produce huge changes in the later. This exceptional nonlinear magnification has been recently demonstrated experimentally for a niobium/sapphire interface [5].

Mechanics models of the toughness of structural metals and metal/ceramic interfaces must be divided into at least two classes: those which are apply to the ductile fracture process of void growth and coalescence, and those for which the atomic separation is the fracture process. The important distinction between two classes arises because of the widely different scales of the fracture processes. Void growth and coalescence takes place on a scale ranging typically from ten to a hundred microns. Conventional continuum plasticity theory is sufficiently accurate at this scale to connect the region where the loads are applied down to the crack tip where the fracture process occurs. Models [1,2,6] which embed a traction-separation relation characterizing the fracture process within a elastic-plastic continuum representation of the solid are capable of capturing the most important features of interaction between fracture process and plastic deformation. By contrast, conventional plasticity theory cannot be used at the very small scales bridging down to the tip when atomic separation is the fracture process. According to continuum plasticity crack solutions, the maximum stresses which can be attained in the vicinity of a crack tip are never more than about five times the tensile flow stress of the metal. Such levels are generally well below the stresses needed to bring about atomic separation of a strong interface. The problem of bridging from the macroscopic level down to the atomistic level at the crack tip through a plastic zone constitutes a major challenge.

In this paper, attention is directed to metal/ceramic interfaces which fail by atomic separation. The problem just alluded to will be side-stepped with a model proposed by Suo, Shih and Varias [7], which will hereafter be designated as the SSV model. This model has been proposed for metals containing at least moderate densities of pre-existing dislocations such that plastic deformation is associated with motion of these dislocations. It is assumed that no dislocations are emitted from the crack tip such that the tip remains atomistically sharp. The essential modeling step is the imposition of an elastic strip of thickness t between the interface and the plastically deforming metal layer, as shown in Fig. 1. In a crude way, the elastic strip is intended to: (i) account for the high flow stress of metal which is expected to develop in the vicinity tip where strain gradients are very large, and (ii) represent any dislocation-free region at the tip. An elaboration on the SSV model [8] proposes a self-consistent way to identify t by matching the dislocation density and effective flow stress a distance t from the tip. In the present paper, the thickness of the elastic strip t is regarded as a parameter in the model whose value, for example, can be chosen to fit one set of experimental data.

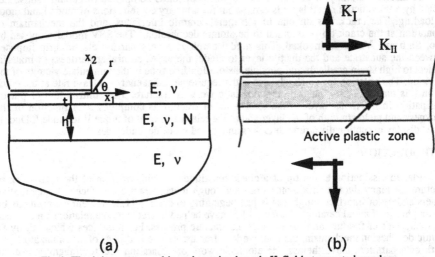

(a)   **(b)**

Fig.1 The joint system subjected to mixed mode K-field at remote boundary

Specifically, we model the plane strain system with a thin metal layer sandwiched between two ceramic substrates. A long crack lies along one of the interfaces. As depicted in Fig. 1, the crack is loaded by combinations of the mode I and mode II stress intensity factors, K_I and K_{II}. The tensile yield stress of the metal is σ_Y and its strain hardening exponent is N. To limit the number of material parameters in the study, the substrates and metal layer are assigned the same Young's modulus E and Poisson's ratio ν. The object of the study is to predict the total work of fracture Γ_{ss} for the steady-state propagation of the crack down the interface in terms of the following parameters of the model: the atomistic work of separation Γ_0, the relative amount of mode II to mode I, as measured by $\psi=\tan^{-1}(K_{II}/K_I)$, the thickness of the metal layer h and its yield stress and hardening exponent, σ_Y and N, and E and ν. The thickness t of the thin metal strip imposed between interface and the elastic-plastic layer is taken to be small compared to the metal layer thickness h.

FORMULATION OF THE STEADY-STATE CRACK GROWTH PROBLEM

Each ceramic substrate in Fig. 1 is assumed to be semi-infinite. The remote loading is specified by the mode I and II stress intensity fields:

$$\sigma_{ij} = \frac{K}{\sqrt{2\pi r}}\left[\tilde{\sigma}_{ij}^I(\theta)\cos(\psi) + \tilde{\sigma}_{ij}^{II}(\theta)\sin(\psi)\right] \quad (r \to \infty)$$

$$\psi = \tan^{-1}(K_{II}/K_I), \quad K = \left[K_I^2 + K_{II}^2\right]^{1/2}$$

(1)

The functions depending on θ can be found in any text book on fracture. The remote loading is characterized by K_I and K_{II} or, equivalently, by G and ψ, where Irwin's result for the energy release rate is

$$G = \frac{(1-v^2)}{E} K^2 = \frac{(1-v^2)}{E}\left[K_I{}^2 + K_{II}{}^2\right] \tag{2}$$

This quantity is interpreted as the remote, or applied, energy release rate. It will be identified with the total work of fracture for steady-state growth, Γ_{ss}.

As the crack tip is fully surrounded by elastic material, it is characterized by local stress intensity factors, K_I^{tip} and K_{II}^{tip}. Alternatively the tip deformation can be characterized by G_{tip} and ψ_{tip}, where

$$G_{tip} = \frac{(1-v^2)}{E}\left[K_I^{tip2} + K_{II}^{tip2}\right] \quad \text{and} \quad \psi_{tip} = \tan^{-1}(K_{II}/K_I) \tag{3}$$

These quantities will be computed. In the application of the crack solution, G_{tip} will be identified with the atomistic work of fracture Γ_0.

The tensile stress-strain curve of the metal in the layer is taken to be

$$\varepsilon = \frac{\sigma}{E} \qquad\qquad \sigma < \sigma_Y$$

$$= \frac{\sigma_Y}{E}\left(\frac{\sigma}{\sigma_Y}\right)^{1/N} \qquad \sigma > \sigma_Y \tag{4}$$

This relation is generalized to multi-axial stress states by J_2 flow theory for small strain incremental plasticity (von Mises theory). For plastic loading, increments in stress and strain are related by

$$\dot{\varepsilon}_{ij} = \frac{1+v}{E}\dot{\sigma}_{ij} - \frac{v}{E}\dot{\sigma}_{kk}\delta_{ij} + \frac{3\dot{\sigma}_e}{2\sigma_e}\left[\frac{1}{E_t(\sigma_e)} - \frac{1}{E}\right]s_{ij} \tag{5}$$

where s_{ij} is the stress deviator, $\sigma_e = \sqrt{3s_{ij}s_{ij}/2}$ and E_t is the tangent modulus of the tensile stress-strain curve at σ_e from (4).

The emphasis here is on steady-state growth wherein the crack has advanced sufficiently far from initiation such that stresses and strains no longer change from the vantage point of a observer translating with the crack tip. The crack problem is posed for steady-state crack growth under constant G and ψ. A zone of active plasticity moves with the tip, and a wake of plastically deformed, but elastically unloaded, material extends behind the tip, as depicted in Fig. 1. The steady-state condition for any quantity such as a Cartesian component of incremental stress is

$$\dot{\sigma}_{ij} = -\dot{a}\frac{\partial\sigma_{ij}}{\partial x_1} \tag{6}$$

where the crack is advancing in the x_1 direction and $\dot{a}$ is the increment of crack advance. A numerical method [9] which employs iteration to satisfy conditions (6) is used to directly obtain the steady-state solution. A similar method was employed in [7] and [8]. A finite element procedure with a grid especially designed to cope with the steady-state wake has been used to carry out the calculations, as will be discussed further in the next section.

The main results of interest from the solution are the functional relations between the local crack tip fields, as characterized by G_{tip} and ψ_{tip}, and the corresponding applied quantities, G and ψ. The non-dimensional forms of these relations are

$$\frac{G}{G_{tip}} = f_1\left[\psi, \bar{h}, \bar{t}, N, \sigma_Y/E, v\right] \quad \text{and} \quad \psi_{tip} = f_2\left[\psi, \bar{h}, \bar{t}, N, \sigma_Y/E, v\right] \tag{7}$$

where the two dimensionless thickness quantities are

$$\bar{h} = \frac{h}{R_0} \text{ and } \bar{t} = \frac{t}{R_0} \text{ where } R_0 = \frac{EG_{tip}}{3\pi(1-v^2)\sigma_Y^2} \qquad (8)$$

The reference length, R_0, can be regarded as an estimate of the small scale yielding plastic zone size in plane strain for a crack loaded by G_{tip}. It will be seen to serve as a basic material length scale against which to gage whether the metal layer is thick or thin. In the limit when $\bar{h}$ is very small, plasticity is essentially eliminated and G_{tip} and ψ_{tip} approach their applied counterparts, G and ψ. When plastic dissipation is not negligible, the steady-state work balance is precisely

$$G = G_{tip} + G_{plastic} \qquad (9)$$

where $G_{plastic}$ denotes the plastic dissipation rate (energy per unit length of crack edge per unit length of crack advance). Thus the steady-state model permits a clear-cut partitioning of the work of crack advance into plastic dissipation and the portion available for the local fracture process, G_{tip}.

In applying the solution to establish the relation of the total work of fracture for steady-state crack advance along the joint, Γ_{ss}, to the atomistic work of separation of the interface, Γ_0, Γ_{ss} is identified with G and Γ_0 with G_{tip}. Thus, results will be presented for the ratio Γ_{ss}/Γ_0, reflecting the magnification of toughness due to plastic dissipation. The nondimensional thickness quantities are still defined in (8), but with R_0 evaluated using $G_{tip}=\Gamma_0$.

NUMERICAL METHOD

As described in [9], the steady-state solution procedure requires stress histories to be integrated along lines parallel to the x_1 axis within the active plastic zone and the wake. This is most effectively implemented in a finite element framework if the plastic zone and wake lie within a horizontal grid whose elements at any distance above the crack line all have the same height in the x_2 direction. The grid used in the present work is shown in Fig. 2. The horizontal mesh near the tip which contains the metal layer transitions to a radial mesh in the outer field. Tractions obtained from (1) are prescribed on a circular boundary of radius R. In all the numerical results presented below, the computations have been performed with $R/R_0=1012$. This choice ensures that layer thickness and the plastic zone extent ahead of the tip will always be small compared to R in the examples considered here. Four-noded quadrilateral elements were employed and four Gauss points are used for integration. A mesh with 3664 elements was used in all the calculations reported below.

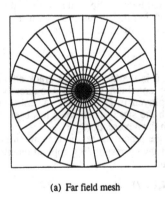

(a) Far field mesh

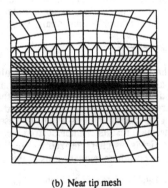

(b) Near tip mesh

Fig.2 Finite element mesh

The local energy release rate, G_{tip}, was computed using the J-integral such that the contour below the crack line was chosen to lie within the thin elastic strip, i.e. $x_2 > t$, ensuring path-independence. The local measure of mode mix, ψ_{tip}, was identified with $\tan^{-1}(\sigma_{12}/\sigma_{22})$ as evaluated on $x_2 = 0$ just ahead of the tip.

NUMERICAL RESULTS

All the dimensionless variables in (7) except σ_Y/E and ν are important in determining Γ_{ss}/Γ_0. Numerical results will now be presented which reveal the main trends. Plots of Γ_{ss}/Γ_0 as a function of $\bar{h}$ are given in Fig. 3 for an applied mode I loading, $\psi=0$, for two choices of $\bar{t}$ and three values of the strain hardening exponent N. For dimensionless metal layer thicknesses $\bar{h}$ above a certain level (between 2 and 4 for the examples in Fig. 3), Γ_{ss}/Γ_0 becomes independent of layer thickness. For values of $\bar{h}$ above this level the plastic zone does not spread all the way across the layer to the other substrate, and a further increase of the layer thickness does not alter the plastic dissipation. The layer and the substrates have the same elastic properties in this example, and thus the limiting value of Γ_{ss}/Γ_0 for large $\bar{h}$ represents the small scale yielding toughness enhancement for the metal/ceramic interface. At the other limit, as $\bar{h}$ approaches zero, the amount of metal deforming plastically becomes unimportant and the total work of fracture approaches Γ_0. The qualitative features seen in Fig. 3 are similar to those predicted by the embedded fracture process model [1,2].

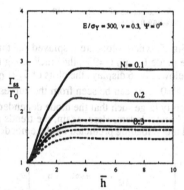

Fig.3 Variation of toughness with ductile layer thickness for different N. Solid lines correspond to $\bar{t}$=0.15, and dashed lines to $\bar{t}$=0.25.

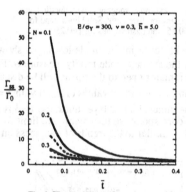

Fig.4 The variation of toughness with elastic strip thickness. Solid lines stand for ψ=90°, and dashed lines for ψ=0°.

The dimensionless parameter $\bar{t}$ characterizing the thickness of the plasticity-free elastic strip in the SSV model is a critical intrinsic parameter in establishing Γ_{ss}/Γ_0. Curves displaying this ratio as a function of $\bar{t}$ for mode I ($\psi=0$) and mode II ($\psi=90°$) are shown in Fig. 4 for several values of N. When $\bar{t}$ exceeds a value of roughly 1/4, there is little plasticity enhancement of toughness. The elastic strip is then sufficiently thick that it eliminates most of the crack tip shielding due to plastic deformation. For values of $\bar{t}$ less than 1/10, the toughness enhancement becomes significant. In this range, shielding is a strong function of $\bar{t}$. One approach to applying the SSV model is to regard $\bar{t}$ as a parameter to be chosen to fit one set of experimental data for a given interface. Alternatively, and more fundamentally, $\bar{t}$ can be assigned its value using an approach such as that suggested in [8] which attempts to estimate the width of the plasticity-free region from first principles. Either way, the value of $\bar{t}$ is associated with the interface and must be regarded as being independent of the metal layer thickness, the mode mixity, and other extrinsic factors, such as residual stress in the metal layer, which are not considered in this paper. The value of the model rests in the fact that once $\bar{t}$ has been set, the dependence of Γ_{ss}/Γ_0 on the extrinsic factors can be computed. There is a correspondence discussed in [2] between the parameter $\bar{t}$ of the SSV model and the ratio of the peak separation stress to the yield stress, $\hat{\sigma}/\sigma_Y$, of the embedded fracture process zone model. For corresponding behaviors, the smaller is $\bar{t}$ the larger is $\hat{\sigma}/\sigma_Y$.

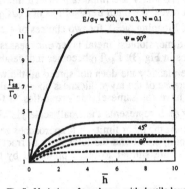

Fig.5 Variation of toughness with ductile layer thickness for different Ψ. Solid lines stand for $\bar{t}$ =0.15, and dashed lines for $\bar{t}$=0.25.

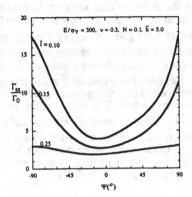

Fig.6 The curves of toughness with applied loading mixity.

The effect of mixed mode loading is shown in Fig. 5, where plots are displayed for three values of the applied mode mixity measure ψ. For pure mode II with ψ=90°, the crack along the interface remains open to the tip, as will be discussed below. Fig. 6 displays the plots of Γ_{ss}/Γ_0 as a function of ψ for three values of $\bar{t}$, all with $\bar{h}$=5 and N=0.1. As can be seen from the curves in Fig. 5, the dimensionless layer thickness $\bar{h}$=5 is sufficiently large such that the mode dependence in Fig. 6 corresponds to the asymptotic limit for "thick" metal layers. Here, again, the trends are remarkably similar to the trends of toughness on mixed mode loading predicted by the embedded

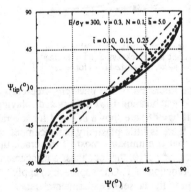

Fig.7 Relations of applied loading mixity with crack tip mixity for three different elastic strip thickness.

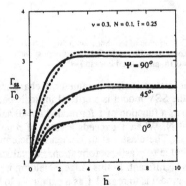

Fig.8 Effect of yielding strain on the toughness. Solid lines stand for E/σ_Y =1000, and dashed lines for E/σ_Y =50.

fracture process model [1]. The lowest total work of fracture Γ_{ss} is attained for near mode I loadings. Large enhancements of toughness arise for loadings which have a major component of mode II. Of course, these enhancements are lost when the layer thicknesses are small ($\bar{h} \ll 1$ in Fig. 3). Companion plots to those in Fig. 6 showing the relation of the local mode mixity ψ_{tip} to the applied mode mixity ψ are given in Fig. 7. The asymmetry in the relation is due to the fact that the metal layer lies below the plane of the cracked interface. Note that pure mode I at the local level corresponds roughly to $\psi = 15^\circ$ at the applied level. The crack tip is open under $\psi = 90^\circ$, but is closed at values of ψ more negative than about -85°. There is a range of applied mixed mode loadings ψ, from roughly -45° to 30°, over which the local mode mixity is only weakly dependent on the applied mode mixity with $\psi_{tip} \cong -15^\circ$

For completeness, curves displaying the effect of two choices of σ_Y/E are shown in Fig. 8. The effect of this parameter is very small. Note, however, that σ_Y has a strong influence on the toughening enhancement through $\bar{h}$ and $\bar{t}$ by virtue of their dependence on R_0. Calculations with different values of Poisson's ratio ν show that this parameter also has a very weak effect on Γ_{ss}/Γ_0.

DISCUSSION

The SSV model predicts a strong dependence of the interface toughness on the relative amount of mode II to mode I applied to the joint when plastic dissipation makes a major contribution to the macroscopic toughness. This effect is due to the more effective crack tip shielding by plastic deformation when the loading has a large component of mode II. The version of the SSV model invoked here assumes that the atomistic work of fracture Γ_0 is independent of the local crack tip mixity, ψ_{tip}. Any local dependence on ψ_{tip} would be further magnified at the macroscopic level. Qualitatively, the mode dependence of the SSV model is similar to that displayed by the embedded fracture process zone model [1]. Both models also display a similar trend of joint toughness with the thickness of the metal layer. When the thickness of the metal layer becomes comparable to the characteristic plastic zone size, R_0, plastic dissipation is reduced. The plasticity contribution to the total work of fracture becomes nearly inconsequential for layer thicknesses below about $R_0/4$.

As discussed in the Introduction, the imposition of the thin elastic strip of thickness t between the interface and the metal layer in the SSV model is a device for circumventing the limitations of conventional plasticity at scales below about several microns. In this spirit, t can be regarded as a parameter in the model to be chosen to fit one set of experimental data for a given interface. More rigorous models for bridging down to the crack tip will require a plasticity theory capable of characterizing behavior at the micron to sub-micron scale when large strain gradients are present. The following numerical example illustrates the inconsistency inherent in applying the SSV model to a metal/ceramic interface undergoing atomic separation. Take the following representative values for the interface system: $E=200$GPa, $\sigma_Y=200$MPa, and $\Gamma_0=1$Jm^{-2}. Then, by (8), $R_0=0.5\mu$m. From the numerical results (cf. Fig. 4), values of $\bar{t}=t/R_0$ needed to give rise to significant levels of plastic dissipation must be less than about 0.2. Thus, the thickness of the plasticity-free elastic strip must satisfy $t<0.1\mu$m. While such values are reasonable, it is not reasonable to expect conventional plasticity will remain valid down to the strip. In other words, even with the introduction of the elastic strip, there remains a significant domain of the model within which the rationale for using conventional plasticity is highly questionable.

ACKNOWLEDGMENT

This work was supported in part by the Natural Science Foundation under Grants MSS-92-02141 and DMR-94-00396, in part by Young Scientist Award for Y.W. from the Chinese Academy of Sciences, and by the Division of Applied Sciences, Harvard University.

REFERENCES

1. V. Tvergaard and J. W. Hutchinson, J. Mech. Phys. Solids, **41**, p.1119(1993).

2. V. Tvergaard and J. W. Hutchinson, Phil. Mag., **70**, p.641(1994).

3. I. E. Reimanis, B. J. Dalgleish, M. Brahy, M. Ruhle, M. and A. G. Evans, Acta Metall. Mater., **38**, p.2645(1990).

4. I. E. Reimanis, B. J. Dalgleish, and A. G. Evans, Acta Metall. Mater., **39**, p.3133(1991).

5. G. Elssner, D. Korn and M. Ruhle, Scripta Metallurgica *et* Materialia, **31**, p.1037(1994).

6. A. Needleman, J. Appl. Mech., **54**, p.525(1987).

7. Z. Suo, C. F. Shih and A. G. Varias, Acta Metall. Mater., **41**, p.1551(1993).

8. G. E. Beltz, J. R. Rice, C. F. Shih and L. Xia, A Self-Consistent Model for Cleavage in the Presence of Plastic Flow, Submitted to Acta Metall. Mater., (1995).

9. R. H. Dean and J. W. Hutchinson, Quasi-static steady crack growth in small-scale yielding, Fracture Mechanics: Twelfth Conference, ASTM STP700, p.383(1980), Philadelphia, PA.

MEASURING INTERFACE FRACTURE RESISTANCE IN THIN HARD FILMS

H. GUO, B. C. HENDRIX, X.D. ZHU, K. W. XU, J. W. HE
State Key Laboratory for Mechanical Behavior of Materials, Xi'an Jiaotong University,
Xi'an, 710049 China, jwhe@xjtu.edu.cn

ABSTRACT

Measuring the resistance of an interface to fracture becomes quite difficult for practical, wear-resistant coatings of the TiN/TiC family deposited on hardened steel. Because of the relatively high bonding strengths, most methodologies currently in use depend on a significant amount of substrate plastic deformation to induce interface fracture, yet practical failures relate more directly to interface roughness and chemistry as well as film thickness rather than small changes in substrate plasticity. Previous work has shown clearly that high cycle (>10^6 cycles) contact fatigue is quite sensitive to interfacial condition and very insensitive to the plastic properties of both substrate and film. The current work uses spherical contact with the film, rather than previously used cylindrical contact, and has thus eliminated the stress concentration at the edge of the contact area, greatly simplifying the required stress analysis. Spherical contact also allows the test to be performed more quickly and on flat substrates. The shear stress range at the interface is calculated as a function of coating thickness and elastic properties of the coating and substrate. In this way, the resistance to fracture of the interface is evaluated independent of substrate plasticity and film thickness.

INTRODUCTION

Thin hard films have found increasing applications in various sectors of industry to improve the tribological performance, however, failure of a hard film-substrate system is seldom caused by conventional wear, but by film fracture or delamination from the substrate (interfacial adhesive failure) [1,2]. The importance of the film-substrate fracture resistance has been emphasized by many investigators and several techniques have been developed to assess the interfacial fracture resistance in thin hard films [3-5]. Uniaxial tests and crack growth tests have well-defined stress states but are limited to the testing of interfaces which are weaker than the strongest adhesives whose upper limit is much lower than the thin hard film-substrate adhesion [6]. The scratch test is well known and widely used because it meets the requirement of load range, but, it always involves static or quasi-static elastoplastic loading. The critical load, which is used to evaluate the fracture resistance is determined not only by the coating-substrate adhesion strength but also by the intrinsic parameters including scratch speed, loading rate and diamond tip radius, and the extrinsic parameters including coating thickness, coating and substrate hardness, surface roughness, friction coefficient and humidity of the test environment [5]. In addition, the complexity of the stress field induced during elastoplastic loading makes it difficult to calculate the stress at the interface and thus determine the critical stress value for coating debonding. Accordingly, it is difficult to quantitatively compare the fracture strength of different kinds of coatings of varying thicknesses on various substrates. The critical load measured by the scratch test is a measure of the load bearing capacity of the coating-substrate system, rather than the interfacial fracture resistance [7].

From the point of view of application, where the coated tools or components are rarely under plastic deformation conditions, the fracture or delamination of coating films usually occurs after a

171

large number of cycles. These facts indicated that the coating failure by debonding during service is a fatigue failure process. So a cylindrical rolling contact fatigue test technique has been developed to investigate the interfacial fatigue behaviour and the fracture resistance of thin hard coatings under cyclic elastic loading conditions [8]. The experimental results show that the method is very sensitive to changes of the interface conditions but not sensitive to non-interfacial factors such as substrate or coating hardness. This result is consistent with Auger spectroscopy that shows no residual Ti or N on the substrate after film spallation [9]. Studies of crack initiation and growth under similar conditions [10] also show crack initiation along the interface as a precursor to crack growth through the film and subsequent spallation.

However, this method also has some shortcomings. To get to the fatigue limit 5×10^6 cycles, each test takes about 9 days. The specimen must be coated on the outside circumference of a cylinder which is inconvenient for line-of-sight coating technique. Also, there is some stress concentration at the edge of the contact area, and during experiments the delamination usually occurs on the edges of the contact area confirming the controlling effect of the stress concentration. If a round counterpart with no stress concentration is used (such as [10]), then it has been shown that the overall contact stress decreases 20% or 30% of the initial stress with even a few micron of the counterpart wear [11].

A spherical rolling test with a series of balls rolling on a flat coated specimen has been proposed to solve these problems [12]. It can eliminate the stress concentration as well as shorten the testing time to about 13 hours. In the present paper, the basic principles of the technique are outlined and typical experimental results are presented. Emphasis is given to the theoretical analysis of the inhomogeneous stress field and to the comparison with experimental results from cylindrical rolling tests under the same interfacial conditions.

MECHANICAL STATE OF THE EXPERIMENT

During the test, the maximum normal stress on the specimens' surface ranges from 500MPa to 2000MPa. In such a load range, bulk plastic deformation does not occur in most of the specimens according to contact stress analysis using the classical Hertzian theory. The elastic nature of the loading in the spherical rolling contact fatigue test allows detailed contact stress analysis using the theory of elasticity. A sketch of a pressure over a circular area of the top of surface and coordinate systems of both coating and substrate is shown in Figure 1. In order to compare tests which are performed on coatings of different thicknesses or of different elastic properties, we should calculate the additional stress from the strain mismatch between the coating and the substrate. Using the integral transforms of the stress fields written by Chen [13], the stresses in the coating and the substrate are determined. Because the contact radius is much larger than the coating thickness, a Hertzian elliptical pressure distribution on the surface is used.

We compiled a general computer program to calculate stresses in a layered medium so that the stresses can be calculated to a high degree of accuracy and the computational labor is reduced significantly. With minor changes in a few subroutines, this computer programme has been used for the plane strain situations of cylindrical rolling test.

Since failure of coatings occurs at the interface, the stress state at the interface is examined. Our previous stress analysis for cylindrical contact shows that the

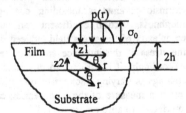

Figure 1. Sketch of Contact Half-space With Ball

shear stress range, $\Delta \tau_{rz}$, at the interface is the appropriate parameter that should be used to characterize the interfacial fracture. Using the above numerical method based on the contact between inhomogeneous materials, the criterion of debonding, $\Delta \tau_{rz}$, at the interface, can be calculated taking into account the influence of elastic properties of the coating and substrate, and thickness of coating.

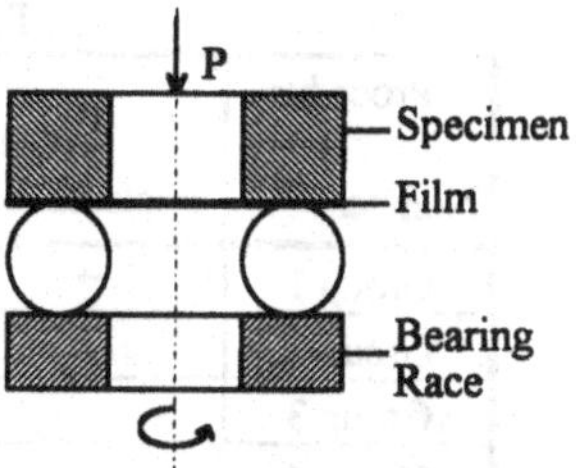

EXPERIMENTAL

Figure 2. Scheme of Spherical Rolling Test

Test Procedure

The spherical rolling test is being carried out on a self-made machine designed for contact fatigue test. The test configuration is illustrated in Figure 2. The specimen is coated with the investigated coating on its flat surface while the balls made of uncoated hardened bearing steel are rolling on the specimen surface. The whole testing system is emerged into the lubrication oil so wear is minimized.

The test was interrupted periodically and the tested surface was examined. The coating was regarded as failed if 5% of the coated contact area had been exfoliated. In this way, under a specific load, the number of cycles at which the coating begins to fail could be determined.

Specimen Preparation

Two hardened and tempered alloy steels, i.e. M2 high speed steel and AISI 52100 ball bearing steel were used as substrate materials for this study because they are commonly used and microstructurally distinct materials.

The films TiN and TiCN with C:N approximately 1:4 are produced by the plasma enhanced chemical vapour deposition (PCVD) method. Prior to coating, the substrate surface was prepared by grinding, blasting or plasma nitriding respectively. It is evident that the state of the substrate surface prior to coating has a strong influence on the bonding strength of hard coatings [14,15]. Different surface pretreatments will lead to a substantial difference in the resultant adhesion strength. In order to investigate the sensitivity of the test to the hardness of the substrate, two testing conditions were used giving substrate steels the hardness of HRC 45 and HRC 63. The thickness of the coatings is about 2.5-3 μ m. Some 4.2 μ m thicker TiCN films are prepared in order to investigate the influence of coating thickness on the coating's fracture resistance. Table I summarizes the relevant details of the tested coating-substrate systems.

TiN films were also produced by ion beam enhanced physical vapor deposition (IBED) method. Details of different ion bombardment procedures are given in Table II . The substrate surface was prepared by polishing. The thickness of the coatings is about 1 μ m.

Table I. Different PCVD Procedures

Substrate / Coating	M2(HRC63)			M2 HRC45 Ground
	Ground	Blasted	Nitrided	
TiN2.5um	+	+	+	
TiCN2.7um	+			+
TiCN4.2um	+			

Table II. Different IBED Procedures

Procedure Number	Ion Implantation	No bombardment Deposition	Ion Interface Mixing	IBED	No bombardment Deposition
Group 1	→		→	→	
Group 2		→	→	→	
Group 3			→	→	
Group 4			→		→

Experimental Results

Typical curves of load $\Delta \tau_{rz}$ which is the dominant stress component for the interfacial fracture versus the number of cycles for coating failure of different coating-substrate systems are given in Figure 3 and Figure 4 respectively. It can be seen that the critical number of cycles increase with decreasing applied stress.

From Figure 3, it is evident that surface preparation by blasting and plasma nitriding can improve the interfacial fracture resistance significantly. Also, the spherical rolling test is very sensitive to the chemical composition at the interface since the fracture resistance limit of TiCN film is nearly two times higher than that of TiN film. On the other hand, the characteristic of the test also makes it independent of substrate hardness.

From Figure 4, when different ion bombardment procedures were used, different interfacial microstructures could be obtained which could induce different values of interfacial fracture resistance. It also confirms that the test is very sensitive to the interfacial conditions.

DISCUSSION

Comparison With Cylindrical Rolling Test

For comparison purpose, the shear stress range $\Delta \tau_{rz}$ has been calculated for some coating-substrate systems evaluated by the cylindrical rolling test as given in Figure 5. The stresses in the spherical rolling test are always 1.8-2.0 times higher than those in the cylindrical rolling test under the same interfacial conditions. This result indicates that the stress concentration factor at the edge of contact area in cylindrical rolling tests is about 1.8-2.0. The fracture resistance limits of different coating-substrate systems have the same trend in the cylindrical test as the spherical test which shows that the two rolling contact fatigue tests are consistent with each other.

Effect of the Film Thickness

The shear stress range $\Delta \tau_{rz}$ at the interface is calculated by varying the coating thickness. From Figure 6, the ratio of $\Delta \tau_{rz}$ to σ_0 (normal stress amplitude on the surface) will increase with increasing thickness. This calculation indicates that under the same external load the thicker coating is more susceptible to delamination-type failure than the thinner coating. This is in agreement with the experimental results reported in Figure 3. After the same number of testing cycles, the thicker coating tends to delaminate at lower load. However, in numerical calculation, the ratio of $\Delta \tau_{rz}$ to σ_0 of a thicker film is larger than that of the thinner coating, so the absolute value of $\Delta \tau_{rz}$ could be nearly same for both thicker and thinner films if they have the same

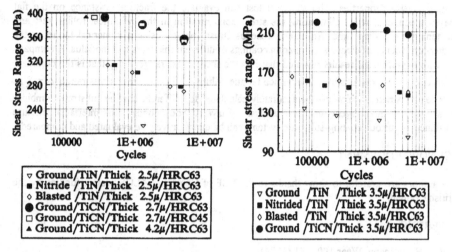

Figure 3. $\Delta \tau_{rz}$ of Different PCVD Films by Spherical Rolling Test

Figure 5. $\Delta \tau_{rz}$ of Different PCVD Films by Cylindrical Rolling Test

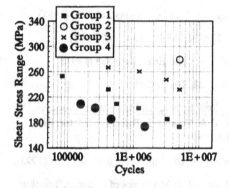

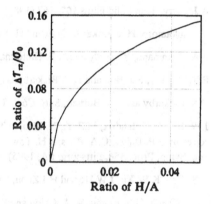

Figure 4. $\Delta \tau_{rz}$ of Different IBED Films by Spherical Rolling Test

Figure 6. $\Delta \tau_{rz}$ as A Function of Film Thickness

interfacial conditions. It thus can be concluded that it is the shear stress range, $\Delta \tau_{rz}$, that can serve as an index to characterize the interfacial fracture resistance of hard films.

SUMMARY

The rolling contact test including both the cylindrical rolling test and the spherical rolling test is sensitive to the interfacial fracture resistance because in the high cycle region the interface, being the weakest point, is not only the point of failure, but also the only point which sees significant deformation. This characteristic of the test also makes it insensitive to substrate and coating plastic properties. The spherical test can evaluate the fracture resistance on the flat coated specimens, while eliminating the stress concentration in the cylindrical rolling test and shortening the testing time. An inhomogeneous stress analysis using a numerical model in the layered system is sufficient for comparing coatings of different thickness and modulus. A comparison with experimental results revealed that it is the shear stress range $\Delta \tau_{rz}$ that is the dominant stress component responsible for interface fatigue failure. Using the shear stress range $\Delta \tau_{rz}$ as an index parameter, which can be accurately calculated for different coating-substrate combinations from the inhomogeneous stress analysis of layered body contact, the interfacial fracture resistance of various coating-substrate systems can be quantitatively be evaluated and compared.

ACKNOWLEDGEMENT

The authors are grateful to show thanks to the NSF of China and ALCS project funding by the British Council.

REFERENCES:

1. H.E. Hintermann, Wear 100, 381 (1984).

2. E. Bergnann, J. Vogel and L. Simmen, Thin Solid Films 153, 219 (1987).

3. A.J. Perry, Thin Solid Films 107, 167 (1983).

4. Y. Tsukamoto, H. Kuroka, A. Sato and H. Yamaguchi, Thin Solid Films 213, 220 (1987).

5. P.A. Steinmann, Y. Tardy and H.E. Hintermann, Thin Solid Films 154, 333 (1987).

6. P.R. Chalker, S.J. Bull and D.S. Rickerby, Materials Sci. and Eng. A140, 583 (1991).

7. D.S. Rickerby and P.J. Burnelt, Surf. Coat. Technol. 33, 191 (1987).

8. J.W. He, B.C. Hendrix, M.Z.Yi, N.S.Hu, in Thin films: Stresses and Mechanical Properties V, edited by S.P. Baker, C.A. Ross, P.H. Townsend, C.A. Volkert, P. B ϕrgesen (Mater. Res. Soc. Symp. Proc. 356, Pittsburg, PA, 1995) pp.881-885.

9. R.S. Gao, K.W.Xu, J.W.He and H.J Zhou, Chinese J. Mech. Eng. (in English) 6(1), 39 (1993).

10. T.P. Cheng, H.S. Cheng, W.A. Chiou and W.D. Sproul, in Mechanics of Coatings, Tribology Series 17, edited by D. Dowson, C.M. Taylor and M. Godet (Elsevier, Amsterdam, 1990) pp. 81-88.

11. H. Guo, B.C. Hendrix, J.W. He, unpublished results.

12. H.Y. Liu, M.S. Thesis, Xi'an Jiaotong University, 1995.

13. W.T. Chen, Int. J. Engng. Sci. 9, 775 (1971).

14. Y. Sun and T. Bell, Trans. Inst. Metal Finish. 70, 38 (1992).

15. Hua Chen, M.Z. Yi, K.W. Xu and J.W. He, Surf. Coat. Technol. 74-75, 253 (1995).

ADHESION IN NiAl–Cr FROM FIRST PRINCIPLES

James E. Raynolds*, John R. Smith**, David J. Srolovitz*, and G.-L. Zhao***
*Department of Materials Science and Engineering, University of Michigan, Ann Arbor, Michigan 48109-2136
**Physics Department, General Motors Research Laboratories, Warren, Michigan, 48090
***Dept. of Physics and Astronomy, Louisiana State University, Baton Rouge, LA 70803.

ABSTRACT

We have calculated the work of adhesion (i.e. energy for rigid fracture) and peak interfacial stress for NiAl, Cr, and NiAl–Cr using self-consistent density functional calculations to obtain the complete energy vs. separation curve for each system. Our calculations indicate that the work of adhesion is largest for Cr and smallest for NiAl while those for interfaces of NiAl with Cr are intermediate. We have also estimated that segregation processes could alter the work of adhesion for the AlNi/Cr interface by up to 20% since Al tends to segregate to the free NiAl surface while Ni tends to segregate to the AlNi/Cr interface.

INTRODUCTION

There is considerable interest in intermetallics such as nickel aluminide (NiAl) and intermetallic composites since these materials show great promise for a wide variety of applications.[1,2] Nickel aluminide, in particular, is a prototypical intermetallic for which much experimental data exists and thus serves as a natural starting point for studies of intermetallics. There is also interest in NiAl for use as a structural material for high-temperature applications in the aerospace industry.[3,4,5] Unfortunately NiAl is brittle at room temperature and has relatively poor high-temperature strength.

The fact that pure NiAl is brittle at room temperature has prompted researchers to search for composites with improved mechanical properties by alloying NiAl with refractory metals such as Cr to improve the fracture toughness. In particular, we will focus on eutectic composites formed by reinforcing NiAl with Cr.[3,4,6] The alloy NiAl–34Cr forms a eutectic composite in which fibers (rods) of Cr form in the NiAl matrix. By adding small amounts (< 1%) of other refractory metals such as Mo, V, or W, the morphology changes from fibrous to lamellar.[7] These composites are formed by a process of directional solidification which tends to align the fibers (or lamellae) along the solidification direction. The microstructure of the lamellar composites has a cube on cube orientation meaning that the crystallographic axes of the reinforcing phase are aligned with those of the NiAl matrix.

Experimental evidence suggests that the fracture toughness of a material has a sensitive dependence on the ideal work of adhesion, W_{ad}, which is defined as the energy needed to rigidly cleave an interface. By studying W_{ad} we hope to obtain a better understand of the relationship between observed mechanical properties and chemical bonding.

The objective of the present study is to gain a detailed knowledge of NiAl–Cr interfacial adhesion and bonding characteristics by using first-principles quantum-mechanical calculations to determine W_{ad} as well as changes in electron charge-density distributions which occur upon forming interfaces. We use first-principles methods since this is the only way to accurately describe energetics in these interfaces which contain a variety of types of atoms and types of bonds (covalent, metallic, ionic). A more complete presentation of the present work has been presented elsewhere.[8]

METHODS

We form an interface between two thin slabs and the adhesion curve is determined by computing the total energy vs. spacing as the slabs are rigidly separated. The arrangement of atoms in each slab is determined by the experimental crystal structure. Figure 1 shows the computational unit cell used to represent the NiAl–Cr interface. The lattice constant is fixed at the experimental value of 2.88Å. This unit cell was chosen to have a mirror symmetry plane through the lowest layer shown so as to reduce the computational effort. A computation using the unit cell shown in Fig. 1 thus represents a larger system which is obtained from the one of Fig. 1 by reflecting through the mirror symmetry plane. We therefore calculate the total energy to rigidly separate two three-layer Cr slabs from a central five-layer NiAl slab. We choose to study the {001} interface which is a typical interface between a lamella of Cr-rich phases and the NiAl matrix. We expect that impurity effects will be important for understanding the energetics of the lamellar microstructure. In the present study we confine ourself to clean interfaces.

For a number of values of the interfacial separation we calculate the total energy using the Self-Consistent Local-Orbital (SCLO) method.[9] The details of this method have been presented elsewhere.[9] By using this first-principles technique we are able to obtain an accurate description of the wide range of bonding situations (metallic, covalent, ionic) which can arise in the systems which we consider.

Experience has shown that the use of thin slabs to represent the interface does not present any problems due to finite size effects.[10,11,12] Tests show that as the thickness of the slabs is increased there is very little change in the computed value of W_{ad} for slabs thicker than three layers. Physically this reflects the fact that a disturbance of the electronic charge density is screened within a very short distance, as is typical of metallic systems.

As in all of our previous studies of interfacial adhesion, we find that it is possible to express the total energy vs. interfacial separation using a simple analytic form that has been called the universal-binding-energy (UBER) relation.[10] This relation is given by:

$$E = -\frac{W_{ad}}{2}(1 + a^*)e^{-a^*}, \tag{1}$$

where

$$a^* = (d - d_0)/l, \tag{2}$$

E is the total energy per unit surface area and d is the interfacial separation. The parameter l is a scaling length. The UBER has been found to apply to a large class of metallic and covalently bonded systems.[13]

In practice, we determine W_{ad} by fitting our calculated adhesion curve with the function of Eq. (1). The interfacial stress vs. separation is obtained by differentiating Eq. (1) with respect the separation. In this way, the stress vs. interfacial separation is given by:

$$\sigma = \sigma_{max}a^*e^{(1-a^*)}. \tag{3}$$

In Eq. (3), the peak interfacial stress σ_{max} is related to the work of adhesion W_{ad} and the scaling length l by

$$\sigma_{max} = \frac{W_{ad}}{le}, \tag{4}$$

RESULTS

We find that Cr has the largest value of W_{ad}, NiAl has the smallest, and W_{ad} for the interface between NiAl and Cr is of intermediate value. For NiAl, {001} planes contain either Al or Ni so fracture occurs between Ni and Al planes. Figure 2 displays our calculated results for the total energy vs. separation for the systems NiAl, Cr, and the two arrangements of the interface which we consider. The two possibilities are (1) to have a Ni layer in contact with the Cr (Ni terminated interface), which we will denote as AlNi/Cr, and (2) to have an Al layer in contact with Cr (Al terminated interface) which we denote as NiAl/Cr. The solid curves are the UBER fits. The work of adhesion W_{ad} is given as twice the depth of the adhesion energy curve (see Eq. (1)). Since the lower energy situation for the interface is the one in which Ni is in contact with Cr, we assume that the interface is Ni terminated from now on. Note the similarity in the value of W_{ad} for NiAl and NiAl/Cr. This reflects the fact that bonds between Al and Cr are very similar to bonds between Al and Ni.

The interfacial stress vs. separation curves for the systems involving Cr are displayed in Fig. 3. As mentioned above, these curves are obtained by differentiating the UBER (see Eq. (3)). There are no data points in Fig. 3; the symbols merely indicate which curve is which. The conclusions to be drawn from these stress curves are consistent with those corresponding to the adhesion curves. In other words, Cr is the strongest material, NiAl is the weakest and the strength of the interfacial bonds is intermediate. The peak interfacial stress for the AlNi/Cr (Ni terminated) interface is larger than for NiAl/Cr which indicates that the interface should be Ni terminated.

By studying how charge density distributions change upon forming interfaces we have determined that the bonding character is metallic with some evidence of covalent bonding. To see how the electronic charge density changes to form interfacial bonds we subtract from the charge density distribution for the AlNi/Cr interface the distribution corresponding to isolated crystals and plot this difference as a contour plot in Figure 4. The plane of the figure is an {110} crystal plane. Curves corresponding to charge accumulation are denoted by solid lines and dashed lines represent charge depletion. The continuous band of charge which forms at the interface is indicative of metallic bonding. In addition we see significant charge accumulation along the lines which join the atomic centers across the interface. This is indicative of covalent bonding and is a result of the directional character of the localized d-orbitals of Ni and Cr. Note also that the disturbance in the charge density is confined to a small region within roughly two atomic layers of the interface.

It is possible that segregation effects might lead to a significant correction to the work of adhesion since the AlNi/Cr interface is Ni terminated while the free NiAl surface tends to be Al terminated.[14] If segregation turns out to be important, we assume its effect to be largest on the free NiAl surface and smallest on an equilibrium interface between NiAl and Cr. We can obtain a crude estimate of the effect of segregation on the work of adhesion by lowering the energy at large separation by the difference in surface energy for pure Ni and pure Al. The surface energy of pure Ni is 2664 mJ/m^2,[15,16] while the Al surface energy is 1170 mJ/m^2.[17,12] The difference of these two surface energies, 1494 mJ/m^2, gives a rough estimate of the change to the AlNi/Cr work of adhesion: $W_{ad}^{AlNi/Cr} = 7388\ mJ/m^2 \longrightarrow W_{ad}(\text{segregation}) \approx 5894\ mJ/m^2$. The solid curve in Fig. 5 is the UBER curve for AlNi/Cr. We lower the point at the largest separation by the difference in surface energy between Ni and Al, keep fixed the three points at the smallest separation, and delete the remaining three points. We then construct a fit to the resulting points using the UBER which is plotted in Fig. 5 as the dotted curve which is labelled "correction". The resulting value of W_{ad} is lower by roughly 20%. We thus conclude that segregation processes could give rise to a significant correction to the work of adhesion.

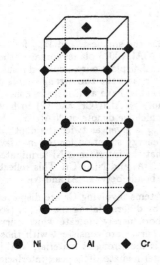

Energy vs. separation (d)

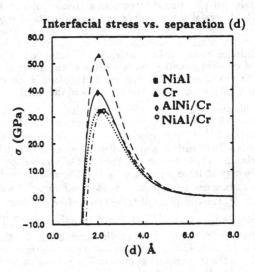

Figure 1. Half unit cell used to study interfacial adhesion between NiAl and Cr.

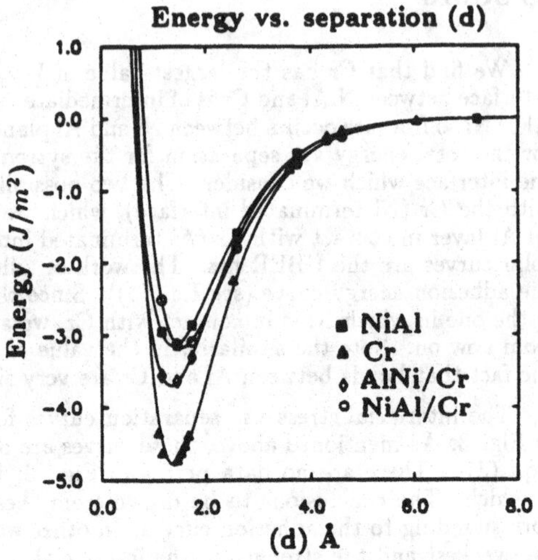

Figure 2. Calculated total energy vs. interfacial separation (d) for the systems NiAl, Cr, and NiAl–Cr.

Interfacial stress vs. separation (d)

Figure 3. Calculated ideal interfacial stress vs. separation for the systems NiAl, Cr, NiAl–Cr.

Figure 4. Charge density rearrangements which arise upon forming the NiAl–Cr interface from isolated NiAl and Cr crystals.

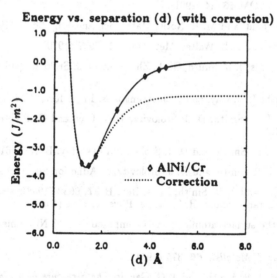

Figure 5. Estimated correction to the adhesion energy curve for the Ni terminated NiAl–Cr (AlNi/Cr) interface due to segregation.

CONCLUSIONS

We have presented results of first-principles studies of adhesion in NiAl and Cr interfaces. The total energy vs separation follows the universal form (UBER).[13] We find that the the work of adhesion is ordered as: $W_{ad}^{Cr} > W_{ad}^{AlNi/Cr} > W_{ad}^{NiAl}$. From the fit to the adhesion energy curve we have obtained the interfacial stress vs. separation by differentiation. The stress curves as well as the adhesion curves lead to the interpretation that the equilibrium interface tends to be Ni terminated. By analyzing the way that the interfacial charge density changes upon forming the interface we conclude that the bonding is predominantly metallic in nature with, however, some covalent character. We also conclude that segregation processes can play a role in modifying the work of adhesion. A crude estimate for AlNi/Cr suggests that this correction could be as large as 20%.

REFERENCES

1. D. B. Miracle, Acta Metall. **41**,649 (1993).

2. R. D. Noebe, R. R. Bowman, and M. V. Nathal, Int. Mater. Rev. **38**, 193 (1993).

3. X. F. Chen, D. R. Johnson, R. D. Noebe, B. F. Oliver, J. Matter. Res. **10**, 1159 (1995).

4. D. R. Johnson, X. F. Chen, B. F. Oliver, R. D. Noebe, and J. D. Whittenberger, Intermetallics **3**, 99 (1995).

5. R. Darolia, JOM **43**, 44 (1991).

6. S. M. Merchant, and M. R. Notis, Mater. Sci. and Eng. **66** 47 (1984).

7. H. E. Cline and J. L. Walter, Met. Trans. **1**, 2907 (1970).

8. J. E. Raynolds, J. R. Smith, G. -L. Zhao, and D. J. Srolovitz (submitted to Phys. Rev. B).

9. J. R. Smith, J. G. Gay, and F. J. Arlinghaus, Phys. Rev. B **21**, 2201 (1980).

10. T. Hong, J. R. Smith, D. J. Srolovitz, J. G. Gay, and R. Richter, Phys. Rev. B **45**, 8775 (1992).

11. T. Hong, J. R. Smith, and D. J. Srolovitz, Phys. Rev. B **47**, 13615 (1993).

12. T. Hong, J. R. Smith, and D. J. Srolovitz, J. Adhesion Sci. Technol. **8**, 837 (1994).

13. A. Banerjea and J. R. Smith, Phys. Rev. B **37**, 6632 (1988); also see P. Vinet, J. H. Rose, J. Ferrante, and J. R. Smith, J. Phys. **1**, 1941 (1989).

14. Actually the surface atomic layer is composed of 35% Ni atoms (see section 3.3.3 of Ref. 1).

15. H. Wawra, *Z. Metallkd.* **66**, 375 (1974).

16. R. Richter, J. R. Smith, and J. G. Gay, in *The Structure of Surfaces*, edited by M. A. Van Hove and S. Y. Tong (Springer-Verlag, New York, 1983), p. 35.

17. H. Wawra, *Z. Metallkd.* **66**, 395-401, 492-498 (1975).

18. J. L. Walter and H. E. Cline, Met. Trans. **1**, 1221 (1970).

FIRST-PRINCIPLES DETERMINATION OF THE EFFECTS OF BORON AND SULFUR ON THE IDEAL CLEAVAGE FRACTURE IN Ni₃Al

SHENG N. SUN,[†] NICHOLAS KIOUSSIS,[†] MIKAEL CIFTAN[‡] AND A. GONIS[††]

[†]Department of Physics, California State University, Northridge, CA 91330-8268
[‡]Physics Division, U.S. Army Research Office, Research Triangle Park, N.C. 27709-2211
[††]Department of Chemistry and Materials Science, Lawrence Livermore National Laboratory, CA 94550

ABSTRACT

The effects of boron and sulfur impurities on the ideal cleavage fracture properties of Ni₃Al under tensile stress are investigated for the first time using the full-potential linear-muffin-tin-orbital (FLMTO) total-energy method, with a repeated slab arrangement of atoms simulating an isolated cleavage plane. Results for the stress-strain relationship, ideal cleavage energies, ideal yield stress and strains with and without impurities are presented, and the electronic mechanism underlying the contrasting effects of boron and sulfur impurities on the ideal cleavage of Ni₃Al is elucidated.

INTRODUCTION

The L1₂-type ordered nickel aluminide, Ni₃Al, exhibits unique mechanical properties that make it attractive for structural applications at elevated temperatures.[1] These are its high melting temperature, low density, resistance to oxidation,[2] and the increase of yield stress with increasing temperature,[3] in contrast to conventional compounds or disordered alloys. However, as with many other intermetallics, an inherent drawback to using polycrystalline ordered stoichiometric Ni₃Al alloys as a structural material is the tendency to undergo brittle intergranular fracture,[4] even though single crystals of Ni₃Al are highly ductile. Weak grain boundary cohesion and low cleavage strength have been suggested[5] to be two of the main causes of brittleness in polycrystalline intermetallic systems. Microalloying studies have shown that doping with boron, which strongly segregates to the grain boundaries, can significantly improve the ductility of polycrystalline Ni₃Al,[5] whereas sulfur reduces ductility and causes embrittlement.[6] These contrasting behaviors of the added impurities have been interpreted[7] as caused by modification of the local electronic structure induced by these defects near and within the grain boundary in specific ways; "embrittling elements", such as sulfur, draw charge from neighboring metal atoms onto themselves, thereby weakening the metal-metal bond charge network, whereas "cohesive enhancers" such as boron, do not draw charge from their neighboring metal atoms, form homopolar bonds, and maintain the cohesiveness of the metal-metal atom charge network. [7]

While significant progress is currently being made in atomistic simulation calculations of fracture processes, [8] inclusion of realistic materials characteristics (grain boundaries, dislocations, etc.) in first-principles electronic structure calculations remains to be very difficult. Since in the case of the Ni₃Al the crystalline structure of the grains remain ordered up to very close to the relatively sharp grain boundary with or without boron addition,[9] it is reasonable to expect that a first-principles study of the bonding characteristics of atoms in a supercell with and without boron/sulfur impurities will provide a first order insight into

183

the cohesive forces that control the ideal cleavage fracture. The next level of study would then to include the relaxation of the crystalline atomic positions (emulating the partial disorder that has been revealed very close to and at the grain boundary), and to determine the effect of such perturbations on the cleavage fracture properties. This also indicates why experimental studies of single Ni_3Al crystals with and without these impurities, which will per force be at specific interstitial sites, are important not only in providing an upper bound of tensile yield stresses, the "ideal yield stress" that is used in phenomenological theories of propagation of cracks, but also most significantly in these Ni_3Al systems where the region in and around the grain boundaries are rather well ordered. More recent experiments[10] have shown that boron also improves the ductility of single Ni_3Al crystals, suggesting that a "bulk effect" should be considered in addition to the grain boundary strengthening of boron when explaining the improvement in ductility of polycrystalline Ni_3Al due to B additions. Moreover, the ideal cleavage energy (which is equal to the total surface energy γ_s of the two cleaved surface planes) is a quantity which appears in the phenomenological theory recently proposed by Rice [11] for determining the intrinsic ductile versus brittle behavior of materials. Rice proposed that a simple rule to measure the brittle/ductile behavior of materials is the ratio γ_{us}/γ_s, which determines the competition between dislocation emission from a crack tip and crack cleavage. Dislocation nucleation is characterized by the unstable stacking fault energy γ_{us}, which is the maximum energy barrier encountered in sliding one half of a crystal relative to another along a slip plane. It is for these reasons that we present below, our results of *ab initio* total energy calculations restricted to the effect of boron and sulfur impurities on the ideal cleavage properties of single Ni_3Al crystals. A much more appropriate set of calculations should involve the geometry of a grain boundary and such calculations are currently being contemplated.

Our method of calculation is the full-potential linear-muffin-tin-orbital (FLMTO) total-energy electronic structure method, and has been described elsewhere.[12] Though the method, as used here, is capable of dealing with very open systems, such as surfaces,[12] it is not constrained to surface situations, so that a continuous transition from a bulk, closely packed arrangement of atoms to an open arrangement, such as when cleavage planes are created, can also be treated. These calculations differ from the pioneering electronic structure calculations of Messmer and Briant using small clusters,[7] in that they are self-consistent full-potential *total-energy* calculations with no shape approximation to the potential and the charge density, and they can provide for the first time valuable information of the effect of impurities on the ideal cleavage energy and the ideal yield stress, and on the tensorial/directional properties of the bonding forces across and parallel to the cleavage plane in terms of the anisotropy of the detailed bonding charge characteristics.

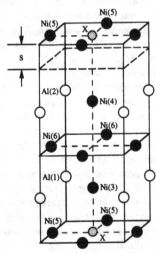

Figure 1. Supercell geometry used in the cleavage calculation. The filled, empty, and gray shaded circles represent sites for Ni, Al, and impurity (X) atoms, respectively.

We shall find that the bonding charge has both metallic and directional characteristics. That is why a full potential treatment, which includes the non-spherical corrections to the spherical potential within the muffin-tin spheres is essential. We present results for the stress-strain relationship, cleavage energies, ideal yield stress and ideal strains with and without impurities, and discuss the origins, in the electronic structure, of the bonding behavior induced by boron and sulfur impurities across the cleavage plane and the resulting effect on resistance to cleavage.

RESULTS AND DISCUSSION

The creation of a cleavage plane is achieved in this calculation by using a repeated slab construction. Each cell, shown in Fig. 1, contains one "slab" and each slab consists of four layers of (001) planes of Ni_3Al. Inequivalent atoms in the cell are denoted by numerical labels (enclosed in parenthesis), depending on their point group symmetry. The simulation of the separation of two neighboring planes (cleaved surfaces) is accomplished by varying the separation, s, between the slabs. Thus, for s=0 we simply have bulk Ni_3Al, and as s is increased the slab separation increases, until, we have a repeated slab calculation of isolated surfaces at large separations. The impurity is placed at the octahedral interstitial site labeled by X, where the impurity has four $Ni(5)$, one $Ni(4)$, and one $Ni(3)$ nearest neighbors. Such an arrangement corresponds to an impurity concentration of 11.1 at.% . In order to minimize the effect of interactions between impurities within the (001) planes, we are currently carrying out electronic structure calculations employing a slab which contains twice as large number of Ni and Al atoms as that in Fig. 1. Such an arrangement, which corresponds to an impurity concentration of 5.88 at.%, will allow us to investigate the effect of impurity concentration on cleavage properties. Even though the impurity concentration in the calculations is high relative to the experimental values, [1] we believe that these calculations serve as a first step in understanding the effects of impurities on the cleavage properties in Ni_3Al.

In Fig. 2(a) we have plotted the difference in total energy per unit area, ΔE, of the four-layer unit cell versus the separation (s in Fig. 1) between neighboring slabs, for the pure Ni_3Al system (solid squares), the $Ni_3AlB_{\frac{1}{2}}$ system (open squares), and the $Ni_3AlS_{\frac{1}{2}}$ system (solid circles). Here, ΔE refers to the total energy of the slab at separation s minus the total energy at separation s_0, where s_0 represents the slab separation at the minimum of the total energy. Atomic relaxation between the (001) layers within the slab was ignored and the lattice constant was held fixed at the value of 3.568Å for pure Ni_3Al.[13] The difference in total energy ΔE converges to the ideal cleavage energy (Griffith energy) as the separation between neighboring slabs increases (the ideal cleavage energy is the total surface energy of the two cleaved surface planes). We find that sulfur reduces the ideal cleavage energy of the pure Ni_3Al system by 35%, while boron reduces it only by 5%. This small reduction in total energy and the small increase in the maximum tensile stress of Ni_3Al when doped with boron is due to the impurity-induced charge density which is presented below. These results are consistent with the experimental findings that single crystals show *reduced* cleavage strength when the boron concentration is more than 0.8 at.%. [10]

The nonzero values of s_0 for the $Ni_3AlB_{\frac{1}{2}}$ and $Ni_3AlS_{\frac{1}{2}}$ systems indicate an outward (001) surface relaxation induced by the impurities. By fitting ΔE to a simple function (cubic polynomial times an exponential) by the nonlinear least-squares method, we have determined the variation of tensile stress with cleavage separation. The tensile stress, a quantity of central interest for the cleavage calculation, is simply $\sigma = \frac{1}{a^2}\frac{d\Delta E}{ds}$, where a is the lattice constant. Figure 2(b) shows the variation of tensile stress with separation between neighboring slabs for pure Ni$_3$Al (solid curve), for Ni$_3$AlB$_{\frac{1}{2}}$ (dotted curve), and for Ni$_3$AlS$_{\frac{1}{2}}$ (dashed curve). We find an ideal cleavage energy of 6.3 Jm^{-2} for pure Ni$_3$Al, which compares well with the value of 5.8 Jm^{-2} found by Fu and Yoo.[14] It is important to notice, that while boron increases the maximum cleavage stress (ideal yield stress) by about 2% compared to that of the pure Ni$_3$Al system (3.8 X 10^{10} N/m^2) , sulfur causes a significant reduction of σ_{max} by 29 %. The contrasting effects of boron and sulfur are already evident. It remains to be seen if the maximum yield stress induced by boron is even higher for lower concentration levels as alluded above.

In order to understand the breaking and rearrangement of electronic bonds induced by the impurities within and across the cleavage plane, we have calculated the *impurity-induced* charge density, $\Delta\rho_{ind} = \rho_{solid}(\text{Ni}_3\text{AlX}_{\frac{1}{2}}) - \rho_{solid}(\text{Ni}_3\text{Al}) - \rho_{atom}(\text{X}_{\frac{1}{2}})$. (The first two terms are the self-consistent charge density of the slab with and without the impurity, and the last term is the superposition of atomic charge densities centered on impurity sites).

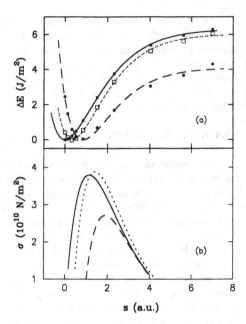

Figure 2. (a) The calculated total energy difference per unit area, ΔE, versus cleavage separation for Ni$_3$Al (solid squares), Ni$_3$AlB$_{\frac{1}{2}}$ (open squares), and Ni$_3$AlS$_{\frac{1}{2}}$ (solid circles). The solid, short- and long-dashed lines are the least-squares fits. (b) The cleavage stress versus cleavage separation for Ni$_3$Al (solid line), Ni$_3$AlB$_{\frac{1}{2}}$ (short-dashed line), and Ni$_3$AlS$_{\frac{1}{2}}$ (long-dashed line).

In Figs. 3a and 3b we show the boron- and sulfur-induced charge density on the vertical (200) plane (the plane containing just Ni atoms and the impurity atom in Fig. 1), respectively, for several values of the slab separation. The solid (dotted) contours represent accumulation (depletion) of electronic charge. We find a large charge accumulation between the boron and Ni(4) atoms across the cleavage plane for Ni$_3$AlB$_{\frac{1}{2}}$, while there is a large depletion of charge in the region between the sulfur and the Ni(4) atom for Ni$_3$AlS$_{\frac{1}{2}}$. The strong accumulation of bonding charge which "hangs on" across the interface much longer, explains the fact that the ideal cleavage strength for Ni$_3$AlB$_{\frac{1}{2}}$ is larger than that of Ni$_3$AlS$_{\frac{1}{2}}$, and it is the microscopic basis for the resistance of the boron-doped system to cleavage. On the other hand, both boron and sulfur impurities induce an accumulation of interstitial bonding charge between the nearest-neighbor Ni(5) atoms in the (001) plane, indicating clearly the anisotropic character of the bonding charge and hence the tensorial nature of the forces along and across the cleavage plane. Examination of the density of states indicates that the electronic bonding mechanism underlying the contrasting effects of the boron and sulfur impurities seems to be strong d-p hybridization for the boron impurity and a more embedded-

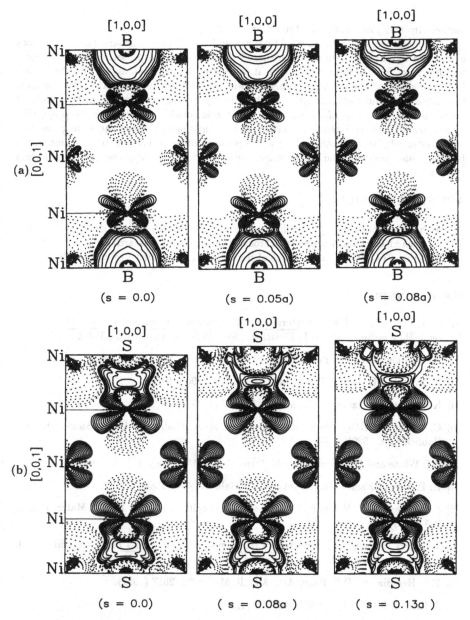

Figure 3. (a) Boron-induced charge density on the (200) plane for various slab separation ratios s/a. (b) Sulfur-induced charge density on the (200) plane for various slab separation ratios s/a. Solid (dotted) contours denote contours of increased (decreased) charge density. Contours start from $\pm 8.0 \times 10^{-4}$ e/(a.u.)^3 and increase successively by a factor of root 2.

like (electrostatic) interaction for the sulfur impurity.

To summarize, we have presented total-energy FLMTO electronic structure calculations of the role of boron and sulfur impurities on the ideal cleavage properties of Ni_3Al under tensile stress. For the impurity-concentration used in the calculations, we find that boron has little effect on the cleavage properties, whereas sulfur reduces greatly both the ideal cleavage energy and yield strength. Work currently in progress is aimed at the effects of boron concentration and of relaxation of the atomic positions on the cleavage properties. In the near future we intend to proceed with calculations involving the sliding one half of a crystal relative to another along a slip plane in order to investigate the competition between dislocation emission and cleavage, and calculations involving the geometry of grain boundaries.

ACKNOWLEDGMENTS

We have benefited greatly from discussions with Dr. Ruqian Wu and Dr. Say Peng Lim. The research at CSUN was supported through the U.S. Army Research Office under Grant No. DAAH04-93-G-0427 and the Office of Research and Sponsored Projects at CSUN.

References

[1] C. T. Liu and D.P. Pope, in Intermetallic Compounds - Principles and Practices, edited by J. H. Westbrook and R.L. Fleischer, (John Wiley & Sons, 1995), Vol. 2, p 17.

[2] G. H. Meier and F. S. Petit, Mater. Sci. Eng. A, **153**, 548 (1992).

[3] K. Aoki and O. Izumi, J. Jap. Inst. Metals **43**, 1190 (1970).

[4] K. Aoki and O. Izumi, Acta. Metall. **27**, 807 (1979).

[5] C. T. Liu, in *Alloy Phase Stability*, edited by A. Gonis and G. M. Stocks (Kluwer Publications, 1989), p 7.

[6] C.L White and D.F. Stein, Metall. Trans. A **9A**, 13 (1978).

[7] R. P. Messmer and C. L. Briant, Acta Metall. **30**, 457 (1983).

[8] S. P. Chen, A. F. Voter, R. C. Albers, A. M. Boring, and P. J. Hay, J. Mater. Res. **5**, 955 (1990).

[9] Takayuki Takasugi, in Intermetallic Compounds - Principles and Practices, edited by J. H. Westbrook and R.L. Fleischer, (John Wiley & Sons, 1995), Vol. 1, p 585.

[10] F.E. Heredia and D.P. Pope, Acta Metall. Mater. **39**, 2017 (1991).

[11] J. Rice, J. Mech. Phys. Solids **40**, 239 (1992).

[12] D. L. Price, J. M. Wills, and B. R. Cooper, Phys. Rev. B **48**, 15301 (1993); D. L. Price and B. R. Cooper, Phys. Rev. B **39**, 4945 (1989).

[13] D. Hackenbracht and J. Kubler, J. Phys. F **13**, L179 (1983).

[14] C.L. Fu and M.H. Yoo, Mat. Chem. and Phys. Vol.**32**, 32 (1992).

DYNAMIC EMBRITTLEMENT OF BERYLLIUM-STRENGTHENED COPPER IN AIR

Ranjani C. Muthiah[1], C. J. McMahon, Jr.[1], and Amitava Guha[2]

[1] Department of Materials Science and Engineering,
University of Pennsylvania, Philadelphia, PA 19104, USA

[2]Brush Wellman Inc.
17876 St. Clair Avenue, Cleveland, OH 44110, USA.

ABSTRACT

A precipitation-hardened Cu-0.26%Be alloy is used as a low-temperature model material for "dynamic embrittlement", or quasi-static diffusion-controlled intergranular brittle fracture. This alloy is shown to undergo intergranular cracking in air at 150°C and to be almost free of this cracking in 2×10^{-6} Torr vacuum at 200°C. The time to failure is highly stress dependent. The temperature dependence of cracking was found to be 30 kcal/mole. This is about 50% greater than the activation energy for oxygen diffusion in copper, but the present experiments also include an unknown incubation time.

INTRODUCTION

Dynamic embrittlement refers to the quasi-static diffusion-controlled decohesion, usually along grain boundaries, that can occur when a high-strength alloy is stressed in the presence of surface-adsorbed impurities. It has been applied [1,2] to the sulfur-induced cracking of alloy steels in what is known as stress-relief cracking and to the tin-induced cracking of a Cu-Sn alloy, both of which involve an embrittling element emanating from the bulk alloy. It has also been applied to the cracking of (boron- and hafnium-doped) Ni_3Al polycrystals tensile tested in air at 600°C [3].

We are studying the effects of oxygen on the cracking of precipitation-hardened Cu-Be alloys as a low-temperature model for nickel-based alloys. The advantage of the Cu-Be alloys is that they can be heat treated to very high strengths, with a wide range of strengths, thus enabling the effects of stress to be readily investigated. It was postulated that, at a few hundred degree Celsius in air, this alloy would exhibit intergranular cracking due to the stress-driven diffusion of oxygen into the grain boundaries. The results we report here support that hypothesis.

The postulated mechanism of dynamic embrittlement is as follows [2]: In any solid subjected to a tensile stress, there is a driving force for inward diffusion of surface atoms equal to the stress, σ, times the atomic volume, Ω, of the surface element. At elevated temperatures this can lead to Herring-Nabarro creep in a single crystal or Coble creep in a polycrystalline material. In the case of a high-strength material at lower homologous temperatures (where matrix-atom mobility is negligible), a similar process can occur if the surface contains adsorbed atoms of high mobility, such as low-melting-temperature impurities or solute elements. When such atoms become concentrated in grain boundaries, the usual result is intergranular embrittlement. Thus, if the applied stress is sufficiently high, one should get incremental, quasi-static crack growth by decohesion in the diffusion zone ahead of a crack.

In principle, the rate of such crack growth can be calculated with the aid of the relevant diffusion equation, (1), which includes a stress term in addition to the usual Fick's-law term for mixing by random jumping of atoms:

$$\frac{\partial C}{\partial t} = D_b \frac{\partial^2 C}{\partial x^2} - \frac{D_b \Omega}{kT} \frac{\partial}{\partial x} \left[C \frac{\partial \sigma}{\partial x} \right] \qquad (1)$$

To calculate a crack-growth rate with this equation, it would be necessary to know three things:

189

(1) the dependence of the decohesion stress on the intergranular concentration of the impurity

(2) the intergranular diffusion coefficient, and

(3) the stress profile at the crack tip.

EXPERIMENTAL PROCEDURE

The alloy tested here contains 0.26% Be, 0.52% Co and 0.05% Ni. Static-load tests were carried out on circumferentially notched tensile specimens. Prior to machining, these specimens had been hot extruded, solution annealed at 900°C for 0.5h and aged at 400°C for 4h. In order to get a homogenous grain-size, the as-machined specimens were further annealed at 925°C for 6.5h, rapidly quenched to room temperature and aged at 400°C for 4h, in vacuum. In the aged condition, the alloy has a nominal 0.2% yield stress of 585 MPa, a grain size of ~75μm and a hardness of DPH 175.

The heat-treated specimens were chemically polished in a solution of 60% acetic acid, 35% phosphoric acid and 5% nitric acid at 65°C. The notch of the test specimens was further cleaned with 600 grit SiC paper, to remove any oxide that might have formed. Prior to the test, the specimen was cleaned in the ultrasonic cleaner using acetone, and the root of the notch examined in the stereo microscope for the presence of flaws.

One test was conducted in a vacuum of 2×10^{-6} Torr at 200°C, while all the other tests were carried out in air. At 200°C, the specimens were tested at various percentages of the nominal yield load. At 63% of the yield load, tests were also done at 150°C and 175°C.

For the tests in air, the cleaned specimen was heated to about 70°C above the test temperature in a rough vacuum (approx. 50 millitorr), and dry air from a cylinder (10ppm H_2O) was passed through the chamber. After the temperature had dropped and stabilized, the static load was applied to the specimen.

The time to failure in air was plotted as a function of the applied stress, and the fracture surface was examined in a scanning electron microscope (SEM). The variation of the critical crack size with the applied load was also determined. At an applied load of 63% of the yield load, the failure time as a function of the test temperature was plotted to determine the temperature dependence of the cracking process.

RESULTS

After 148h, the specimen tested at 2×10^{-6} Torr at 96% of the yield load had not failed. A longitudinal section of this specimen revealed the presence of a fine intergranular crack about 400μm (20 grains) in length at the root of the notch, as seen in Fig. 1. The size of the crack would indicate that the crack propagation was very slow, presumably due to the extremely low partial pressure of oxygen in the atmosphere.[4]

The failure time under static-load in air at 200°C decreased with increase in the applied stress, as can be seen in Fig. 2. The curve is drawn through the minimum failure times. These represent tests for which the incubation times were minimized. The longer failure times presumably result from longer incubation times, which are postulated to be caused by an unknown and variable surface reaction that precedes oxygen entry into the specimen All efforts were taken to prepare the root of the notch in the same manner for all the specimens. However, this did not eliminate the scatter. To eliminate this variable, measurements of crack-growth rate at various temperatures and oxygen pressures are planned.

Examination of the fracture surfaces revealed that failure was by intergranular cracking; an example from the fracture surface of the specimen that failed in 0.5h at 63% of the yield stress is shown in Fig.3(a). An image of the fracture surface at a higher magnification, Fig.3(b), shows the presence of smooth grain facets, indicating that the cracking was by intergranular decohesion.

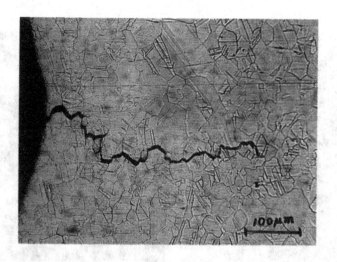

Fig.1 : Longitudinal section of the specimen held at 2×10^{-6} Torr at 96% of the yield stress for 148h. An intergranular crack of ~400 μm in length can be seen at the root of the notch.

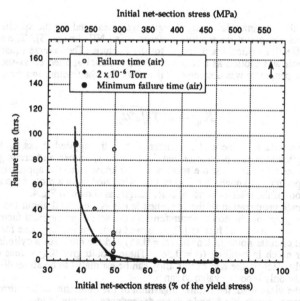

Fig.2 : Dependence of the time to failure on the applied stress during static-load tests at 200°C.

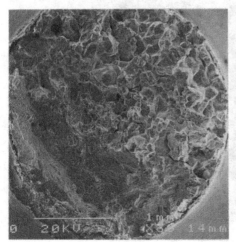

Fig.3 : (a) SEM fractograph of the specimen tested at 200°C in air under a load of 63% of the yield load. (b) A higher magnification, reveals the smoothness of grain boundary facets.

The area of the intergranular fracture region was measured for the specimens tested at 200°C in air on digital images of the fracture surfaces, using NIH Image [5]. The square root of this area gives an effective critical crack length for the fracture. The inverse square root of this effective crack length was plotted as a function of the nominal applied cross-sectional stress, Fig.4, and a linear dependence was obtained. It is known that the fracture toughness is given by the equation:

$$K_{IC} = \alpha\sigma\sqrt{\pi a_c}$$ (2)

where, K_{IC} is the fracture toughness of the material, σ is the applied stress, a_c is the critical effective crack length and α is a parameter that depends on specimen and crack geometry [6]. From (2), it can be seen that if α for the specimen were known and the appropriate geometric factor for the test specimen is known, the K_{IC} can be calculated from a plot of the applied stress vs. inverse square root of the effective critical crack length, as shown in Fig.4.

Since the crack propagation in the specimen is not symmetrical about the loading axis, but originates at one point on the notch circumference, the actual geometrical factor has not yet been calculated. Two approximate factors [7] were used to determine K_{IC}: one for a cylindrical bar with an external circular notch in tension ($\alpha = 0.48$) and the other for a cylindrical bar with an external circular notch in bending ($\alpha = 0.38$). These would give K_{IC} values of 44 and 35 MPa√m, respectively. Hence, the K_{IC} of this alloy can be estimated to be about 40 ± 5 MPa√m. This appears to be the only reported value so far.

At 63% of the yield load, the natural logarithm of the minimum failure time was plotted vs. the inverse test temperature, Fig.5, and a linear dependence was obtained.

The activation energy for oxygen diffusion in copper is reported to be: (i) 600 - 1000°C, Q= 15 kcal/mole [8] and (ii) 800 - 1030°C, Q= 16 kcal/mole [9]. From Fig.5, the temperature dependence of the cracking process, Q, was found to be = 30 kcal/mole. From Fig.2 it is obvious that the failure time of the present specimens includes some variable incubation time. Therefore,

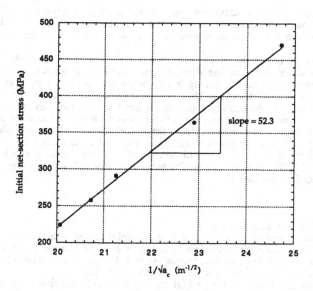

Fig.4 : Variation of the effective critical crack length of the specimen with nominal applied stress for specimens tested at 200°C. The slope of the plot is used to calculate the fracture toughness of the material.

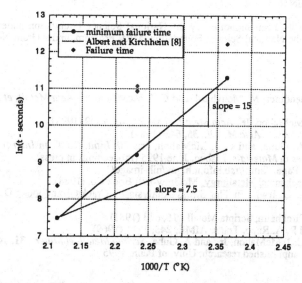

Fig.5 : Temperature dependence of the time to failure for the specimens tested in air. The temperature dependence of the diffusion coefficient of oxygen in copper [8] is also shown.

the plot in Fig.5 cannot give the activation energy of the crack-growth process. All that can be said at present is that the temperature dependence is not inconsistent with the postulated rate-controlling step for crack growth: stress-induced diffusion of oxygen from the surface into the crack-tip region.

The results are obviously consistent with the initial postulate. An earlier preliminary experiment [10] showed that a similar alloy to which had been added 400ppm zirconium is essentially immune to this kind of cracking in air, and, in fact, the commercial version of this alloy contains this amount of zirconium. Thus, this alloy system illustrates both the phenomenon of dynamic embrittlement, which can be anticipated in any high-strength alloy stressed in air, and the inhibition of such embrittlement by the addition of a small amount of an innocuous element which segregates to grain boundaries and blocks the inward diffusion of the embrittling element, which presumably occurs in the case of zirconium. We suggest that this is analogous to the cracking of nickel-based alloys in air and to the recently observed cracking of a high-strength ferritic steel in air [11]. We would expect similar behavior in any high-strength alloy containing a surface-adsorbed embrittling element.

CONCLUSIONS

From the results obtained it can be seen that:
(1) Beryllium-strengthened copper undergoes dynamic embrittlement in air at temperatures as low as 150°C. The failure time drops precipitously as the applied stress is increased.
(2) The time to failure includes a variable incubation time that presumably arises due to some sort of reaction that prevents the entry of oxygen into the specimen.
(3) The temperature dependence of the total failure time is about 30 kcal/mole, which is about double that for oxygen diffusion. However, the temperature dependence includes both the incubation process and the crack-growth process.
(4) The K_{IC} of this alloy was estimated to be 40 ± 5 MPa√m from the variation of critical crack area with applied net section stress.

ACKNOWLEDGMENTS

The research has been supported by the National Science Foundation under Grant no. CMS 95-03980 and the National Science Foundation, MRL Program, under Grant No. DMR91-20668.

REFERENCES

1. D. Bika, J. A. Pfaendtner, M. Menyhard, and C. J. McMahon, Jr. *Acta Metall. et Mater.*, **43**, 1895 (1995).
2. D. Bika and C. J. McMahon, Jr., *Acta Metall. et Mater.*, **43**, 1909 (1995).
3. C. T. Liu and C. L. White, *Acta Metall.*, **35**, 643 (1987).
4. R. C. Muthiah, A. Guha, and C. J. McMahon, Jr., *7th Intnl. Conf. on Intergranular and Interphase Boundaries in Materials*, Lisbon, June 1995 (Proceedings in press).
5. NIH Image Home Page, "http://rsb.info.nih.gov/nih-image/".
6. G. E. Dieter, in Mechanical Metallurgy, McGraw-Hill Book Co. (1988).
7. H. Tada, P. Paris, G. Irwin, in The Stress Analysis of Cracks Handbook, Del Research Corporation (1973)
8. E. Albert and R. Kirchheim, Scripta Metall., **15**, 673 (1981).
9. R. L. Pastorek and R. A. Rapp, Trans. AIME, **245**, 1711 (1969).
10. R. D. K. Misra, C. J. McMahon, Jr., and A. Guha, *Scripta Metall. et Mater.*, **31**, 1471 (1995).
11. J. A. Pfaendtner, unpublished research, Univ. of Penn. 1995.

EXPERIMENTAL EVIDENCE FOR SULFUR INDUCED LOSS OF DUCTILITY IN COPPER SHAPED-CHARGE JETS

D. K. CHAN,* D. H. LASSILA,* W. E. KING,* and E. L. BAKER**
*Lawrence Livermore National Laboratory, P.O. Box 808, Livermore, CA 94551
**U. S. Army, ARDEC, Picatinny, NJ 07806

ABSTRACT

We have observed that a change in the bulk sulfur content of oxygen-free electronic copper markedly affects its high temperature (400-1000°C), high strain-rate (> 10^3 s^{-1}) deformation and fracture behavior. These conditions are typical of those found in "jets" formed from the explosive deformation of copper shaped-charge liners. Specifically, an increase in the bulk sulfur concentration from 4 ppm to 8 ppm shortens the breakup time, t_b, of the copper jets by nearly 20% as measured using flash x-ray radiographs recorded during breakup of the jets. At bulk concentrations of 4 ppm, the jet was observed to be uniform and axisymmetric with a breakup time of 186 μs. Jet particles exhibited length-to-diameter ratios of roughly 8:1. The addition of sulfur transformed the jet breakup behavior to non-uniform, non-axisymmetric rupture and reduced the breakup time to 147 μs. The length-to-diameter ratios decreased to roughly 5:1 in the sulfur-doped samples. Previously measured sulfur solubilities and diffusivities in copper at the temperatures where this material was processed indicates nearly all of the sulfur was localized to grain boundaries. Therefore, we infer that the increase in sulfur content at grain boundaries is directly responsible for the change in breakup performance of the shaped-charge jets.

INTRODUCTION

The material parameters controlling the breakup behavior of copper jets formed from shaped-charge liners are not well understood. Previous studies [1-3] have shown that the grain size, texture, and impurity content are important factors which affect the breakup behavior of copper jets undergoing high strain rate (> 10^3 s^{-1}), high temperature (400-1000°C) deformation. In general, previous experimental observations indicate that copper liners with relatively small grain size and low impurity concentrations resulted in jets that ruptured ductilely, while large grain size, high impurity copper resulted in jets which exhibit brittle grain boundary fracture. For example, in a study of liners made from ETP (electrolytic tough pitch) copper (99.90% copper) [2], an increase in the average grain size from 15 μm to 120 μm resulted in a change from "semi-ductile" jet breakup to "particulated" break up. Similar observations [3] were reported for OFE (oxygen-free electronic) 99.99% copper. However, no experimental data exists in which a specific impurity was isolated from changes in grain size and texture to determine its effects on breakup behavior.

This paper presents the first results on the effects of the sulfur concentration on copper shaped-charge jet ductility, and analyzes its role on changing the high strain rate, high temperature mechanical behavior of copper. Copper liners with different sulfur concentrations were explosively deformed and examined for changes in breakup behavior. Flash x-ray radiographs were obtained of the deforming jets and analyzed for differences in the time to the onset of jet breakup and shape of the resulting jet particles. The jet breakup time, t_b, is an indication of the amount of ductile rupture that occurs, i.e. the longer the breakup time, the greater the amount of ductile rupture.

195

Increases in the sulfur concentration of the liners resulted in shorter breakup times of the jets, indicating that the jets underwent a loss of ductility with increasing sulfur concentrations.

EXPERIMENTAL PROCEDURES

Copper shaped-charge liners, in the shape of hollow cones (base inner diameter = 81 mm, apex angle = 42°), were prepared using a standard cold-forge process [4]. The starting material was OFE 99.99% copper, Hitachi C10100 bar stock. Table 1 lists the measured impurities in the copper liners [5]. The liners were annealed at 315°C for one hour and 400°C for 10 minutes and 100 hours in order to stabilize the grain structure. The grain size of the liners was measured using standard metallographic techniques. Electron micrographs were obtained from 3 mm disk samples machined from the cones which were jet polished to electron transparency. The breakup times were measured from the flash x-ray radiographs of the jets formed after the explosive deformation of the copper liners.

Table 1. The measured impurity concentrations in OFE copper, Hitachi C10100 stock. All other elements had concentrations of < 0.1 ppm.

	Concentration (ppm)	Impurity	Concentration (ppm)
H	0.9	Ni	1.0
C	5.0	As	0.4
N	<0.1	Se	0.3
O	6.0	Ag	6.4
Si	0.2	Sb	0.3
P	0.4	Pb	0.2
S	4.0	Bi	0.2
Fe	2.0		

Several liners were doped with sulfur without altering the grain size and texture using a Cu_2S powder pack method. This method consisted of placing the copper liners in an evacuated ($< 10^{-5}$ Torr) pyrex container containing Cu_2S powder. The interior of the pyrex container was lined with graphite to getter any residual oxygen. The liners were heated at 250°C for six hours or 310°C for 48 hours. Under these annealing conditions, no grain growth was observed. The partial pressure of sulfur that was obtained in the closed environment was equal to the partial pressure required to maintain equilibrium between Cu_2S with copper and sulfur [6] according to equation 1.

$$Cu_2S \xleftrightarrow{T} 2Cu + S \qquad (1)$$

After annealing, the surfaces of the liners had a slight tarnish which was easily removed by etching in a dilute nitric acid solution. The sulfur concentration of the doped liners was measured using standard titration methods.

RESULTS AND DISCUSSION

Chemical analysis of the sulfur doped liners showed that the sulfur concentration increased, on average, from 4 ppm to 8 ppm, while the remaining impurity concentrations remained unchanged.

The grain size of the liners remained unchanged at 15-25 μm and 30-50 μm after heating at 310 and 400°C, respectively. Optical micrographs of the copper liners shown in Fig. 1a revealed no second phases, inclusions, or porosity. Similar features were observed in the sulfur doped liners (Fig. 1b-d), indicating that no changes in grain size had occurred during the sulphidation process. This result was corroborated by conventional transmission electron microscope results, in which no additional features in the sulfur doped liners were observed as compared with the original liners.

Fig. 1. Optical micrographs showing the grain structure of the copper liners with different sulfur concentrations, c_S. (a) $c_S = 4$ ppm; (b) $c_S = 9$ ppm; (c) $c_S = 7$ ppm; and (d) $c_S = 8$ ppm.

Comparison of the early time (at 140 μs) x-ray radiographs, Fig. 2, of the shaped-charge jets containing 3 ppm and 9 ppm sulfur revealed an earlier onset of necking in the higher sulfur concentration jet. X-ray radiographs (at 260 μs) of the shaped-charge jet with 3 ppm sulfur, shown in Fig. 3a, had jet particles with axisymmetric, uniform breakup with length-to-diameter ratios of greater than 8:1 and breakup times greater than 180 μs. This breakup behavior was typical of jets formed from liners with similar purities. A dramatic difference in breakup behavior was observed in the sulfur doped liners as compared to the undoped liners. X-ray radiographs of the shaped-charge jets formed from higher sulfur concentration liners, shown in Fig. 3b-d, had jet particles with irregular particulation (non-axisymmetric), fragmentation, and off-axis tumbling. The length-to-diameter ratios decreased to approximately 5:1, and the breakup times decreased by approximately 30 μs to 150 μs. These results are summarized in Fig. 4 of breakup time as a function of sulfur concentration. The breakup times measured by Lichtenberger [1] for copper bearing 13 ppm and 20 ppm of sulfur (grain size unknown) were included in this plot to demonstrate the negative relationship between sulfur concentration in the liner and jet breakup time.

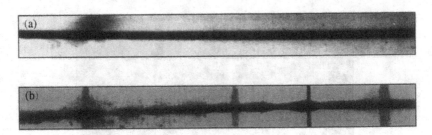

Fig. 2. X-ray radiographs (at 140 μs) showing the early onset of necking of the copper shaped-charge jet with 9 ppm sulfur. The dark contrast is the copper jet, and the vertical lines are fiducials for timing. (a) c_S = 3 ppm; (b) c_S = 9 ppm.

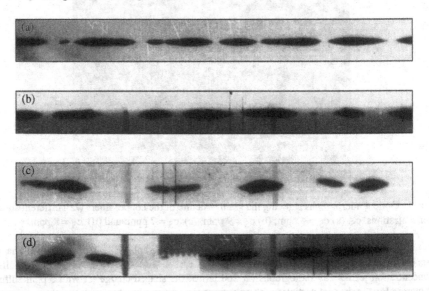

Fig. 3. X-ray radiographs (at 260 μs) showing the breakup behavior of the copper shaped-charge jets. (a) c_S = 3 ppm, t_b = 186 μs, length-to-diameter ratio (L/D) = 8:1; (b) c_S = 9 ppm, t_b = 152 μs, L/D = 5:1; (c) c_S = 7 ppm, t_b = 147 μs, L/D = 5:1; and (d) c_S = 8 ppm, t_b = 148 μs, L/D = 5:1.

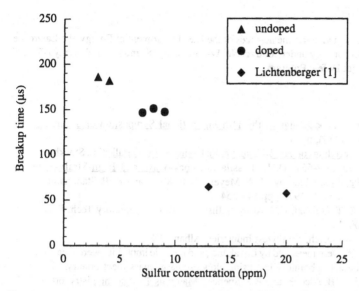

Fig. 4. A graph of the effect of sulfur concentration on jet breakup time.

An increase in the sulfur grain boundary concentration during the doping procedure is believed to be the underlying cause for the change in breakup behavior. The pack-powder doping established a sulfur concentration in the copper liner such that the activity of sulfur at the surface was equal to the activity of sulfur in the grain boundary and in the bulk. However, the establishment of an equilibrium distribution in the copper was diffusion limited. The root mean square diffusion distances of sulfur in bulk copper at 250°C for six hours and 310°C for 48 hours are 4.8×10^{-7} cm and 7.4×10^{-6} cm [7], respectively, while the respective grain boundary diffusion distances are 0.1 cm and 0.7 cm [8]. Therefore, most of the sulfur was dissolved into the grain boundaries, and very little sulfur was dissolved into the bulk. We infer from the sulfur kinetics that the measured bulk increase in the sulfur concentration was due to an increase in the grain boundary sulfur concentration. This increase was calculated as approximately 6 wt. % sulfur if we assume that the grains are cubes with a 3×10^{-3} cm length and an estimated grain boundary width of 1×10^{-7} cm. We believe that this change in the grain boundary sulfur concentration contributed to an increase in high temperature brittle breakup of the jets.

CONCLUSIONS

The change in sulfur concentration in copper, from 4 ppm to 8 ppm, resulted in a change in jet breakup behavior at high strain rates and high temperatures. The measured breakup times of the jets decreased 20%, from 186 µs to 147 µs. We infer from the sulfur kinetics in copper that the sulfur increase in copper is localized to the grain boundaries, resulting in a calculated increase of 6 wt.% at the grain boundaries. Although the total concentration of 8 ppm is under the OFE copper limit of 15 ppm sulfur [9], the 4 ppm increase in sulfur resulted in a loss of the shaped-charge jet ductility.

ACKNOWLEDGMENTS

This work performed under the auspices of the U.S. Department of Energy, the Lawrence Livermore National Laboratory under contract No. W-7405-Eng-48, and the U. S. Army ARDEC Technology Base Target Defeat Program.

REFERENCES

1. A. Lichtenberger, in Proceedings of the 11th International Symposium on Ballistics, Brussels, Belgium (1989), pp. 5-11.
2. D. H. Lassila, in Proceedings of the 13th International Symposium on Ballistics, Stockholm, Sweden (1992), pp. 549-557; D. H. Lassila in Shock-Wave and High-Strain-Rate Phenomena in Materials, edited by M. A. Meyers, L. E. Murr, and K. P. Staudhammer (Marcel Dekker, New York, 1972), pp 543-554.
3. M. L. Duffy and S. T. Golaski, U.S. Army Ballistic Research Laboratory Tech. Report No. BRL-TR-2800, 1987.
4. Prepared by F. W. Kuebrich, Northwest Industries, Albany, OR.
5. Interstitial impurities were measured by Luvak Inc., and metallic impurities were measured by Northern Analytical Laboratory Inc. using glow discharge mass spectrometry.
6. O. Kubaschewski, C. B. Alcock, and P. J. Spencer, Materials Thermochemistry, 6th ed. (Pergamon Press Inc., New York, 1993), pp. 1-27.
7. K. Fueki and Y. Ouchi, Bull. Chem. Soc. Jap. 51, 2234 (1978).
8. G. E. Moya-Gontier and F. Moya, Scripta Metall. 8, 153 (1974); G. E. Moya-Gontier and F. Moya, Scripta Metall. 9, 307 (1975).
9. 1995 Annual Book of ASTM Standards, edited by N. C. Furcola et al. (ASTM 2.01, Philadelphia, PA, 1995) pp. 821, 837.

ATOMIC FORCE AND SCANNING ELECTRON MICROSCOPY OF CORROSION AND FATIGUE OF AN ALUMINUM-COPPER ALLOY

K.Kowal[*], J.DeLuccia[*], J.Y.Josefowicz[**], C.Laird[*], and G.C.Farrington[*]
[*]MSE Department, University of Pennsylvania, Philadelphia, PA 19104
farringt@eniac.seas.upenn.edu
[**]Visiting professor from Hughes Research Laboratory, Malibu, CA

ABSTRACT

The morphological features of 2024-T3 aluminum alloy were delineated using atomic force microscopy (AFM) during separate and combined actions of corrosion and fatigue.

In-situ AFM corrosion studies in hydrochloric acid environments without mechanical deformation showed accelerated dissolution in the vicinity of second phase precipitates leading to intergranular corrosion. During fatigue in air, AFM images revealed steps along grain boundaries, as well as parallel extrusions and intrusions during the early stages of fatigue life. At later stages of mechanical deformation persistent slip bands (PSBs) were observed on the sample's surface. Cracks were observed to nucleate and propagate along PSBs. For experiments where samples were subjected to the simultaneous action of a corrosive environment and mechanical deformation, intergranular cracking was observed during the early stages of fatigue life. The corrosive environment was observed to accelerate the crack nucleation process.

INTRODUCTION

Since the development of scanning tunneling microscopy (STM) [1], which was first used only to obtain high resolution images of conducive surfaces in vacuum environments, scanning probe microscopy techniques have evolved to overcome the restriction of high vacuum environments, as well as the need for high conductivity of the imaged surfaces [2]. Among these techniques atomic force microscopy (AFM) [2] allows for obtaining high resolution surface images regardless of the electronic properties of the imaged surface. Furthermore, AFM also provides a opportunity to image the structure of a surface immersed in liquid environments [3]. This offers a revolutionary approach for the *in-situ* characterization of electrochemical processes. The combination of AFM with various galvanostatic and potentiostatic techniques makes it possible to study electrode surface morphology while electrochemical reactions are underway [4-8]. This work is the first attempt at imaging of corrosion fatigue processes, during which electrochemical (environmental) and mechanical deformation factors are simultaneously applied.

Stress corrosion and corrosion fatigue of aluminum alloys have been problems since the first aluminum alloys were produced in early 1900s. Pure aluminum is very corrosion resistant, but unfortunately, it is also soft and has little strength. Alloying aluminum with Cu, Mg, Zn, or Li, and then heat treating it can substantially increase the metal's strength, but also increases the likelihood of cracking when the metal is exposed under static or cyclic stress to a corrosive environment [9].

In most aluminum alloys such as 1xxx (commercially pure aluminum), 5xxx (aluminum magnesium based alloys), and 7xxx (aluminum zinc based alloys) corrosion fatigue and stress corrosion cracking are believed to be hydrogen induced [9,10]. The hydrogen embrittlement model postulates that atomic hydrogen is absorbed and somehow weakens the grain boundaries, which allows cracking [9]. However, for aluminum-copper based alloys (2xxx series), most researchers accept an anodic dissolution model of stress corrosion and corrosion fatigue cracking [9,11,12]. This model assumes that cracking is due to the preferential dissolution along grain boundaries [12]. However, it has been recently reported that hydrogen embrittlement might play a significant role during stress corrosion of aluminum copper alloys [13,14]. Since these alloys are widely used for aerospace purposes and are known to undergo acidic corrosion in the media containing chloride ions [15,16], it is important to understand their corrosion mechanism and the influence of corrosive environments on their mechanical strength.

Mat. Res. Soc. Symp. Proc. Vol. 409 © 1996 Materials Research Society

EXPERIMENTAL

Aluminum (2024-T3) specimens were prepared from 1/8" thick sheet in the form of cantilever beam samples for fatigue and corrosion fatigue studies (fig.1), or circular samples with a 1/2" diameter for corrosion investigations. Specimens were mechanically and electrochemically polished prior to experiments. Aqueous solutions of 0.01M and 0.1M HCl were prepared by diluting a standard solution of 1.005-0.995N HCl (Fisher Scientific) with deionized ultra-filtered water (Fisher Scientific).

Mechanical deformation tests were performed in an alternate bending geometry. A unique mechanically driven deformation stage has been designed and constructed for this purpose. The specially designed holders eliminate all tensile and compression stresses except those due to bending. For the cantilever beam samples (fig.1), uniform strains in the range of 0.01% - 0.6% may be applied. Mechanical cyclic frequency varies from 0.01 Hz to 1Hz. A special liquid cell has been designed to perform fatigue experiments in corrosive environments. Fig.1 shows schematic drawings of the specimen, the liquid cell (environmental chamber), and the gripping mechanism of the mechanical stage. It has not been found possible to apply stress while the AFM is actually in observational mode because bending amplitudes are too large. However imaging can be carried out between strain excursions without specimen removal from the solution.

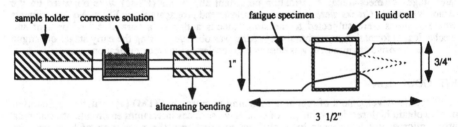

Fig.1. A schematic drawing of the specimen, the liquid cell, and the gripping mechanism of the mechanical stage.

In-situ corrosion AFM experiments were performed with a Nanoscope II (Digital Instruments, Inc.) using a 120 µm scanner. A standard fluid cell was used. Additional description of the fluid cell has been presented elsewhere [4].

Fatigue and corrosion fatigue AFM studies were performed with a Dimension 3000 (Digital Instruments, Inc.) using a 120 µm scanner. Typical experiments lasted between 4 to 60 hours. AFM morphological investigations were performed in the same, well defined cycle intervals of the fatigue lives i.e. before mechanical deformation, after 100 cycles, 200 cycles, 500 cycles, and then in 1,000 or 5,000 cycle intervals depending on the strain amplitude, and at the end of the cyclic life. All AFM images were obtained after stopping the mechanical deformation. The computer controlled stage of the Dimension 3000 microscope allows automatic localization of the surface region within an error less than 10 µm. This makes it possible to observe the progressive changes in the same region of the specimen surface .

After termination of the *in-situ* experiments, samples were removed from the wet cell, washed in deionized water, dried, and subjected to scanning electron microscopy (SEM, JELO 3000) analysis. SEM and *ex-situ* AFM were used to verify the reliability of the observed morphology for the other surface regions. Furthermore, energy dispersive X-ray analysis (EDX) was used for qualitative compositional analysis of the different surface regions during SEM observations.

RESULTS

a) Corrosion

Corrosion of 2024 aluminum alloy was investigated in 1M, 0.1 M HCl, and 0.01 M HCl. The processes observed in all three concentrations were very similar. First the local morphology was transformed from the morphology of an electropolished surface to a rough hill-like morphology. Once the hill-like morphology formed, dissolution of the grain interiors and formation of new pits began to occur [4].

The series of images in fig.2 shows processes occurring along grain boundaries which led to intergranular corrosion. The morphology of a grain boundary region prior to solution injection is shown in fig.2a. After 15 minutes in solution pits began to develop in one of the grain interiors (fig.2b). After one hour, fine particles were observed on the sample surface. They appeared to be a result of the redeposition of the dissolved material at local cathodes. Pits which developed upon immersion in solution did not show any significant changes in their dimensions for immersion times shorter than 4 hours.

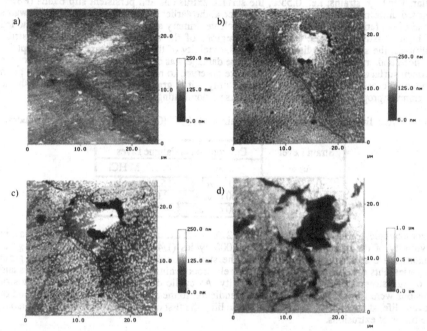

Fig.2. AFM images of a 2024-T3 sample immersed in 0.1 M HCl; different immersion times: a) before injecting solution, b) 15 minutes, c) 4 hours, and d) 8 hours.

The protruding region observed in fig.2b and fig.2c in between the pits might be a large second phase precipitate. Similar particles were determined to be copper rich constituent precipitates by EDX analysis after the termination of the corrosion experiments [4]. As proposed by Dix [17] the region adjacent to copper-rich second phase precipitates would be copper depleted and therefore more vulnerable to corrosion. This hypothesis could explain the faster dissolution of the regions in the vicinity of the second phase precipitates and the formation of pits due to the lower pitting potential of the copper depleted regions [18].

After 4 hours in solution, the local morphology of the surface completely transformed to a rough hill-like morphology, and the beginning of a preferential dissolution along grain boundaries was observed (fig.2c). At this point, hydrogen evolution began to occur. For longer immersion times, the sample surface roughened rapidly due to preferential dissolution along the grain boundaries and the formation of a large cavity in the center of the imaged area (fig.2d). This, in turn, led to extensive intergranular damage along grain boundaries for immersion times between 8 to 10 hours. After 8 hours in solution the second phase precipitate disappeared. The second phase precipitate probably lost contact with the matrix and was swept away as a result of the fast dissolution of the copper depleted regions surrounding it.

This series of images clearly shows the morphological changes from the beginning of passivity breakdown to corrosion along grain boundaries. Although such events have been postulated previously [17,18],this is the first time those events have been directly observed as they actually happened.

b) Fatigue in air

Fatigue experiments were performed with three different strain amplitudes: 0.55%, 0.4%, and 0.27%. The cyclic life, defined as the number of cycles from beginning of the experiment to fracture are shown in table 1. For all strain amplitudes investigated, the surface damage was very similar. For high strains, i.e., 0.55%, the surface extrusions and persistent slip bands (PSBs) nucleated uniformly all over the sample surface at the earlier stages of the fatigue life. Towards the end of the fatigue life, cracks tended to nucleate primary along PSBs and extrusions, and led to a high density of fatigue cracks in all regions of the sample surface. For lower strain amplitudes, the surface was limited only to the vicinity of the final fracture crack. Except for extrusions and cracking in this area, no surface damage was detected in the other regions of the specimen surface. Although many cracks were observed to nucleate at pits at or near the grain boundaries, cross-sectional analyses showed only initial grain boundary penetration with the transgranular propagation mode of all the cracks predominating.

Table 1. Cyclic lives of the samples cycled in air and 0.1 M HCl with different strain amplitudes.

Strain Level	Environmental Fatigue Lives	
$\Delta\varepsilon$	in air	in 0.01M HCl
0.27%	78,000 cycles	55,000 cycles
0.40%	9,700 cycles	8,000 cycles
0.55%	4,500 cycles	4,100 cycles

For example, for the strain amplitude of 0.27% small extrusions were observed to nucleate in the vicinity of the grain boundary after 20,000 cycles (1/4 of the fatigue life). Since grain boundaries are softer than grain interiors, the strain would be highly localized at the grain boundaries. Pits observed on the surface after electropolishing acted as stress concentrators and led to high stress localization in their vicinity. As cyclic deformation progressed, bands of extrusions were formed in the direction perpendicular to the straining axis. Towards the end of the cyclic life, at about 50,000 (2/3 of the cyclic life), the first cracks were observed to open along these bands of extrusions.

c) Corrosion fatigue

A significant reduction in cyclic life was observed for the specimens fatigued in an aggressive medium as compared to those observed in air. Typical values of cyclic lives for specimens cycled in 0.01 M HCl are given in table 1. As the straining amplitude decreased the reduction in the cyclic life became more severe.

Different surface and cracking morphology were observed for the samples fatigued in the aggressive aqueous environment as compared to those observed in air. Fig.3 shows the typical series of events on the specimen surface fatigued in 0.01 M HCl with strain amplitude of 0.27% and cyclic frequency of 0.5Hz. After the injection of the corrosive solution no changes were observed during the first 5,000 cycles. Typical morphological features observed up to 5,000 cycles is shown in fig.3.a. The first crack was observed after 10,000 cycles, 6 hours 45 minutes in solution (fig.3.b). This kind of damage resembled preferential corrosion attack along grain boundaries. This preferential dissolution occurred faster, i.e. after about 6 hours, than for experiments where the specimens were subjected to a corrosive environment alone, e.g. after 24 hours. Furthermore, cracking along grain boundaries seemed to propagate faster in the direction perpendicular to the straining axis. These observations support the hypothesis that dissolution at grain boundary regions is strongly enhanced by mechanical straining [9,11].

After 30,000 cycles (fig.3.c), the observed crack appeared to change its propagation mode. Intergranular corrosion assisted paths transformed into a transgranular crack propagating along the direction of maximum shear stress (fig.3.d). This path showed characteristic extrusions similar to those observed in air. During the rest of the life, the crack propagated transgranularly and strong intergranular corrosion damaged was limited to the nucleation site. The final fracture crack was thus a result of the interconnection of small cracks nucleated intergranularly.

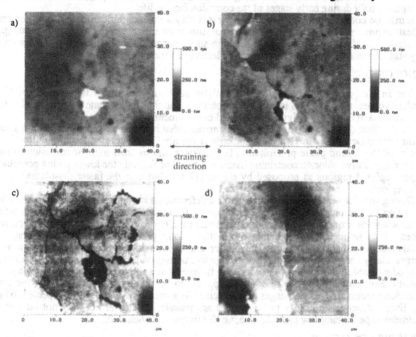

Fig.3. AFM images of a 2024-T3 sample fatigued in 0.1 M HCl with strain amplitude 0.27% and cyclic frequency 0.5Hz; a) before injecting solution, and after 500, 1000, and 5000 cycles b) after 10,000 cycles, c) after 20,000 cycles, and d) transgranular cracking after 30,000 cycles.

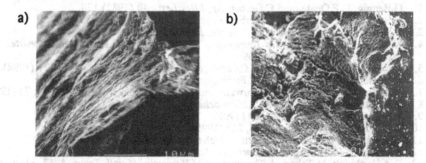

Fig.4. Fracture surfaces of the samples cycled in: a) in air, and b) 0.01 M HCl solution.

The fracture surface for a crack propagating in air showed features typical for a transgranular mode of crack propagation throughout the whole fatigue life (fig.4a). A typical fracture micrograph obtained at the crack nucleation site for a specimen fatigued in 0.01 M hydrochloric acid is shown in fig.4b. It can be seen that intergranular fracture was predominant during crack nucleation and the early stages of propagation in the aggressive environment. In the later stages of the cyclic life, transgranular fracture was observed for cracks propagating in 0.01M HCl. These results were consistent with AFM investigations which showed the intergranular mode of crack propagation only during early stages of the corrosion fatigue life.

It may be concluded that corrosion processes play a most significant role during the crack nucleation process during which aggressive environments enhance crack nucleation along grain boundaries in this alloy. This leads to the high stress concentration at these regions during the early stages of corrosion fatigue and therefore leads to a shortening of fatigue lives.

CONCLUSIONS

The *in-situ* corrosion and corrosion fatigue AFM studies make it possible to follow progressive, real time changes in surface morphology and thus to delineate detailed mechanisms of corrosion and corrosion fatigue in controlled environments.

In-situ corrosion investigations have demonstrated that intergranular corrosion of heat treated aluminum copper alloys in HCl environments is mainly due to accelerated dissolution in the vicinity of the second phase precipitates [4]. Therefore, severe intergranular corrosion of these alloys in aggressive chloride containing media is a result of not only the lower pitting potential of copper depleted regions as proposed by others [17,18], but also the faster dissolution rate of these regions.

During simultaneous exposure to mechanical deformation and an acidic environment, a strong interaction between mechanical and electrochemical factors was observed. A reduction in fatigue life in acidic environments appears to be attributed mainly to the accelerated crack nucleation process. It is important to note that dissolution along grain boundaries is accelerated by mechanical deformation. Therefore the surface cracks start out as intergranular ones, proceed intergranularly for a few grains and then change to a transgranular mode. Such behavior supports the anodic dissolution model for the corrosion fatigue of aluminum copper alloys [9]. However, in acidic environments anodic dissolution appears to be predominant only during the crack nucleation process. In the later stages of the cyclic life, a corrosive environment appeared to have very little influence on cracks propagating transgranularly. This effect may not be true for specimens experiencing low cycle fatigue, i.e. for frequencies below 0.5 Hz.

ACKNOWLEDGMENTS

The authors wish to thank the US Air Force Office of Scientific Research and National Science Foundation MRL Program for the financial support of this research.

REFERENCES

1. G.Bennig, C.F.Quate, and C.Gerber, *Appl.Rev.Lett.*, **40** (1982) 178.
2. G.Bennig, C.F.Quate, and C.Gerber, *Phys.Rev.Lett.*, **56** (1986) 930.
2. R.Sonnenfeld and P.K.Hansma, *Science*, **232**, 211 (1986).
4. K.Kowal, J.J.DeLuccia, J.Y.Josefowicz, C.Laird, and G.C.Farrington, *submitted to J. Electrochem. Soc.*, November 1995.
5. K.Kowal, L.Xie, R.Huq, and G.C.Farrington, *J.Electrochem. Soc.*, **141**, 121 (1994).
6. J.Y.Josefowicz, L.Xie, and G.C.Farrington, *J.Phys.Chem.*, **97**, 11995 (1993).
7. J.Josefowicz, J.DeLuccia, V.Agarwala, and G.C.Farrington, *Mat.Charact.*, **34**, 73 (1995).
8. R.M.Rynders and R.C.Alkire, *J.Electrochem. Soc.*, **141**, 1166 (1994).
9. T.D.Burleigh, *Corrosion*, **47**, 89 (1991).
10. T.Magnin, *ISIJ International*, **36**, 223 (1995).
11. M.R.Bayoum, *Engineering Fracture Mechanics*, **45**, 297 (1993).
12. K.Urushino and K.Sugimoto, *Corrosion Science*, **19**, 225 (1979).
13. D.A.Hardwick, M.Taheri, A.Thompson, I.M.Bernstein, *Metall. Trans. A*, **13**, 811 (1982).
14. F.Zeides and I.Roman, *Materials Science and Engineering*, **A125**, 21 (1990)
15. H.Kaesche in Localized Corrosion. Houston, TX, NACE 1974, p.516.
16. E.H.Hollingsworth and H.Y.Hunsicker in Metals Handbook volume 13 Corrosion. American Society of Metals, Metals Park, OH (1987).
17. E.H.Dix, Jr., *Trans. AIME*, **137**, 11 (1940).
18. J.R.Galvele and S.M. de DeMicheli, *Corrosion Science*, **10**, 795, 1970.

FILM EFFECTS ON DUCTILE/BRITTLE BEHAVIOR
IN STRESS-CORROSION CRACKING

Tong-Yi Zhang*, Wu-Yang Chu** and Ji-Mei Xiao**
* Department of Mechanical Engineering, Center for Advanced Engineering Materials, Hong Kong University of Science and Technology, Kowloon, Hong Kong
** Department of Materials Physics, Beijing University of Science and Technology, Beijing 100083, P.R. China

ABSTRACT

The present work analyzes the effects of a passive film formed during stress corrosion cracking on ductile/brittle fracture behavior, considering the interaction of a screw dislocation with a thin film-covered mode III crack under an applied remote load. Exact solutions are derived, and the results show that the crack stress field due to the applied load is enhanced by a harder film or abated by a softer film. The critical stress intensity factor for dislocation emission from the crack tip is greatly influenced by both the stiffness and thickness of the film. A dislocation is more easily to be emitted from the crack tip if the covered film has a shear modulus larger than that of the substrate. The opposite is also true, i.e., a softer film makes dislocation emission more difficult. Both phenomena become more significant when the film thickness is smaller.

INTRODUCTION

Stress corrosion cracking (SCC) can cause catastrophic failures of engineering structures and components due to resultant acting of applied loads and environment. All SCC failures have in common a macroscopic appearance of brittleness, in the engineering sense that the ductility of the material is impaired. However, how the macroscopic ductility is faded at microscopic levels still remains unclear. Many different mechanisms have been proposed to explain the synergistic stress-corrosion interactions that occur at a crack tip, and there are several processes that can contribute to SCC, including anodic dissolution, film growth and cracking, and hydrogen-induced cracking [1-2].

Sieradzki [3] studied, using angularly resolved X-ray Photoelectron Spectroscopy, embrittlement of High Strength Low Alloy (HSLA) steel in gaseous chlorine. His results indicated that a thin film of $FeCl_2$, forming above a threshold pressure of chlorine gas, is responsible for this SCC. There is evidence from Transmission Electron Microscopy that an oxidized thin film or a de-alloyed "sponge" is formed in austentic stainless steels, as they are susceptible to chlorine-induced cracking [4]. According to the experimental results, Sieradzki [3], and Sieradzki and Newman [5,6] proposed a mechanism of film-induced cleavage for SCC based on dislocation-crack interaction. The dislocation emission from a crack tip may become difficult if a thin film with a thickness of several hundred nanometers is formed around the crack tip owing to an anodic process. Also, a thin ductile film may initiate cleavage if the interface between the film and the substrate is coherent and if appropriate mismatch strains exist. Their analysis, however, relies on a major assumption: the stress field in the vicinity of the crack tip due to the applied loads remains the same whether film forms or not. They also assume that the image force of a dislocation near the crack tip covered with a film is the same as that near a flat film-substrate

207

surface [3,6]. To our knowledge, nobody has examined the validity of these two assumptions. Therefore, one purpose of the present work is to examine the film-crack system using linear fracture mechanics. For simplicity, a mode III crack and a screw dislocation are considered as simple demonstrations of the effects of a thin film on ductile/brittle fracture behavior. This note reports highlights of the modeling and results; detailed analysis will appear elsewhere [7]. Furthermore, interaction of a mode I or a mixed mode I + mode II + mode III crack with dislocations having arbitrary Burgers vectors is under preparation [8]. The derived exact solutions indicate that the local stress field around the crack tip may be dramatically changed due to the formation of a passive thin film at the tip, and consequently, the energy barrier for dislocation emission may be reduced or increased depending on the nature of the film.

ANALYSIS

In order to model a crack covered by a thin film, we assume that the film has an elliptical shape of $x_1^2 / a^2 + x_2^2 / h^2 = 1$. Inside the film there is a crack occupying from -c to c along the x_1 axis and $c^2 = a^2 - h^2$. The crack covered with the film is embedded in an infinite medium under remote loading, as shown in Fig. 1.

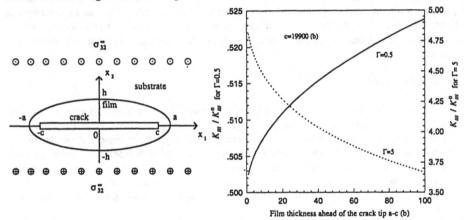

Fig. 1. Schematic illustration of a thin film-covered mode III crack under an applied remote load in the z plane.

Fig. 2. The normalized stress intensity factors of K_{III} / K_{III}^0 as a function of the film thickness for $\Gamma=0.5$ and $\Gamma=5$.

Stress Fields Due to an Applied Load

The stress fields inside the film and the substrate due to an applied remote load, σ_{32}^- are formulated, and only the stress field inside the film is given here by

$$\sigma^* = \frac{\Gamma \sigma_{32}^-}{1+\Gamma+m(1-\Gamma)} \left(z+\sqrt{z^2-c^2} + \frac{c^2}{z+\sqrt{z^2-c^2}} \right) \frac{1}{\sqrt{z^2-c^2}}, \tag{1}$$

where complex stress $\sigma = \sigma_{32} + i\sigma_{31}$, $\Gamma = \mu^* / \mu$ is a ratio of the shear modulus of the film to that of the substrate, μ is the shear modulus of the material, $z = x_1 + ix_2$, $R = (a + h)/2$, $m = (a - h)/(a + h)$, and the asterisk denotes the film. From the stress field, the stress intensity factor at the right crack tip is calculated. It is

$$K_{III} = \frac{2\Gamma\sigma_{32}^\infty\sqrt{\pi c}}{1 + \Gamma + m(1 - \Gamma)} = \frac{2\Gamma}{\Gamma(1 - m) + 1 + m}K_{III}^0,$$ (2)

where K_{III}^0 is the nominal stress intensity factor without any film. Since m<1, the stress intensity factor is enhanced if the shear modulus of the thin film is larger than that of the substrate. Being a function of m, the stress intensity factor depends on the film thickness. Figure 2 indicates the normalized stress intensity factor as a function of the film thickness of a-c, respectively, for $\Gamma = 0.5$ and $\Gamma = 5$. The normalized stress intensity factor decreases with increasing film thickness if the film is harder than the substrate, as shown in Fig. 2 for $\Gamma = 5$. In this case, the normalized stress intensity factor varies from 5 to 1 as the film thickness changes from infinity to zero. When the film is softer than the substrate, the normalized stress intensity factor increases with increasing film thickness, as shown also in Fig. 2. For $\Gamma = 0.5$, the normalized stress intensity factor ranges from 0.5 to 1 as the film thickness increases from zero to infinity. When the crack length is much larger than the film thickness, Eq. (2) can be reduced to

$$K_{III} = \Gamma K_{III}^0, \quad \text{for h<<c.}$$ (3)

In this case, the stress intensity factor is enlarged by Γ times.

<u>Stress Fields of a Dislocation Located Inside The Film</u>

For a screw dislocation inside the film, stress fields within the film and substrate are evaluated. The stress field in the film has the following form:

$$\sigma^* = \left[\frac{\mu^* b}{2\pi}\left(\frac{1}{\zeta - \zeta_d} - \frac{1}{\zeta - m/\bar{\zeta}_d}\right) + \mu^*\sum_{n=1}^{\infty} n(\bar{A}_n^* m^n \zeta^{-n} + A_n^* \zeta^n)\right]\frac{z + \sqrt{z^2 - c^2}}{2R\sqrt{z^2 - c^2}},$$ (4)

where the overbar means the complex conjugate, b is the Burgers vector, and both A_{-n} and A_n^* are constants given in [7]. The image force exerted on a dislocation is calculated by the Peach-Koehler formula. When the dislocation is located on the x_1 axis, the image force is evaluated as:

$$f_1 = \frac{\mu^* b^2}{2\pi}\left(-\frac{z_d}{z_d^2 - c^2}\right) + \mu^* b\sum_{n=1}^{\infty} nA_n^*\left[m^n\left(\frac{z_d + \sqrt{z_d^2 - c^2}}{2R}\right)^{-n} + \left(\frac{z_d + \sqrt{z_d^2 - c^2}}{2R}\right)^n\right]\frac{z_d + \sqrt{z_d^2 - c^2}}{2R\sqrt{z_d^2 - c^2}}.$$ (5)

Figure 3 shows the image force as a function of the distance from the crack tip, where the film is 100 b thick. When the film is softer than the substrate, the image force is always negative. In this case, the crack attracts the dislocation, while the interface (or the harder substrate) also pushes the dislocation towards the crack, as shown in Fig.3 for $\Gamma = 0.5$. However, if the film is harder than the substrate, both the crack and the interface attract the dislocation. Therefore there is an equilibrium point at which the total image force acting on the dislocation is zero. If the distance from the crack tip is smaller than the equilibrium distance, the dislocation will be attracted by the crack; otherwise, it will be attracted by the interface, as shown in Fig.3 for $\Gamma = 5$. As a reference, Fig.3 also indicates the image force for $\Gamma = 1$.

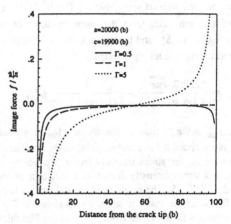

Fig. 3. The image forces of a screw dislocation in the film as a function of the distance from the crack tip for Γ=0.5, Γ=1 and Γ=5.

DISLOCATION EMISSION FROM THE CRACK

Recently, Beltz and Rice [9] re-solved the emission problem using the Peierls-Nabarro model. The critical stress intensity factor for dislocation emission from a crack tip is determined by the unstable stacking fault energy. However, the unstable stacking fault energy is not available in the literature for most materials, especially for passive films. In the present work, we emphasize how a hard or soft passive film can influence dislocation emission from the crack tip. Therefore, we determine the critical stress intensity factor for dislocation emission from the Rice-Thomson model [10], wherein we assume that the size of the dislocation core in the film is the same as that in the substrate. Thus, the effect of the shear modulus ratio and the film thickness can be clearly demonstrated. The critical stress intensity factor in a two-dimensional calculation of the Rice-Thomson model is given by [11]

$$K_{III,e} = -\frac{\sqrt{2\pi r_0}}{b} f(r_0) , \qquad (6)$$

where r_0 is the radius of the dislocation core. Substituting Eqs. (2) and (5) into Eq.(6) leads to

$$K_{III,e}^0 = \frac{\mu b}{2\sqrt{2\pi r_0}} - \frac{\mu(\Gamma-1)b\sqrt{c}}{4R\sqrt{\pi}} \sum_{n=1}^{\infty} \left[\left(\frac{c+\sqrt{2cr_0}}{2R} \right)^{2n} - \left(\frac{2Rm}{c+\sqrt{2cr_0}} \right)^{2n} \right] \frac{1}{\Gamma(1-m^n)+1+m^n}. \qquad (7)$$

The radius of the dislocation core is chosen as one Burgers vector in plotting the critical stress intensity factor for dislocation emission. Figure 4 shows that the critical stress intensity factor decreases as the shear modulus ratio increases. This phenomenon becomes more significant when the film thickness is smaller, as shown in Fig.4 for the film thicknesses of 2 b, 5 b and 10 b. The film thickness effect is illustrated in Fig.5 for $\Gamma = 0.5$ and $\Gamma = 5$. As can be seen in Fig.5, for both cases of $\Gamma = 0.5$ and $\Gamma = 5$, the influence of the film on the critical stress intensity factor becomes smaller and smaller as the film thickness gets larger and larger. There are two reasons for this phenomenon. Firstly, the normalized applied stress intensity factor increases (or decreases) for $\Gamma = 5$ (or $\Gamma = 0.5$) as the film thickness decreases. This means the driving force for dislocation emission becomes stronger (or weaker) for $\Gamma = 5$ (or $\Gamma = 0.5$) as the film is thinner. Secondly, if the film is harder (or softer) than the substrate the image force induced by the interface attracts (or pushes) the dislocation towards to the interface (or crack) and consequently, the total image force acting on the dislocation is smaller (or larger). As a result, the critical stress intensity factor for dislocation emission is smaller for a harder film or larger for a softer film, and the changes are substantial for thin films with having thicknesses less than 20 b, as shown in Fig.5.

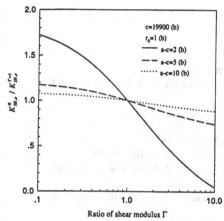
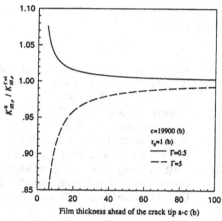

Fig. 4. The normalized critical stress intensity factors of $K_{III,e}^0 / K_{III,e}^{\Gamma=1}$ for dislocation emission from the crack tip as a function of the shear modulus ratio Γ for the film thickness a-c=2 b, 5 b and 10 b, where $K_{III,e}^{\Gamma=1}$ is the critical stress intensity factor for dislocation emission from a crack tip in a homogeneous solid having the same shear modulus as the substrate.

Fig.5. The normalized critical stress intensity factors of $K_{III,e}^0 / K_{III,e}^{\Gamma=1}$ for dislocation emission from the crack tip as a function of the film thickness for Γ =0.5 and Γ =5, where $K_{III,e}^{\Gamma=1}$ is the critical stress intensity factor for dislocation emission from a crack tip in a homogeneous solid having the same shear modulus as the substrate.

CONCLUSIONS

For simplicity, a stress corrosion crack is modeled in the present work as a thin film-covered mode III crack. The exact solutions are obtained for both the applied stress field and the dislocation stress field. Consequently, the dislocation emission from the thin film-covered crack tip is studied using the Rice-Thomson model. The results indicate that the crack stress field due to the applied load is enhanced by a harder film or abated by a softer film. If the film thickness is much smaller than the crack length, then, the stress intensity factor can be simply expressed as the product of the nominal stress intensity factor times the shear modulus ratio. The critical stress intensity factor for dislocation emission from the crack tip is greatly influenced by both the stiffness and thickness of the film. A dislocation is more easily to be emitted from the crack tip if the covered film has a shear modulus larger than that of the substrate, and a softer film makes dislocation emission more difficult. Both phenomena become more significant when the film thickness is smaller.

ACKNOWLEDGMENTS

The authors thank Mr. Cai-Fu Qian for his help in plotting the figures. Financial support from both the Hong Kong Research Grants Council through an RGC grant and the National Natural Science Foundation of China is gratefully acknowledged.

REFERENCES

1. R.N. Parkins, JOM, December issue (1992) p.12-19.
2. R.H. Jones and R.E. Ricker, in "Stress-Corrosion Cracking" ed. R.H. Jones, ASM International (1992) p.1-40.
3. K. Sieradzki Acta Metall. **30**, p.973 (1982).
4. G.M. Scanmans and P.R. Swann, Corros. Sci. **18**, p. 983 (1978).
5. K. Sieradzki and R.C. Newman, Phil. Mag. **A51**, p. 95 (1985).
6. K. Sieradzki and R.C. Newman, J. Phys. Chem. Solids **48**, p.1101 (1987).
7. Tong-Yi Zhang and Cai-Fu Qian, Interaction of a screw dislocation with a thin film-covered mode III crack, unpublished work.
8. Cai-Fu Qian and Tong-Yi Zhang, Interaction of a dislocation with a thin film-covered crack, unpublished work.
9. G.E. Beltz and J.R. Rice, Acta Metall. Mater. **40**, p. S321 (1992).
10. J.R. Rice and R. Thomson, Phil. Mag. **29**, p. 73 (1974).
11. I.H. Lin and R. Thomson, Acta Metall. **34**, p. 187 (1986).

Part IV

Fracture in Ceramics and Composites

CRITERIA FOR PROGRESSIVE INTERFACIAL DEBONDING WITH FRICTION IN FIBER-REINFORCED CERAMIC COMPOSITES

CHUN-HWAY HSUEH
Metals and Ceramics Division, Oak Ridge National Laboratory, Oak Ridge, TN 37831

ABSTRACT

Criteria for progressive debonding at the fiber/matrix interface with friction along the debonded interface are considered for fiber-reinforced ceramic composites. The energy-based criterion is adopted to analyze the debond length, the crack-opening displacement, and the displacement of the composite due to interfacial debonding. The analytical solutions are identical to those obtained from the mismatch-strain criterion, in which interfacial debonding is assumed to occur when the mismatch in the axial strain between the fiber and the matrix reaches a critical value. Furthermore, the mismatch-strain criterion is found to bear the same physical meaning as the strength-based criterion.

INTRODUCTION

Bridging of matrix cracks by fibers, which debond from and slip frictionally against the matrix, is an important toughening mechanism in fiber-reinforced ceramic composites [1,2]. To analyze the toughening effect, a criterion for progressive debonding at the fiber/matrix interface accompanied by friction along the debonded interface is required. The loading stress on the fiber to initiate debonding (or the debond stress for a frictionless interface), σ_d, has been analyzed by using either the energy-based [3-6] or the strength-based criterion [7-9]. The effect of constant friction along the debonded interface on progressive debonding was analyzed recently by Nair [10] using the energy-based criterion and by Budiansky et al. [11] using the strength-based criterion. It is noted that refinement is required in Nair's analysis regarding the work done by load. An alternative debonding criterion was proposed recently in which debonding is assumed to occur when the mismatch in the axial strain between the fiber and the matrix reaches a critical value [12]. Based on this assumption, the solutions for progressive debonding have been obtained [13]. A question is raised as to whether the solutions obtained from the three debonding criteria mentioned above agree with each other.

The purpose of the present study is to address the above question. First, using the energy-based criterion, solutions for progressive debonding with a constant friction along the debonded interface are obtained by modifying Nair's analysis [10]. These solutions are then compared to those obtained from the mismatch-strain criterion. Finally, the physical meaning of the approach using the strength-based criterion is examined and compared to the mismatch-strain criterion.

THE ENERGY-BASED CRITERION

A unidirectional composite subjected to a tensile load in the direction parallel to the fiber axis is considered. Matrix cracking occurs perpendicular to the loading direction and is bridged by intact fibers, which exert a bridging stress, σ_o, to oppose crack-opening. This problem can be modeled by using a representative volume element shown in Fig. 1. A fiber with a radius, a, is located at the center of a coaxial cylindrical shell of matrix with an outer radius, b, such that a^2/b^2 corresponds to the volume fraction of fibers, V_f, in the composite (Fig. 1a). When the interface remains bonded, the composite is subjected to a tensile stress, $V_f\sigma_o$, and has a displacement, u_{bonded}, in the axial direction (Fig. 1b). In the presence of interfacial debonding, the bridging fiber is subjected to a tensile stress, σ_o, and the matrix is stress-free at the crack surface (Fig. 1c). Interfacial debonding and sliding occur along a length, h, with a frictional

215

stress, τ, and the end of the debonding zone and the crack surface are located at $z=0$ and $z=h$, respectively. The half crack-opening displacement, u_0, is defined by the relative displacement between the fiber and the matrix at the crack surface (Fig. 1c). Also, compared to the composite without interfacial debonding (Fig. 1b), the composite with interfacial debonding has an additional displacement, u_{debond}, in the loading direction (Fig. 1c).

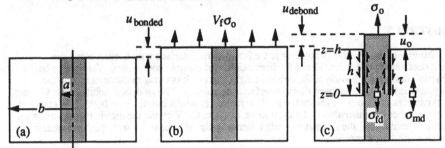

Fig. 2. A representative volume element for the fiber bridging problem: (a) prior to loading, (b) loading without interfacial debonding, and (c) loading with interfacial debonding. The half cracking opening displacement, u_0, and the displacement of the composite due to interfacial debonding, u_{debond}, are also shown.

<u>Stresses in the fiber and the matrix</u>

When the interface is bonded, the equilibrium axial stresses in the fiber and the matrix, σ_f and σ_m, satisfy both the equilibrium and the continuity conditions, such that

$$V_f \sigma_f + V_m \sigma_m = V_f \sigma_0 \qquad (1)$$

$$\frac{\sigma_f}{E_f} = \frac{\sigma_m}{E_m} \qquad (2)$$

where V_m ($=1-V_f$) is the volume fraction of the matrix, and E_f and E_m are Young's moduli of the fiber and the matrix, respectively. Combination of Eqs. (1) and (2) yields

$$\sigma_f = \frac{V_f E_f \sigma_0}{E_c} \qquad \text{(for bonded interface)} \qquad (3a)$$

$$\sigma_m = \frac{V_f E_m \sigma_0}{E_c} \qquad \text{(for bonded interface)} \qquad (3b)$$

where $E_c = V_f E_f + V_m E_m$.

For a frictional interface, both σ_f and σ_m can be approximated to be independent of the radial coordinate [4,5], and Eq. (1) is satisfied. The axial stresses in the fiber and the matrix at the end of the debond length, σ_{fd} and σ_{md}, can be obtained from the stress transfer equation, such that

$$\sigma_{fd} = \sigma_0 - \frac{2h\tau}{a} \qquad (4a)$$

$$\sigma_{md} = \frac{2h V_f \tau}{a V_m} \qquad (4b)$$

Solutions of σ_{fd} and σ_{md} are contingent upon the determination of h. With constant friction, the axial stress distributions in the fiber and the matrix, σ_f and σ_m, along the debond length are

$$\sigma_f = \sigma_{fd} + \frac{z(\sigma_o - \sigma_{fd})}{h} \qquad (0 \leq z \leq h) \qquad (5a)$$

$$\sigma_m = \left(1 - \frac{z}{h}\right)\sigma_{md} \qquad (0 \leq z \leq h) \qquad (5b)$$

Displacements

In the debonded region, the axial displacements resulting from the axial stresses described by Eqs. (5a) and (5b) are

$$w_f = \frac{z\sigma_{fd}}{E_f} + \frac{z^2(\sigma_o - \sigma_{fd})}{2hE_f} \qquad (0 \leq z \leq h) \qquad (6a)$$

$$w_m = \left(z - \frac{z^2}{2h}\right)\frac{\sigma_{md}}{E_m} \qquad (0 \leq z \leq h) \qquad (6b)$$

The half crack opening displacement, u_o (=w_f-w_m at z=h), becomes (Fig. 1c)

$$u_o = \frac{h\sigma_o}{E_f} - \frac{h^2 \tau E_c}{aV_m E_f E_m} \qquad (7)$$

In the absence of interfacial debonding, the axial displacement in the composite, w_c, within a length, h, is (Fig. 1b)

$$w_c(h) = \frac{hV_f \sigma_o}{E_c} \qquad (8)$$

Hence, the additional axial displacement of the composite due to debonding, u_{debond} (=$w_f(h)$-$w_c(h)$), becomes (Fig. 1c)

$$u_{debond} = \frac{hV_m E_m \sigma_o}{E_f E_c} - \frac{h^2 \tau}{aE_f} \qquad (9)$$

Solutions of u_o and u_{debond} are also contingent upon the determination of the debond length, h, which is solved using the energy-based criterion as follows.

The debond length and related solutions

Based on the energy-based criterion, the following energy terms are involved: (1) U_e, the elastic strain energy in the composite, (2) U_s, the energy due to sliding at the debonded interface, (3) G_i, the energy release rate for interfacial debonding, and (4) W, the work done by the applied stress. The equilibrium debond length, h, can be determined by using the energy balance condition when the fiber is subjected to a loading stress, σ_o, the debond length is assumed to advance a distance dh, and the corresponding energy changes are dU_e, dU_s, dG_i and dW. The energy balance condition requires that

$$dW = dU_e + dU_s + dG_i \qquad (10)$$

The above condition has been used by Nair [10] to derive the debond length; however, refinement of the derivation of dW is required. To determine the debond length, the present study summarizes the results for dU_e, dU_s and dG_i, and derives dW. However, a complete analysis of the debond length can be found elsewhere [14].

217

The results for dU_e, dU_s and dG_i are [10,14]:

$$dU_e = \frac{\pi a^2 V_m E_m}{2 E_f E_C} \left(\sigma_0 - \frac{2h\tau E_C}{a V_m E_m} \right)^2 dh \qquad (11)$$

$$dU_s = 2\pi a\tau \left(\frac{h\sigma_0}{E_f} - \frac{2h^2 \tau E_C}{a V_m E_f E_m} \right) dh \qquad (12)$$

$$dG_i = 2\pi a G_i dh \qquad (13)$$

With the bridging stress, σ_0, on the fiber, the work done due to interfacial debonding is $W = \pi a^2 \sigma_0 u_{debond}$. The change in the work done is hence

$$dW = \pi a^2 \sigma_0 du_{debond} \qquad (14)$$

It is noted that instead of using u_{debond}, u_0 was incorrectly used in Nair's analysis in deriving dW. Substitution of Eq. (9) into Eq. (14) yields

$$dW = \pi a^2 \sigma_0 \left(\frac{V_m E_m \sigma_0}{E_f E_C} - \frac{2h\tau}{a E_f} \right) dh \qquad (15)$$

Substitution of Eqs. (11), (12), (13), and (15) into Eq. (10) yields

$$h = \frac{a V_m E_m}{2\tau E_C} \left[\sigma_0 - 2 \left(\frac{E_f E_C G_i}{a V_m E_m} \right)^{1/2} \right] \qquad (16)$$

The stress required for initial debonding, σ_d, can be obtained from Eq. (16) by letting $h=0$, such that

$$\sigma_d = 2 \left(\frac{E_f E_C G_i}{a V_m E_m} \right)^{1/2} \qquad (17)$$

The solutions of u_0 and u_{debond} can be obtained by substituting Eq. (16) into Eqs. (7) and (9), such that

$$u_0 = \frac{a V_m E_m \sigma_0^2}{4 E_f E_C \tau} - \frac{G_i}{\tau} \qquad (18a)$$

$$u_{debond} = \frac{a V_m^2 E_m^2 \sigma_0^2}{4 E_f E_C^2 \tau} - \frac{V_m E_m G_i}{E_C \tau} \qquad (18b)$$

In the absence of interfacial bonding (i.e., $G_i=0$), equations (18a), and (18b) become

$$u_0 = \frac{a V_m E_m \sigma_0^2}{4 E_f E_C \tau} \qquad (19a)$$

$$u_{debond} = \frac{a V_m^2 E_m^2 \sigma_0^2}{4 E_f E_C^2 \tau} \qquad (19b)$$

Equations (19a) and (19b) are identical to the displacements derived in the MCE [15] and the ACK [16] models, respectively. While u_{debond} is considered in the ACK model [16], u_O is considered in the MCE model [15].

The steady-state increase in toughness, ΔG, of the composite due to frictional bridging of the matrix crack by fibers is given by [11,17]

$$\Delta G = 2V_f \int_0^{u^*} \sigma_o du_{debond} \tag{20}$$

where u^* is the displacement of the composite due to interfacial debonding when the loading stress on the fiber, σ_O, reaches the fiber strength, σ_s. Substitution of Eq. (18b) into Eq. (20) yields

$$\Delta G = \frac{aV_f V_m{}^2 E_m{}^2}{3E_f E_c{}^2 \tau} \left(\sigma_s{}^3 - \sigma_d{}^3 \right) \tag{21}$$

Hence, in order to achieve toughening effect (i.e., $\Delta G > 0$), the fiber strength, σ_s, must be greater than the initial debond stress, σ_d.

COMPARISON WITH MISMATCH-STRAIN CRITERION

A simple debonding criterion has been proposed such that debonding occurs when the mismatch in the axial strain between the fiber and the matrix reaches a critical value [12]. Based on this criterion, solutions for progressive debonding with friction along the debonded interface have been derived [13] which are reviewed and compared with the present results as follows.

When the bridging stress reaches the initial debond stress, σ_d, debonding initiates at the crack surface, and the critical mismatch strain, ε_d, is

$$\varepsilon_d = \frac{\sigma_d}{E_f} \tag{22}$$

During subsequent loading (i.e., $\sigma_O > \sigma_d$), debonding extends underneath the surface, and the mismatch strain at the end of the debonding zone remains ε_d, such that

$$\varepsilon_d = \frac{\sigma_{fd}}{E_f} - \frac{\sigma_{md}}{E_m} \tag{23}$$

where σ_{fd} and σ_{md} are the axial stresses in the fiber and the matrix at the end of the debonding zone which satisfy the mechanical equilibrium condition described by Eq. (1). Combination of Eqs. (1), (22), and (23) yields

$$\sigma_{fd} = \frac{V_f E_f \sigma_o + V_m E_m \sigma_d}{E_c} \tag{24}$$

The debond length, h, can be obtained from Eqs. (4a) and (24), such that

$$h = \frac{aV_m E_m (\sigma_o - \sigma_d)}{2E_c \tau} \tag{25}$$

Equation (25) is identical to the results obtained from both the energy-based [Eq. (16)] and the strength-based [11] criteria. Both u_O and u_{debond} have also been derived using the mismatch-strain criterion [13], and they are identical to those obtained in the present study.

THE STRENGTH-BASED CRITERION

For the strength-based criterion, debonding occurs when the interfacial shear strength, τ_s, is reached. A difference has been noted between debonding at the crack surface and debonding underneath the crack surface [18]. Whereas the matrix is stress-free at the crack surface, it is subjected to axial stresses underneath the crack surface due to the stress transfer from the fiber to the matrix. Hence, the magnitude of the interfacial shear stress induced by a loading stress σ_d on the fiber at the crack surface is different from that induced by an axial stress σ_d in the fiber underneath the crack surface. Assuming that the axial stresses at the end of the debonding zone are σ_{fd} and σ_{md} respectively in the fiber and the matrix, the relation between σ_{fd} and σ_d can be derived using the strength-based criterion and this is shown as follows.

At the end of the debonding zone, the interfacial shear stress can be analyzed using the following procedures. First, tractions of $E_f\sigma_{md}/E_m$ and σ_{md} are imposed on the fiber and the matrix, respectively (Fig. 2a). This would result in a uniform axial strain σ_{md}/E_m in the composite, and no interfacial shear stress is induced. Then, a traction of $\sigma_{fd}-E_f\sigma_{md}/E_m$ is imposed on the fiber, and this would induce the interfacial shear stress (Fig. 2b). Combining the above two procedures, the tractions imposed on the fiber and the matrix are σ_{fd} and σ_{md} respectively (Fig. 2c). Hence, the interfacial shear stress at the end of the debonding zone is equivalent to that if a traction of $\sigma_{fd}-E_f\sigma_{md}/E_m$ is imposed on the fiber alone at the crack surface. To satisfy the debonding condition at the end of the debonding zone, the following relation is hence required:

$$\sigma_{fd} - \frac{E_f\sigma_{md}}{E_m} = \sigma_d \tag{26}$$

It is noted that Eq. (26) can also be obtained by combining Eq. (22) with Eq. (23). Hence, the strength-based criterion yields the same results as those using the mismatch-strain criterion.

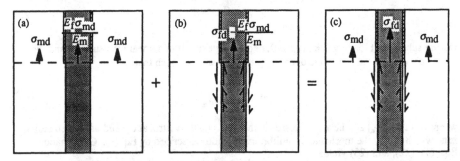

Fig. 2. The procedures in deriving the interfacial shear stress at the end of the debonding zone: (a) tractions of $E_f\sigma_{md}/E_m$ and σ_{md} are imposed on the fiber and the matrix, respectively, at the end of the debonding zone resulting a uniform axial strain in the composite, (b) a traction of $\sigma_{fd}-E_f\sigma_{md}/E_m$ is imposed on the fiber, and the interfacial shear stress is induced, (c) combination of the above two procedures results in the condition of tractions at the end of the debonding zone.

CONCLUSIONS

Using the energy-based criterion, progressive debonding at the fiber/matrix interface with friction along the debonded interface is analyzed for fiber-reinforced ceramic composites. It is noted that the displacement term involved in calculating the work done by load is the displacement of the composite due to interfacial debonding not the crack opening displacement. The present results for progressive debonding are identical to those obtained from the mismatch-strain criterion, in which interfacial debonding is assumed to occur when the mismatch in the axial strain between the fiber and the matrix reaches a critical value. Also, the mismatch-strain criterion is found to have the same physical meaning as the strength-based criterion.

ACKNOWLEDGMENTS

The author thanks Drs. P. F. Becher, E. Lara-Curzio and S. Raghuraman for reviewing the manuscript. Research sponsored jointly by the U.S. Department of Energy, Division of Materials Sciences, Office of Basic Energy Sciences, and Assistant Secretary for Energy Efficiency and Renewable Energy, Office of Industrial Technologies, Industrial Energy Efficiency Division and Continuous Fiber Ceramic Composites Program, under contract DE-AC05-84OR21400 with Lockheed Martin Energy Systems.

REFERENCES

1. A. G. Evans and R. M. McMeeking, *Acta Metall.*, 34, 2435-2441 (1986).
2. P. F. Becher, C. H. Hsueh, P. Angelini and T. N. Tiegs, *J. Am. Ceram. Soc.*, 71, 1050-1061 (1988).
3. C. Gurney and J. Hunt, *Proc. Roy. Soc. Lond.*, A299, 508-524 (1967).
4. Y. C. Gao, Y. W. Mai and B. Cotterell, *J. Appl. Math. and Phys. (ZAMP)*, 39, 550-572 (1988).
5. J. W. Hutchinson and H. M. Jensen, *Mech. Materials*, 9, 139-163 (1990)
6. C. H. Hsueh, *Mater. Sci. and Eng.*, A159, 65-72 (1992).
7. P. Lawrence, *J. Mater. Sci.*, 7, 1-6 (1972).
8. A. Takaku and R. G. C. Arridge, *J. Phys. D: Appl. Phys.*, 6, 2038-2047 (1973).
9. C. H. Hsueh, *Mater. Sci. and Eng.*, A123, 1-11 (1990).
10. S. V. Nair, *J. Am. Ceram. Soc.*, 73, 2839-2847 (1990).
11. B. Budiansky, A. G. Evans, and J. W. Hutchinson, *Int. J. Solids Structures*, 32, 315-328 (1995).
12. N. Shafry, D. G. Brandon and M. Terasaki, *Euro-Ceramics*, 3, 3.453-457 (1989).
13. C. H. Hsueh, *J. Mater. Sci.*, 30, 1781-1789 (1995).
14. C. H. Hsueh, submitted to *Acta Metall. Mater.*
15. D. B. Marshall, B. N. Cox, and A. G. Evans, *Acta Metall.*, 33, 2013 (1985).
16. J. Aveston, G. A. Cooper, and A. Kelly, "The Properties of Fibre Composites," pp.15-26, Conference Proceedings, National Physical Laboratory, Guildford, IPC Science and Technology Press Ltd., (1971).
17. L. N. McCartney, *Proc. R. Soc. Lond.*, A409, 329-350 (1987).
18. J. K. Kim, C. Baillie, and Y. W. Mai, *J. Mater. Sci.*, 27, 3143-3154 (1992).

R-CURVE RESPONSE OF SILICON CARBIDE WHISKER-REINFORCED ALUMINA: MICROSTRUCTURAL INFLUENCE

E. Y. Sun, C. H. Hsueh, and P. F. Becher
Metals and Ceramics Division, Oak Ridge National Laboratory, Oak Ridge, TN 37831-6068

ABSTRACT

Rising fracture resistance with crack extension (*R*-curve response) can lead to improvements in the mechanical reliability of ceramics. To understand how microstructures influence the *R*-curve behavior, direct observations of crack interactions with microstructural features were conducted on SiC whisker-reinforced alumina. The contribution of the dominant toughening mechanisms to the *R*-curve behavior of these composites is discussed using experimental and theoretical studies.

INTRODUCTION

Rising fracture resistance with crack extension (*R*-curve response) has been observed in SiC whisker-reinforced alumina and explained by crack-bridging mechanisms [1-9]. *R*-curve behavior predicted using constitutive models based on crack-bridging mechanisms agrees well with experimental results [4-9]. However, because of the experimental difficulty in measuring several parameters critical to modeling, a certain degree of ambiguity always exists in identifying the important toughening parameters, especially in the short-crack region (< 50 μm). The objective of the present study was to examine the fracture behavior of SiC whisker-reinforced alumina at a micro-scale level, and to obtain a better understanding of the toughening mechanisms in these materials by *in-situ* observations of crack interactions with microstructural features when subjected to an applied stress.

EXPERIMENT PROCEDURE

Materials

SiC whisker-reinforced alumina composites were prepared by mixing the ~ 0.8 μm diameter SiC whiskers (Advanced Composite Materials Corporation, Greer, SC) with alumina powder suspended in hexane in a shear mill, followed by rapid evaporation of the media to form a granulated, friable mixture. The mixtures were hot pressed in graphite dies at a temperature of 1800°C to obtain composites with densities ≥ 98 % of theoretical. Densification was controlled to maintain a grain size of 1-2 μm in final composites containing ~ 20 vol.% whiskers.

R-Curve Measurement and Direct Observation of Fracture Behavior

R-curve responses of SiC whisker-reinforced alumina composites were investigated using an applied moment double cantilever beam (DCB) geometry with the crack plane oriented parallel to the hot-pressing axis. The DCB blanks had a dimension of 10 × 35 × 2.5 mm. One of the wide surfaces (10 × 35 mm) contained a ~1 mm deep by 1.5 mm wide centerline groove parallel to the length while the opposite surface was mechanically polished to facilitate observation of the crack. One end of the specimen was notched and precracked using a TiB$_2$

223

Figure 1. Applied moment DCB test module and the sample. The testing stage can be placed on an optical microscope stage or housed in the chamber of an SEM.

wedge. The wake of the precrack was removed with a low-speed diamond saw. The initial crack lengths employed in the present study were between 100-200 µm. Stainless steel loading arms were attached to the DCB specimens using an epoxy adhesive. Samples were carbon coated.

Samples were tested using a DCB test module, as shown in Figure 1. The testing stage employs a DC drive motor to generate the various applied loads, which are monitored by a semiconductor load cell. The load is applied through a bending moment on the loading arms. With this geometry, the applied stress intensity depends on the load but not on the crack length [10]. During the experiments, the testing stage was either placed on an optical microscope stage or housed in the chamber of a scanning electron microscope to enable both direct observation of the crack interaction with microstructures and measurement of crack opening displacements under applied loads. Details of sample preparation and testing procedures are reported in Reference 9.

RESULTS

Typical R-curve responses for the present SiC whisker-reinforced aluminas are shown in Figure 2(a). (Comparison between the experimental results and the prediction based on different models is discussed in the next section.) The initial crack (typically 100-200 µm in length) started to propagate at ~ 6 MPa√m and a steady-state toughness (~ 8 MPa√m) was approached at crack lengths of ~700 µm. Direct observation of crack propagation revealed that the crack tip was frequently deflected at whisker surfaces and then propagated along the whisker/matrix interface. Crack deflection was an operative process in these composites. If the embedded length of the whisker on one side of the crack plane is short, the crack then continued around the whisker-matrix interface and into the matrix ahead of the whisker. Otherwise, the crack tip appeared to arrest as it encountered the interface and then reappeared in the matrix ahead of the whisker, leaving the whisker intact and functioning as a bridging ligament. An example is shown in Figure 3(a) where the whisker is ~ 5 µm behind the tip. (The crack propagated from right to left in all the pictures). This type of bridging by intact whiskers has been defined as "frictional bridging" in previous studies [7-9]. It has been predicted by the bridging model that the R-curve rises quickly due to frictional bridging during the initial stages of crack extension and the size of the bridging zone is < 50 µm. *In-situ* observations revealed that a large fraction of such whiskers underwent "partial" debonding and then fractured without experiencing any pull-out, as shown in Figures 3(a) and 3(b). In Figure 3(a), debonding along the whisker/matrix interface is only observed at the right-side of the whisker. No debonding appears along the interface on the left-side of the whisker though it may have been covered by the carbon coating deposited on the specimen surface. In Figure

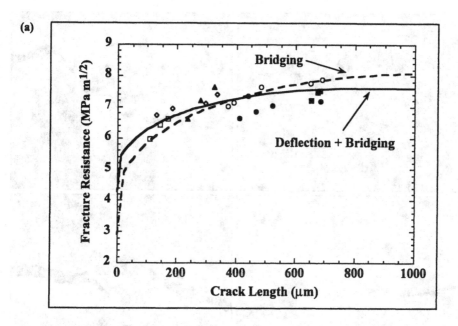

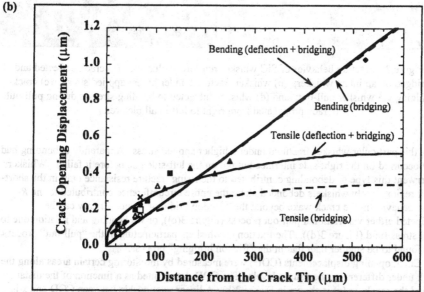

Figure 2 (a) *R*-curve response of SiC whisker-reinforced alumina and (b) crack opening displacements under dynamic loading. (Symbols are experimental data and curves represent calculations based on different toughening mechanisms described in Discussion: dashed lines are for solely bridging processes and solid lines include crack deflection contribution as well.)

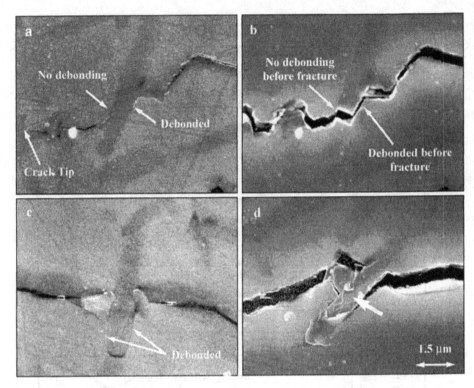

Figure 3. Fracture behavior of SiC whisker-reinforced alumina: (a) crack deflected and bridged by an intact whisker; (b) whisker fractured under higher applied stress; (c) crack-bridging via whisker pull-out; and (d) whisker subjected to bending stresses during pull-out. (Crack propagated from right to left in all pictures.)

3(b), this particular whisker fractured under a higher applied stress. Apparently debonding had only occurred on the right-side interface but not on the left-side one before it failed. Whiskers underwent this type of debonding contributed to the rising fracture resistance only in the short-crack region. Only whiskers debonded along the entire circumference contributed to the R-curve behavior in the crack wake beyond the "friction bridging" zone. Energy can be dissipated either via frictional pull-out process (Figure 3(c)), or via bending and rotation due to local stress field (Figure 3(d)). The fraction of whiskers participating in the "pull-out" process was found to be smaller than that in the "friction bridging" process.

Crack opening displacements (COD) were measured by monitoring certain areas along the crack under different applied stress levels. The COD data plotted as a function of the distance behind the crack tip (x) is shown in Figure 2(b). A linear relationship between COD and x is exhibited. This phenomenon is discussed later. It is also noted that the fracture resistance approached its steady-state value when the COD value at the end of the crack was about twice of the whisker diameter.

DISCUSSION

From the experimental evidence, it can be concluded that crack deflection, "frictional bridging" and pull-out process are active in silicon carbide whisker-reinforced alumina. However, it should be realized that among these three mechanisms, crack deflection increases the "intrinsic" toughness value but has no influence in the crack wake. In other words, crack deflection would shift the entire R-curve up-and-down. Based on this basic interpretation, the previously developed bridging model has been modified to incorporate a crack-deflection contribution to the material's toughness. Specifically, a significant increase in the K_0 value (the toughness at the crack tip) is used to reflect the increase in toughness by crack-tip deflection.

In addition, some other changes have been made in the modeling work based on the *in-situ* observations. First, in the previous calculation [9], debonding was assumed to occur at the interface of a particular whisker on both sides of the crack plane. However, the present experiments indicate that the whiskers more likely only debond on one side of the crack. Secondly, the expression of the crack profile is changed. A $u(x)$ function for simple uniaxial tensile stress condition was used in the previous work, where u is half of the COD value and x is the distance from the crack tip [9]. However, it is found that the COD values calculated using this method are much smaller than that observed experimentally, especially at large distance from the tip, as shown by the curves labeled as "tensile" in Figure 2(b). Thus a new expression of $u(x)$ is derived by analyzing the deformation of a beam under an applied moment, which is more appropriate for the stress condition experienced by a DCB sample. The COD values calculated as a function of x using this new expression is plotted in Figure 2(b) and labeled as "bending".

With these modifications, the R-curve for the present 20 vol.% SiC whisker-reinforced alumina is calculated using both the "deflection + bridging" model and the "bridging" model. The K_0 value used in the "deflection + bridging" model is 4.2 MPa m$^{1/2}$ instead of the 2.8 MPa m$^{1/2}$ used earlier in the "bridging" model [9]. This 50% increase in K_0 is estimated from Faber and Evans' work [11]. In both calculations, it is assumed that 50% whiskers participate the frictional bridging and pull-out processes, considering the random orientation of whiskers in planes perpendicular to the hot-pressing axis. Also, 0.25 is used as the pull-out length to debonding length ratio in both calculations to account for (1) the fact that a bridging whisker can fracture at any point along the debonded portion; and (2) the smaller fraction of whiskers participating in the pull-out as compared to the frictional bridging process. It is found that a higher whisker strength is required in the "bridging" model (9.2 GPa) than that in the "deflection + bridging" model (7.0 GPa) to obtain similar toughness values in the long-crack region. (All the other parameters are remained the same in both calculations except the whisker strength and the K_0 value). Comparing the two curves in Figure 2(a), apparently the major difference lies in the short-crack region. Because of the lack of experimental data in the short-crack region (< 50 μm), it is difficult to justify the actual contribution of crack-deflection to the toughness increase at the crack tip. Future work will focus on development of new techniques to investigate fracture resistance in the short-crack region.

CONCLUSIONS

The R-curve responses of SiC whisker-reinforced aluminas were investigated by direct observation of crack interactions with microstructural features. It is found that crack-deflection and crack-bridging are the operative toughening mechanisms in the composites, with crack-deflection increasing the toughness at the crack tip and crack-bridging contributing to the rising portion of the R-curve. Frictional bridging and pull-out processes are studied in detail at a micro-scale level. Based on experimental observations, the previously developed crack-bridging model is modified to account for the crack-deflection contribution and to obtain an accurate crack profile. In addition, the present work indicates that to verify the contribution of the individual toughening mechanism to the R-curve behavior, one needs to study the fracture behavior in the short-crack region ($< 50 \ \mu m$).

ACKNOWLEDGMENTS

The authors thank Dr. M. K. Ferber and Dr. H. T. Lin for reviewing the manuscript. Research is sponsored by the U.S. Department of Energy, Division of Materials Sciences, office of Basic Energy Sciences, under contract DE-AC05-84OR21400 with Lockheed Martin Energy Systems and by an appointment of E. Y. Sun to the Oak Ridge National Laboratory Postdoctoral Research Associates Program, which is administered jointly by the Oak Ridge Institute for Science and Education and Oak Ridge National Laboratory.

REFERENCES

1. F. Krause, E. R. Fuller, and J. F. Rhodes, J. Am. Ceram. Soc., **73**, 559 (1990).
2. M. G. Jenkins, A. S. Kobayashi, K. W. White, and R. C. Bradt, J. Am. Ceram. Soc., **70**, 393 (1987).
3. J. Homeny and W. L. Vaughn, J. Am. Ceram. Soc., **73**, 2060 (1990).
4. P. F. Becher and G. C. Wei, J. Am. Ceram. Soc., **67**, C-267 (1984).
5. P. F. Becher, T. N. tiegs, J. C. Ogle, and W. H. Warwick in Fracture Mechanics of Ceramics, **7**, edited by R. C. Bradt, A. G. Evans, D. P. H. Hasselman, and F. F. Lange (Plenum Publishing Corp., New York, 1986) pp. 61-73.
6. P. F. Becher, C. H. Hsueh, P. Angelini and T. N. Tiegs, J. Am. Ceram. Soc., **71**, 1050 (1988).
7. C. H. Hsueh and P. F. Becher, Compos. Engng., **1**, 129-143 (1991).
8. P. F. Becher, J. Am. Ceram. Soc., **74**, 255 (1991).
9. P. F. Becher, C. H. Hsueh, K. B. Alexander, and E. Y. Sun, J. Am. Ceram. Soc., in press.
10. S. W. Freiman, D. R. Mulville, and P. W. Mast, J. Mater. Sci., **8**, 1527 (1973).
11. K. T. Faber and A. G. Evans, Acta Metall., **31**, 565 (1983).

FRACTURE OF PARTICULATE COMPOSITES WITH STRONG OR WEAK INTERFACES

A. LEKATOU, S.E. FAIDI, S.B. LYON and R.C. NEWMAN
University of Manchester Institute of Science and Technology
Corrosion and Protection Centre, Manchester, M60 1QD, UK

ABSTRACT

The tensile fracture of glass-bead/epoxy composites has been studied using a large number of glass volume fractions (q) from 0 to 0.25. Absorption of water from a saturated NaCl solution was used to destroy interface adhesion. The strength of the dry composites decreased up to q = 0.15, then became constant. The strength of the wet composites showed a plateau around q = 0.12, then an abrupt drop between 0.12 and 0.15, then became nearly constant. It is possible that the abrupt drop in strength occurs at the percolation threshold. However, electrical measurements on similar composites prepared with conducting (silver-coated) beads showed that the applicable percolation threshold (q_c) was ≥ 0.16, in agreement with the observation of anomalous water uptake only for $q \geq 0.18$. It is concluded that the abrupt drop in strength occurs below q_c, and is due to large finite interface clusters (or rather the more planar parts of such clusters) acting as nuclei for a progressive matrix fracture process. The plateau around q = 0.12 is due to favourable fibre-like interactions of smaller interface clusters, and gives the highest normalized (wet/dry) strength of about 0.9.

INTRODUCTION

Recent interest in the mechanical properties of particulate composites or porous solids has focused on their elastic properties [1,2] and on mainly two-dimensional analyses of fracture [3]. In two dimensions, percolation of weak interfaces or pores has an obvious significance for fracture: the material falls apart at the percolation threshold. In three dimensions this is not the case, and it is not clear that any abrupt transition will be observed. Certainly nothing remarkable occurs in porous inorganic glasses [4], but these are so brittle that single pores or very small pore clusters cause a large reduction in strength (i.e. exceed the intrinsic defect size). In the present work we have used a tougher system (epoxy resin reinforced with glass beads) to examine the effect of particle volume fraction and interface strength on the fracture of smooth tensile specimens. The region around the percolation threshold (0.157 for a continuum) is of particular interest.

EXPERIMENTAL PROCEDURE

Full details have been given elsewhere [5]. The composite system was epoxy resin reinforced with silanized sodalime glass spheres from 1 to 20 μm in diameter with a median of about 7 μm. The volume fractions of glass (in %) were 0, 3, 6, 9, 2, 15, 18, 21 and 25. Similar spheres with a silver coating were used for conductivity measurements to test the applicability of the percolation model to this system. Tensile measurements were made on "I"-section tensile specimens with gauge dimensions 70 x 12 x 3 mm, using a strain rate of 10^{-3}/s. Water uptake and electrical resistance measurements were made on discs of the same thickness. Water was absorbed from a saturated NaCl solution at 40°C (water activity: 0.75), to minimize osmotic interface damage due to glass dissolution. All mechanical and electrical measurements were made at room temperature. The immersion times for the tensile specimens were 18 and 70 days, corresponding to different degrees of osmotic interface damage at the higher glass volume fractions; the similarity of behaviour for these two immersion times was part of the evidence that the trends in composite strength were not due to some special effect associated with water exposure, but were indicative of general laws for particulate composites with this level of matrix toughness.

RESULTS AND DISCUSSION

<u>Electrical Measurements using Silver-Coated Glass Beads</u>

The high resistivity of all the composites (more than 10^{13} ohm cm) showed that there was no particle contact. However a clear threshold in the conductivity was evident at q = 0.16, as shown in Figure 1. This corresponds closely to the percolation threshold for a random biphase continuum [6]. By the time q reaches 0.25, the infinite cluster is dense enough that the conductivity would be about 10% that of bulk silver if there was perfect particle contact. The much smaller measured increase in conductivity between q = 0.16 and q = 0.25 (a factor of 20) may thus be considered to result from conduction through films of resin about 0.03 μm thick between the particles of median diameter 7 μm. This calculated thickness was consistent with the fact that the polymer film between the particles was at the limit of resolution in the SEM.

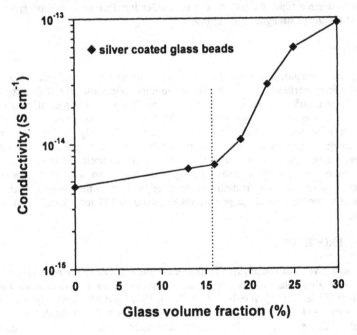

Figure 1   Dry conductivity as a function of glass volume fraction for 3 mm-thick discs of epoxy filled with silver-coated glass beads.

The important aspect of Figure 1 is that it validates the application of percolation concepts to the fracture problem. The surprisingly close agreement between the experimental conductivity threshold and that for a biphase continuum may result in part from the wide particle size distribution in the composite, and also from the high viscosity of the resin during mixing and curing. Many calculations of the elastic properties of random binary media have used a Brownian motion assumption that allows the particles to bounce off one another, leading to a marked ordering of the system [2]. Our situation corresponds to Torquato's alternative 'sticky sphere' assumption.

Water Uptake

Despite the low water activity in the saturated NaCl solution (about 0.75), an even lower one can be created by dissolution of the glass to form sodium silicate. The resulting osmotic pressure is incapable of opening a cavity at an isolated particle-matrix interface, but can help to open a narrow, continuous channel within a connected cluster of interfaces. This allows enhanced water uptake above the percolation threshold, as shown in Figure 2. The reason for including this figure is that no anomalous water uptake occurs at $q = 0.15$, so one cannot ascribe a transition in fracture behaviour at that composition to some special effect of water.

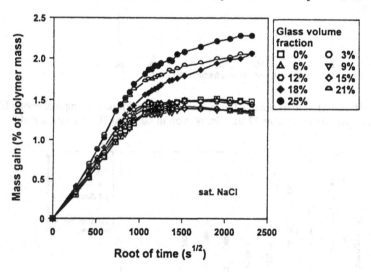

Figure 2   Water uptake for 3 mm-thick discs of epoxy filled with various volume fractions of glass beads, exposed to saturated NaCl solution at 40°C.

Tensile Fracture

The fracture stress of the dry composites (Figure 3) showed a sharp drop between $q = 0.03$ and $q = 0.06$, which may be ascribed to the appearance of specific multiparticle clusters. The defect size represented by a single particle is just too small to cause fracture prior to the propagation of the progressive damage process that leads to failure of the resin (e.g. for a 20

μm particle with a failed interface, the K value cannot be higher than 0.3 MPa m$^{1/2}$ which is lower than reported values of K_{Ic} for epoxies [7]). A likely crack nucleus at q = 0.06 is a cluster of about 7 average particles in a disc-like arrangement, for which we calculate a frequency of about one per cm^3 (the volume of the gauge length is 2.5 cm^3).

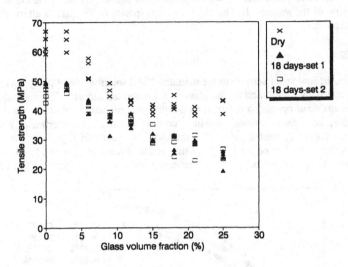

Figure 3   The effect of glass volume fraction on the tensile fracture strength of epoxy filled with glass beads, tested dry or after 18 days immersion in saturated NaCl solution at 40°C.

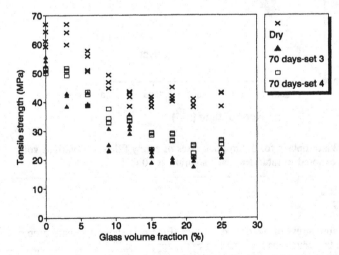

Figure 4   The effect of glass volume fraction on the tensile fracture strength of epoxy filled with glass beads, tested dry or after 70 days immersion in saturated NaCl solution at 40°C.

An interesting feature of the dry behaviour is the constant strength between q = 0.12 and q = 0.25. Apparently the strength is controlled by the interface failure process and not by subsequent crack propagation. Alternatively one may appeal to the relatively slow increase with q of the *crack-like planar features* within the largest particle clusters. In so far as fracture mechanics applies to the fracture stress, one should also remember that the modulus doubles between q = 0.06 and q = 0.25 [5], so a constant value of K_{Ic} would imply a steadily decreasing value of G_{Ic} [7]. Finally, the fracture occurs after a period of damage and must be considered within the context of a system of interacting cracks (though less so than in the case of the wet material).

For the wet material, new detail was observed in the dependence of fracture strength on the particle volume fraction q, as shown in Figures 3-5. This detail was observed for two batches of material at each of two immersion times and for different levels of interface swelling and damage, as mentioned earlier. Once again, there was a sharp drop in strength between q = 0.03 and q = 0.06, except for one sample set (70 days set 3) where the drop occurred between q = 0 and q = 0.03. The really significant behaviour occurred in the range q = 0.09 to q = 0.15, comprising a plateau followed by a sharp fall. Application of Duncan's multiple range test [8] showed that the drop between q = 0.12 and q = 0.15 was significant at better than the 99% level in each case, while the strengths at q = 0.09 and 0.12, and at q = 0.15, 0.18 and 0.21, could not be distinguished statistically. This result implies that the best normalized (wet/dry) strength is achieved at q = 0.12, which is indeed the case:

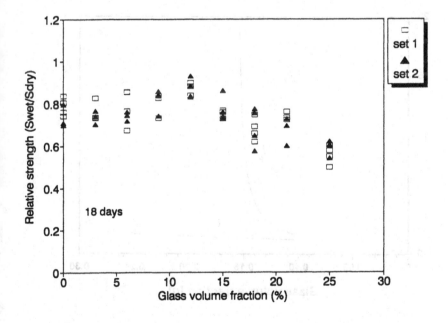

Figure 5   The ratio of wet to dry strength for the data of Figure 3.

The relatively beneficial effect observed at q = 0.12 is readily interpreted using concepts of crack interaction. Particle clusters whose largest planar extension lies parallel to the tensile axis act as fibre-like crack stoppers. One is tempted to ascribe the abrupt fall in strength between q = 0.12 and q = 0.15 to the appearance of a percolating or nearly percolating interface cluster, by analogy with a two-dimensional system [3], but this is not justified without considering the actual cluster geometry in 3 dimensions. Figure 6 shows the calculated dependence of the mean cluster spanning length and the mean blob (high-density particle cluster) diameter on the particle volume fraction q. Clearly the blob size varies so slowly with q that no abrupt transitions are to be expected; if these occur, they are expected at low q values and relate to the extreme value of blob size, which is sample-size dependent as discussed earlier. The spanning length shows the right kind of dependence on q, and simplistically one could rationalize a transition in the range q = 0.12 to q = 0.15 (just below the percolation threshold q_c). However the large clusters are such tenuous objects that it is not permissible to regard them as crack nuclei of characteristic size l_{av}. The main effect of increasing q from 0.12 to 0.15 is to connect small clusters (themselves composed of separated blobs) by 'one-dimensional' strands of particles, but since the particles themselves are not zero-dimensional but have a significant size and mechanical effect in their own right, the situation is not quite as bad as it would be in a true random biphase continuum. We conclude that the sharp increase in l_{av} (and thus the extreme value of l, l_{max}) between q = 0.12 and q = 0.15 causes the fracture transition. However, since not all of a cluster can function as a crack (otherwise the strength would be zero for q > q_c), the critical crack must be assembled from several favourably oriented subunits within a large cluster.

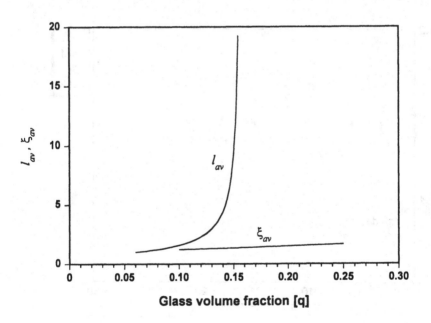

Figure 6   Approximate dependence of the mean cluster spanning length (l_{av}) and the mean blob diameter ξ_{av} (high-density particle cluster diameter) on the particle volume fraction q.

The particle-resin interfaces, though strong in the dry state, do eventually fail and appear on the fracture surface as shown in Figure 7a. The strong interaction of the crack with the particles is evident from the roughening and 'tail' formation behind particles on this surface [7]. The fracture surfaces of the wet composites showed little interaction of matrix cracks with the particles and were much flatter than those seen in dry material - compare Figures 7a and 7b.

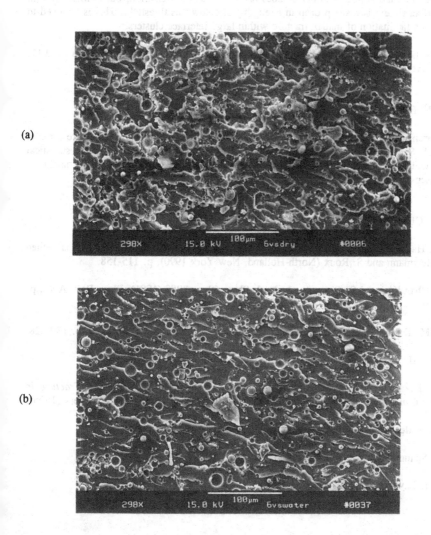

Figure 7   Fracture surfaces of composites with 6% glass: (a) dry; (b) after 18 days immersion in saturated NaCl solution.

CONCLUSIONS

1.   Particulate composites that conform closely to a percolation model have been prepared and characterized both electrically and mechanically.

2.   The percolation threshold *per se* does not set a fracture transition, but in material with weak interfaces there is a sharp drop in strength just below this threshold. This is believed to be due to the facilitation of matrix fracture within large interface clusters.

3.   The best normalized (wet/dry) strength occurs at a particle volume fraction of 0.12, owing to favourable interactions between relatively small interface clusters.

ACKNOWLEDGEMENTS

This research was sponsored by the Office of Naval Research (USA). The authors are grateful to Dr. A.J. Sedriks for his interest and support. Dr. P.J. Laycock carried out the statistical analyses, which will be published in due course. The authors are grateful to Karl Sieradzki for helpful discussions.

REFERENCES

1.   A. Hansen, *Disorder*, in Statistical Models for the Fracture of Disordered Media, edited by H.J. Herrmann and S. Roux (North-Holland, New York 1990), p. 115-158.

2.   I. Chan Kim and S. Torquato, J. Appl. Phys. **69**, p. 2280 (1991); Phys. Rev. A **43**, p. 3198 (1991).

3.   P.M. Duxbury, *Breakdown of Diluted and Hierarchical Systems*, in ref. 1, p. 189-228.

4.   D.P.H. Hasselman and R.M. Fulrath, quoted in ref. 3.

5.   A. Lekatou, S.E. Faidi, S.B. Lyon and R.C. Newman, *Elasticity and Fracture in Particulate Composites with Strong and Degraded Interfaces*, J. Mater. Res., in press (1996).

6.   R. Zallen, Physics of Amorphous Solids, Wiley, New York, 1983.

7.   J. Spanoudakis and R.J. Young, J. Mat. Sci. **19**, pp. 473 and 487 (1984).

8.   P.J. Laycock, unpublished research.

THE NON-LINEAR BEHAVIOUR OF A MODEL RELATED TO THE FRACTURE OF COMPOSITE MATERIALS

Alberto Varone * Franco Meloni,* Francesco Ginesu ** and Francesco Aymerich,**
*INFM-Dipartimento di Scienze Fisiche, Università di Cagliari, I-09124 Cagliari, Italy
** Mechanical Engineering Department, University of Cagliari, Italy

ABSTRACT

The fracture of a fibre reinforced graphite peek polymer is considered when dynamic load is applied. The energy-time diagram shows various critical points during cyclic tensile loading. A quantitative analysis based on the study of the fractal dimension of the chaotic attractor in the representative phase space will allow us to individuate the physical parame ters responsible of the stability range of the material. A new simple model has been derived and successfully tested from a non-linear chaotic scheme. The present approach appears very promising for a future interdisciplinary study for evaluating the physical response of real massive materials under stress.

INTRODUCTION

The change of stiffness of a graphite reinforced fibre when subject to axial periodically applied fatigue has recently been studied by using a non-linear set of equations [1]. The results, mainly qualitative, show that it is possible to enter various apparently different field of research like mechanical engineering and chaotic algorithm in a common physical vision of complex non-conservative energy behaviours [2]. The argument seems to offer a number of soliciting aspects related to the macroscopic response of real massive materials ranging from a comprehensive description of the parameters leading to the fracture and, more important, to the prevision of the future behaviour starting from the analysis of the initial conditions. The parallel solution of the problem as seen by the chaotic theory induces a direct analogy with an order/disorder transition in a bifurcation map [3]. Moreover the theory is able to individuate in a rigorous analytical scheme the degree of freedom the model simulating the real phenomenon in order to take in the right account the responsible parameters and, consequently, to minimize the set of equations to be solved. The reduction of the complexity comes as a direct consequence and allows to individuate the mechanical aspects in the massive material to avoid unexpected fractures in the assembled structures. The aim of the present work is to draw, after a description of the experimental process, a theoretical method able to investigate the mechanical properties in a scenario where the typical terms of the chaos theory are the key players. The theory presently introduced, at difference with the previous initial study [1], must work in the absence of the detailed knowledge of the system dynamics. In this way we shall define the resulting attractor in the phase space by measuring its fractal dimension as well as the topological characteristics resulting from the experimental time-series of data. The final goal will allow the development of a common field of interactive research among engineers and physicists in the area of model building based on signals from non-linear systems [4].

Mat. Res. Soc. Symp. Proc. Vol. 409 ® 1996 Materials Research Society

EXPERIMENT

A graphite fibre reinforced polyetheretherketone (peek) is a composite laminate material made by polymers reinforced with parallel carbon fibres oriented by Θ with respect to the direction of the applied sinusoidal force. The angles may assume different values and the corresponding specimen has the notation $[\Theta_1/\Theta_2]ns$, where ns accounts for the periodic sequence [5]. The material used was ACP2 prepared by Fiberite Europe, consisting of AS4 graphite fibres and peek resin. The specimens tested consisted of angle-ply laminates with 16 layers oriented along the 35° and −35° directions with respect to the load application axis (stacking sequence $[-35°/35°]4s$). Tests were carried out in servo hydraulic $\pm 250kN$ machine under load control. Fatigue loading was sinusoidal tension-tension with $R = s_{min}/s_{max} = 0.1$ and frequency 3 Hz. During the test, signals from the load cell, the extensometers (longitudinal and transverse) and a thermocouple on the surface of the specimen were continuously monitored by means on a 13 bit digital voltmeter. In this way it was possible to evaluate at regular time intervals (100 cycles) some integral parameters such as the stiffness and the hysteresis cycle area, which are reliable indicators of damage progression before fracture. The measure of these parameters, especially of the latter, the most sensitive to damage, is subject to electrical and mechanical noises. Scatter in the results can be reduced by digitally filtering the signals; a low pass filter with a cutoff frequency of 40 Hz was applied in this series of tests.

METHOD

The theoretical study starts with the analysis of the time-series data extracted by the mechanical experiment. This set has been collected in such a way to avoid the presence of the experimental noise due to the characteristics of the measuring instrument by using a low pass filter. The elimination of an extra-perturbing effect is clearly more easy to obtain in linear-response phenomena by using Fourier transform analysis. In a chaotic ambience the separation is more hard to get and the various methods [4] are based on the idea that the signal of interest has a low-dimension chaos with a specific geometric structure in its phase space. The general problem accounts for a description based on the procedure [6, 7, 8, 9] that defines the minimum number of the variables and, as a consequence, of the equations able to describe the physical problem. We have to compute the delay time τ and D, the minimum number of dimensions to define the phase space embedding the attractor.

τ measures the average mutual information that implies that the time between samples is so long that the orbits are non-correlated in an information-theoretical sense. τ is used to write a D-dimensional vector [10]:

$$y(n) = x(n), x(n+T), ..., x(n+(D-1)T \tag{1}$$

whose components describe a coordinate system in which is completely defined the attractor related to the physical problem. T is an integer multiple of τ. Now we must construct a Euclidean space RD large enough to contain the set of points of dimension D unfolded without ambiguity. This means that the fact that two points are close each other does not depend of the value of D. The problem is related to the elimination the false crossing of the orbits with itself for having projected the attractor into a too low dimensional space. We adopt the method of the false-neighbours [11] to get the embedding dimension D_E. Taking

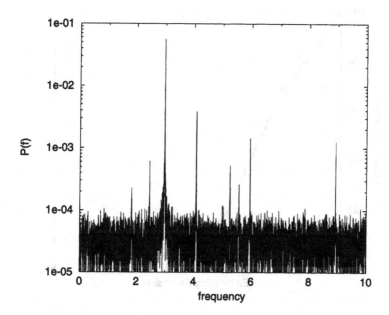

Figure 1: Fourier spectrum of the experimental data series. The highest peak is located at the drive frequency of 3 Hz.

in the right account the possible causes of error coming mainly from the uncontrolled noise in the experimental series of data, we arrive, with the previously obtained information, to compute the dimension D of the problem. Figure 1 reports the experimental Fourier analysis constructed by using a series of 100,000 points. The broad spectrum does not guarantee sensitivity to the initial conditions, but it is a reliable indicator of chaos. To discuss the topological characteristics of the system we calculate the percentage of false neighbours as a function of the embedding dimension. The result of the calculations is reported in Figure 2 from which we assume $D_E = 5$. The dynamical system may then be described by three non-linear equations.

The other mathematical indicator to compute is τ that must not be too large or too small because in both cases the experimental data will be unphysically uncorrelated. Figure 3 reports the curve of $I(\tau)$, the mutual information of the experimental point sequence as a function of time delay τ that makes the data statistically dependent. The first minimum of $I(\tau)$ defines the expected value [4]. Once the time delay τ and the embedding dimension D_E have been calculated, the dimension of the attractor is the last order of business. The numerical evaluation has been subject of many discussions [4]. In this work we use the method suggested by Grassberger and Procaccia [12] which is more general than that used to define the space-phase dimension as the integer number of quantities necessary to fully identify the dynamical state of the system in any instant. In dissipative systems the concepts to be used must consider fractal quantities as invariant describing how the

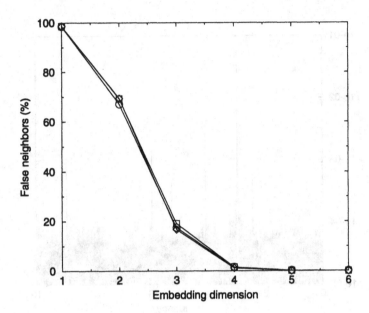

Figure 2: Percentage of global false nearest-neighbour as a function of the embedding dimension D_E. The model requires dimension $D_E = 5$ to unfold the attractor. The picture refers to three equally oriented samples. It is impossible to detect appreciable differences among them.

sample of points along an orbit tends to be spatially distributed. The time evolution of the system is not directly related to the value of the attractor, but some conjectures relating the fractal dimension and the stability of the dynamical properties have been discussed in the last years [13]. We calculate the dimension of the strange attractor $D = 1.8 \pm .1$. This last result ensures the chaotic behaviour of the physical phenomenon and the necessity to use non-linear dynamical scheme to build a model that could describe the dynamical process. To a similar result we may arrive by using a model equation previously presented [1].

$$[M]\ddot{x} = -[G]\dot{x} - [K]x + F_0 \sin(\omega t) + N(\Theta) \qquad (2)$$

where the terms in brackets are related to the global mass, the hardness and to the dislocation function, and the last nonlinear term in the previous equation refers to the orientation of the fibres Θ in the sample and is the responsible of the chaotic dynamical evolution of the model.

$$\begin{aligned}
\dot{x} &= v \\
\dot{v} &= -v - x + F_0 \sin \Phi + N(\Theta) \\
\dot{\Phi} &= \omega
\end{aligned} \qquad (3)$$

The qualitative approach is now supported by the analytical procedure and the next step must consist in the exact definition of the correlation between macroscopic thermodinami-

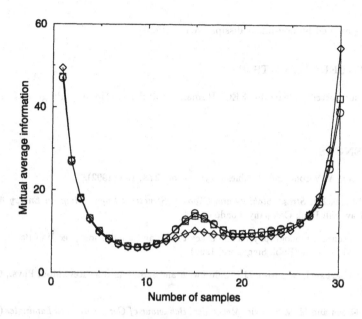

Figure 3: Average mutual information $I(\tau)$ as a function of the time delay. The first minimum indicating τ is at the value of 9 sample steps. As in Fig. 2 we analyze three equally oriented samples for which there is a good agreement in the left side of the curves where the information is taken off. The noise causes the little differences in the region corresponding to longer correlation times.

cally dependent terms in Eq. 2 and the fiber orientation Θ.

CONCLUSIONS

Experimentally controlled phenomena like the deformation and consequent fracture of real materials include a quantity of responsible terms to be taken into account that makes generally impossible to define the dynamical behaviour. The presence of non- linear components increases the difficulty to fit a model into a mathematical framework. In this work we have shown how powerful may be the modelling of a non-linear evolution of a physical system from the reconstruction of a D-dimensional phase space. We started from the analysis of a time-series data obtained by the experiment to determine the number of necessary terms to be taken into account in the analytical description of the phenomenon. The theory is able to define the attractor which, if with fractal dimension, signs the chaotic behaviour of the system. The method may be directly applied to many chaotic processes once the experiment provides a time-sequential array of data. The particular application to real composite materials is a furthermore signal of the interdisciplinary aspect of the study which uses the computational part to give an analytical definition of complex engineering problems governed by non-linear dissipative equations.

ACKNOWLEDGEMENTS

Work supported in part by EEC, Human Capital and Mobility Program CHRX-CT 930153.

REFERENCES

1. F. Meloni, A. Varone and F. Ginesu, MRS Proc. **278**, 135 (1992).

2. M.S. El Naschie, *Stress, Stability and Chaos in Structural Engineering: an Energy Approach* (McGraw-Hill Book Company, London, 1990).

3. For an accurate bibliography on the chaos theory and applications, see Hao Bai-Lin, *Chaos II* (World Scientific Publ.,Singapore, 1990)

4. H.D.I. Abarbanel, R. Brown, J.J. Sidorowich and L.S. Tsimring, Rev. Mod. Phys., **65**, 1331 (1993).

5. W.A. Green and M. Micunovic, *Mechanical Behaviour of Composites and Laminates* (Elsevier Appl. Sc.,London, 1987).

6. J.P. Eckmann and D. Ruelle, Rev. Mod. Phys. **57**, 617 (1985).

7. R. Mae', in *Dynamical Systems and Turbulence*, Warwick 1980 edited by D. Rand and L.S. Young, Lecture Notes in Mathematics Vol. 898 (Springer, Berlin, 1981), p. 230.

8. F. Takens, in Ref [7], p. 366.

9. T. Sauer, J.A. Jorke and M. Casdagli, J. Stat. Phys. **65**, 579 (1991).

10. H.D.I. Abarbanel, T.A. Carroll, L.M. Pecora, J.J. Sidorowich and L.S. Tsimring, Phys. Rev. E **49**, 1840 (1994).

11. M.B. Kennel, R. Brown and H.D.I. Abarbanel, Phys. Rev. A **45**, 3403 (1992).

12. P. Grassberger and I. Procaccia, Phys. Rev. Lett. **50**, 346 (1983); Physica D **9**, 189 (1983).

13. J.L. Kaplan and J.A. Jorke, in *Functional Differential Equations and Approximations of Fixed Points*, edited by H.O. Peitgen and H.O. Walther, Lecture Notes in Mathematics Vol. 730 (Springer, Berlin, 1979), p. 228.

INTERFACIAL FRACTURE BETWEEN BORON NITRIDE AND SILICON NITRIDE AND ITS APPLICATIONS TO THE FAILURE BEHAVIOR OF FIBROUS MONOLITHIC CERAMICS

D. KOVAR*, G.A. BRADY*, M.D. THOULESS**, J.W. HALLORAN*
*Materials Science and Engineering Dept., **Mechanical Engineering and Applied Mechanics Dept., University of Michigan, Ann Arbor, MI 48109-2136

ABSTRACT

Mechanical tests show that fibrous monolithic ceramics can exhibit graceful failure in bending due to marked crack deflection at silicon nitride/boron nitride interfaces. Model sandwich specimens were manufactured from polycrystalline silicon nitride with interphases consisting of mixtures of silicon nitride and boron nitride. The interfacial fracture resistances of the specimens were measured, and the results were correlated to the failure behavior of fibrous monolithic ceramics. It was found that the three modes of failure observed in fibrous monoliths can be correlated with the interfacial fracture resistance. In addition, anisotropy in the properties of the interphases were found to have a profound influence on crack deflection.

INTRODUCTION

Fibrous monolithic ceramics are low cost, ceramic composites that show great promise for use in high-temperature structural applications. Although their architecture can be tailored for specific applications, in this paper we focus on properties of fibrous monoliths with uniaxially aligned filaments of a strong phase surrounded by a coating or interphase of a weak material. The filaments of cell material, in this case, are polycrystalline Si_3N_4 while the weak interphase materials are mixtures of silicon nitride and boron nitride.

EXPERIMENTAL PROCEDURE

Details of the processing of fibrous monoliths are given elsewhere[1] and thus only a brief outline is given here. The cell materials for fibrous monoliths were prepared by mixing silicon nitride (with 6 wt% yttria and 2% alumina additives) with an ethylene co-polymer such that the volume fraction of solids was approximately 51%. The interphase for the filaments was prepared in a similar manner from a combination of silicon nitride, boron nitride, and co-polymer. The ratio of Si_3N_4 to BN was varied from 0-80 vol% of the total solids in the interphase material.

A feedrod, approximately 100 mm long and 20 mm in diameter, was assembled by molding a cladding of the interphase material to a cylindrical core of cell material; the rods were 80% cell material and 20% cladding. The feedrod was then extruded at 160°C through a 320 μm orifice into a fine, flexible filament such that the geometry of the feedrod was maintained but now at a much finer scale. The filament was cut, laid up by hand into a die, and laminated above the softening point of the polymer into rectangular billets 26 mm wide, 52 mm long, and 12 mm deep. The billets were then heated slowly in a vacuum to remove the organics and polymers leaving a ceramic billet which was hot pressed at 1750°C for 1 hour at 25 MPa. Bend specimens were prepared from these billets according to Military Standard 1942b.

Specimens for interfacial fracture tests were prepared from the same mixtures of polymers, organics, and ceramics as the fibrous monoliths. However, rather than mold the materials into a feed rod, the interphase materials were molded into thin sheets and then placed between two blocks of the silicon nitride-polymer mixture and then warm-laminated at 160°C. Binder removal and hot pressing conditions were identical to those used for the fibrous monoliths. The resulting billets consisted of a thin sheet of interphase material containing a mixture of Si_3N_4 and BN (0-80 % Si_3N_4) between thick blocks of Si_3N_4. The billets were cut and ground into specimens approximately 50 mm long, 5 mm thick, and 3 mm wide. One side surface was polished to facilitate the observation of cracking of the interface during testing.

Flexural strengths of the fibrous monoliths were determined with a fully articulating four-point fixture in displacement control at a rate of 0.5 mm/min. The inner and outer spans were 20 mm and 40 mm, respectively.

Considering that the delamination failure observed in fibrous monoliths is driven by a mixed-mode stress intensity, and because the fracture resistance of interfaces often varies as a function of mode mixity, it was important to characterize the interfacial fracture resistance between silicon nitride and the interphase materials at a variety of nominal phase angles.* A schematic summary of the two methods that were used in this study are shown in Figure 1. The first method is a slight modification of the well-known four-point-bending test introduced by Charalambides *et al.*[2] Rather than use the bi-layer specimen described by these authors, however, a sandwich specimen was used. The major advantage of the sandwich specimen is, unlike the bi-layer specimen, residual stress that may be present due to thermal expansion mismatch between materials does not influence the measurement of the fracture resistance.[3]

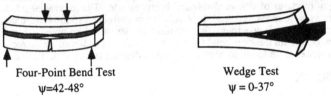

Four-Point Bend Test
$\psi = 42\text{-}48°$

Wedge Test
$\psi = 0\text{-}37°$

Figure 1.   Schematic illustrations of two types of tests conducted on sandwich specimens to measure the interfacial fracture resistance between Si_3N_4 and the interphase.

The test was conducted by applying a load to the specimen and measuring the center-point specimen deflection using an LVDT with a resolution of ± 0.2 μm. The initial portion of the load-deflection response was linear followed by a region where the load was relatively constant. Interrupted tests confirmed that the plateau in the load-deflection response occurred when crack propagation occurred along the interface. This steady-state load was used along with the specimen geometry to calculate the interfacial fracture resistance from:[4]

$$\Gamma_i = \left[\frac{3P^2 L^2 \left(1 - v^2\right)}{2E\left(h_1 + h_2\right)^3 b^2} \right] \left[\left(\frac{1}{n} + 1\right)^3 - 1 \right] \qquad (1)$$

where P is the steady state cracking load, L is distance from the outer loading span to the inner span, E is Young's modulus, n is the ratio of the heights of the lower and upper portions of the beam (h_2 and h_1, respectively), and v is Poisson's ratio.

Because the phase angle can only be varied over a narrow range of values using the bend test (42-48 degrees),[2] wedge tests were also performed to generate a range of interfacial fracture resistance measurements at phase angles varying from 0 (nominally pure mode I) to approximately 37 degrees. Sandwich specimens were machined so that the ratio of h_1 to h_2 was varied from 0.25 to 1.0 and a small notch was cut into one side of the specimen to create an asymmetric double cantilever beam. A silicon nitride wedge was then inserted into the notch to drive a crack out of the notch along the interface. The crack opening displacement at the notch and the crack length after crack arrest were measured using an SEM and used to calculate the steady state fracture resistance from the following equation:

---

* The nominal phase angle, ignoring modulus mismatch between the Si_3N_4 and the interphase, is defined as $\psi = \tan^{-1}(K_{II}/K_I)$ where K_{II} is the mode-two applied stress intensity and K_I is the mode-one applied stress intensity.

$$\Gamma_i = \frac{3En^3\delta^2}{8a^3(1-v^2)(1+n^3)} \qquad (2)$$

where δ is the crack opening displacement and a is the crack length.[5]

RESULTS

An optical micrograph of a cross-section of a fibrous monolith is shown in Figure 2. The Si_3N_4 cell shape is roughly oval in cross section with a width of about 280 μm and a height of about 115 μm. Although the cells are continuous down the length of the specimen, interruptions in the cells are sometimes visible due to imperfections in the alignment of the cells.

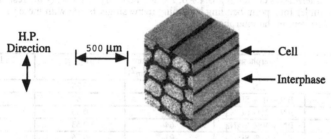

Figure 2.   Three-dimensional optical micrographs of an aligned fibrous monolith showing Si_3N_4 cells and Si_3N_4 + BN interphase.

Typical load-crosshead deflection responses for fibrous monoliths loaded in four-point bending are shown in Figure 3 as a function of Si_3N_4 content in the interphase and a summary of the mechanical properties are presented in Table I. The strength of fibrous monoliths increase with Si_3N_4 content in the interphase from 469±60 MPa for the material with the pure BN interphase to 600 MPa±106 MPa for the material with a 80% Si_3N_4-20% BN interphase. More importantly, three distinct modes of failure were observed. Type I behavior is characterized by non-catastrophic failure with significant retained load following an initial load drop (Figure 3). SEM micrographs show that extensive crack deflection and splintering occur both on the tensile surface and through the thickness of the bar (Figure 4a). Specimens that exhibit Type II behavior are slightly stronger but exhibit a much smaller retained load. Note that Type II specimens exhibit virtually no crack deflection on the tensile surface and more limited crack deflection through the thickness of the bar (Figure 4b). Type III specimens have the highest strength but are brittle and exhibit little or no crack deflection (Figure 4c). Since the composition of the cells is the same in all of the materials, the dramatic differences in the mechanical properties of the fibrous monoliths result from changes to the interfacial properties as the composition of the interphase material is varied.

Turning to the interfacial properties, no systematic variations in the fracture resistance with phase angle were noted over the range of phase angles that were measured (approximately 0-45°) However, the interfacial fracture resistance increases with additions of Si_3N_4 to the interphase. More importantly, it is seen that the qualitative differences between the fracture morphologies of the fibrous monoliths can be correlated to the fracture energy of the interface. For example, extensive crack deflection and delamination (Type I behavior) occurs only in the fibrous monolith with a low interfacial fracture resistance while no crack deflection occurs in the material with a high interfacial fracture resistance (Type III behavior). At intermediate values of

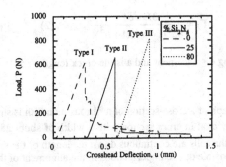

Figure 3.   Load is plotted as a function of crosshead displacement for fibrous monoliths with interphases containing BN + 0, 25, and 80% Si_3N_4 showing the response under four-point bending becoming increasingly brittle with increasing Si_3N_4 content in the interphase.

Cell	Interphase	Interfacial Fracture Resistance, Γ_i (J/m^2)	$\Gamma_i/\Gamma_{Si_3N_4}$	Failure Mode
Si_3N_4	BN	30	0.25	I
Si_3N_4	BN + 10% Si_3N_4	56	0.47	II
Si_3N_4	BN + 25% Si_3N_4	68	0.56	II
Si_3N_4	BN + 40% Si_3N_4	64	0.53	II
Si_3N_4	BN + 80% Si_3N_4	91	0.76	III

Table I.   A summary of the interfacial properties and failure modes for the various fibrous monoliths that were tested

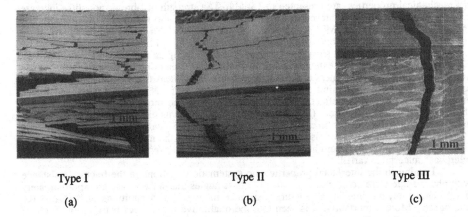

Type I   Type II   Type III

(a)   (b)   (c)

Figure 4.   SEM micrographs of flexural bars after testing with the tensile surface shown on the bottom and the side surface shown on top. Note that crack deflection decreases as the Si_3N_4 content in the interphase is increased from 0 to 80%.

interfacial fracture resistance, only limited crack deflection is observed through the thickness of the specimen.

The conditions for crack deflection at the cell-interphase boundary are usually based on the interfacial fracture resistance and the elastic mismatch between the cell and interphase materials (the Dundur's parameter, α) using the analysis of He and Hutchinson.[6] However, a direct application of this analysis using the measured values of interfacial fracture resistance, mode I fracture resistance of Si_3N_4, and the elastic properties of Si_3N_4 and interphase materials[7] does not properly predict the observed crack deflection; crack deflection occurs in Type II materials when the prediction says it should not. This discrepancy can be rationalized in terms of the large anisotropy that is observed in the fracture resistance of the BN interphase materials in directions parallel and perpendicular to the cell alignment direction. Figure 5a is an SEM micrograph of a crack impinging on a Si_3N_4/BN interface at 90°. The crack initiated from a saw-cut notch in the Si_3N_4 layer and grew unstably into the BN interphase at which point it was arrested. A closer examination of the crack in the BN (Figure 5b) reveals that the crack deflects numerous times perpendicular to the initial crack trajectory. In addition, profuse microcracking is evident in the vicinity of the crack. After the crack was arrested and the load was reapplied, the crack eventually turned in the BN layer and grew parallel to the interface, but always within in the interphase material rather than at one of the interfaces. The fact that the crack turned and grew in the interphase is a clear indication that the fracture resistance of the BN layer is highly anisotropic; the fracture resistance of the BN layer parallel to the interface must be substantially lower than perpendicular to the interface. Given that the BN powder morphology consists of platelets with an aspect ratio of approximately 30 and that these materials are hot-pressed, it is almost certain that the BN platelets in these layers are highly aligned which would explain the anisotropy in toughness that is observed in these materials.

Because crack deflection occurs in the interphase rather than at one of the interfaces, the important material properties that govern crack deflection are the fracture resistance of the BN interphase in directions parallel and perpendicular to cell axis (ignoring elastic anisotropy). However, measurement of the fracture resistance perpendicular to the cell axis is extremely difficult since cracks always tend to deflect and run parallel to the cell axis. A lower-bound estimate of the fracture resistance of BN perpendicular to the cell axis can be determined from the arrest condition for cracks propagating from the Si_3N_4 into the BN interphase (Figure 5a) which gives a value of 120 J/m^2 compared to 30 J/m^2 for the direction parallel to the cell axis. This strong anisotropy in the fracture resistance is necessary for crack deflection in these materials.

CONCLUSIONS

Aligned fibrous monoliths were manufactured with Si_3N_4 cells and interphases consisting of a mixture of BN and Si_3N_4. It was found that the failure behavior of fibrous monoliths are directly influenced by the fracture resistance of the interface; crack deflection decreased markedly as the interfacial fracture resistance was increased by adding Si_3N_4 to the interphase. Three distinct fracture modes were identified based on the measured value of the fracture resistance. Lastly, it was shown that the conditions for crack deflection in this material system are influenced by the large anisotropy in the fracture resistance of the interphase materials. Because crack deflection occurs in the interphase rather than at the interface, the conditions for crack deflection are governed by the ratio of the fracture resistance of the interphase in directions parallel and perpendicular to the initial crack trajectory.

ACKNOWLEDGMENTS

The authors thank Advanced Ceramics Research, Inc. for providing some of the raw materials used in this study. Financial support was provided by ARPA through ONR.

(a)   (b)

Figure 5.   a) A crack impinging on the interface between Si$_3$N$_4$ and the interphase material showing that the crack is arrested in the interphase. b) Higher magnification view of cracking in BN interphase showing tortuous crack path and profuse microcracking.

REFERENCES

[1] G. Hilmas, A. Brady, U. Abdali, G. Zywicki, and J. Halloran, "Fibrous Monoliths: Non-Brittle Fracture from Powder Processed Ceramics," Mat. Sci. Eng. A **A915**, 263 (1995).

[2] P.G. Charalambides, J. Lund, A.G. Evans, and R.M. McMeeking, "A Test Specimen for Determining the Fracture Resistance of Bimaterial Interfaces," J. App. Mech. **56** (3), 77 (1989).

[3] Z. Suo and J.W. Hutchinson, "Sandwich Test Specimen for Measuring Interface Crack Toughness," Mat. Sci. Eng. **A107**, 135 (1989).

[4] J.W. Hutchinson and Z. Suo in *Advances in Applied Mechanics, Vol 29*, edited by J.W. Hutchinson and T.Y. Wu (Academic Press, Inc., San Diego, CA, 1992), p. 64.

[5] M.D. Thouless, "Fracture of a Model Interface Under Mixed-Mode Loading," Acta Metall. Mater. **38** (6), 1135 (1990).

[6] M.-Y. He and J.W. Hutchinson, "Crack Deflection at an Interface Between Dissimilar Elastic Materials," Int. J. Solids Structures, **25** (9) 1053 (1989).

[7] J.H. Edgar in *Properties of Group III Nitrides* edited by J.H. Edgar. (INSPEC, London, U.K., 1994), p. 7

HIGH TOUGHNESS ALUMINA/ALUMINATE: THE ROLE OF HETERO-INTERFACES

M. E. BRITO*, M. YASUOKA and S. KANZAKI
National Industrial Research Institute of Nagoya,
Nagoya 462, JAPAN
*brito@nirin.go.jp

ABSTRACT

Silica doped alumina/aluminate materials present a combination of high strength and high toughness not achieved before in other alumina systems, except for transformation toughened alumina. We have associated the increase in toughness to crack bridging by anisotropically grown alumina grains with concurrent interfacial debonding of these grains. A HREM study of grain boundaries and hetero-interface structures in this material shows the absence of amorphous phases at grain boundaries. Local Auger electron analysis of fractured surfaces revealed the coexistence of Si and La at the grain facets exposed by the noticeable intergranular fracture mode of this material. It is concluded that a certain and important degree of boundaries weakness is related to both, presence of Si at the interfaces and existence of alumina/aluminate hetero-interfaces.

INTRODUCTION

Nowadays, it is generally accepted that fracture toughness of ceramics can be improved by the presence of elongated grains or second phases in the microstructure, which promotes mechanisms such as crack deflection and crack bridging[1]. Nevertheless, this improvement is usually achieved at the expense of strength since second phases act as large flaws. On the other hand, toughness mechanisms heavily rely on the ability for elongated grains to detach to some extent from their matrix, i.e., grain boundary weakness. For instance, impressive mechanical behavior of recently developed silicon nitride based ceramics[2] should be indeed related to the conspicuous presence of a thin glassy film surrounding silicon nitride prismatic grains[3], the result of using a transient liquid phase to densify these materials. The question still remains on whether a "weak" grain boundary, condition necessary for high toughness, should be always related to a glassy film presence. An unprecedented combination of properties, high strength of 600 MPa and a high fracture toughness of 6 MPa•m$^{1/2}$, has been achieved in an Al_2O_3-20 vol% $LaAl_{11}O_{18}$ composite doped with silica (300 ppm)[4]. We present here an evaluation of grain boundary structures for this material.

EXPERIMENT

Material Preparation

Commercially available high purity Al_2O_3 (α-alumina, TM-DAR, purity>99.99%, Taimei Chemical Co. Ltd., Japan), synthesized lanthanum aluminate ($LaAlO_3$) and silica sol were used as starting material. The amount of $LaAlO_3$ addition was adjusted to form 20 vol. % of $LaAl_{11}O_{18}$ and the amount of silica dopant was 300 ppm referred to alumina. The compact was sintered at 1600 ºC for 2h in air. Details on the material processing are given elsewhere[4].

Material Characterization

X-ray diffraction analysis (XRD; RAD-RB, Rigaku Ltd., Japan) was conducted for phase identification under radiation conditions of 40 kV, 100 mA. Silica content in sintered specimens was analyzed by inductively coupled plasma atomic emission spectrometry (ICP-AES; ICAP-1000S, Nippon Jarrell-Ash Ltd., Japan). Microstructural features were revealed by scanning electron microscopy (SEM; JSM-T330AS, JEOL Ltd., Japan). Fractured surfaces were locally analyzed by Auger electron analysis (AES; JAMP-7800, JEOL Ltd., Japan). Specimens for transmission electron microscopy (TEM) observations were prepared by standard techniques. Conventional and high resolution (HREM) observation was performed using an analytical electron microscope (JEM-4000FX) operated at 400 kV with attached energy dispersive x-ray analysis (EDX) system and a high resolution dedicated microscope(JEM-2000EX) operated at 200kV, being the point to point image resolution of each microsocope better than 0.26 nm and 0.20 nm, respectively.

RESULTS

XRD analysis of the specimens indicated the unambiguous presence of α-Al_2O_3 and $LaAl_{11}O_{18}$ as crystalline phases . It is important to notice the total absence of reflections due to minor phases. Silica content of the Al_2O_3-aluminate composite with silica dopants was about 100 ppm higher than the added amount of dopant. The presence of excess silica could be mainly due to silica impurity in the raw powders. The silica doping produces rather uniform changes in Al_2O_3 morphology, from equiaxied to plate-like grains, that could be associated to the presence of a liquid phase at sintering temperatures (Fig. 1). The $LaAl_{11}O_{18}$ appears, in the figure, as the lighter phase. The development of aluminate platelets inhibits the grain growth of the alumina matrix.

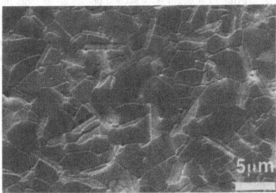

Fig. 1 SEM micrograph of silica doped alumina/aluminate composite.

Conventional TEM and electron diffraction analysis provided further crystallographic information on habit planes observed in plate-like grains and on characteristics of grain boundaries. In general, the Al_2O_3 plate-like grains present grain boundaries parallel to dense planes, either basal (001) or rhombohedral (012) planes; being the basal plane, the characteristic boundary of alumina flat-sided grains. A relatively low EDX signal for silicon has been detected in triple (three grain) junction boundaries. The content of silicon at Al_2O_3-Al_2O_3 or Al_2O_3-

aluminate interfaces is beyond the detection level of the equipment here used.

In contrast to profuse literature on the thin film glassy phase at grain boundaries in glass forming systems, such as liquid phase sintered silicon nitride, scarce information is available on alumina; particularly on, currently available sintered high purity materials. Elongated grains with platelet or plate-like shape have been observed in alumina, when glass forming additives such as Na_2O+SiO_2, $MgO+SiO_2$, $CaO+SiO_2$, etc., were added[5-6]. It is reasonable to expect, considering the relatively high sintering temperature, that also in the Al_2O_3-aluminate composite glass formation during sintering induces the anisotropic growth of alumina. Several Al_2O_3-Al_2O_3 grain boundaries were analyzed by HREM, including those involving the Al_2O_3 basal plane, to verify the existence of a glassy phase. As an example, Fig. 2 show a Al_2O_3-Al_2O_3 grain boundary where no evidence for a second phase could be found. The straight boundary of Fig. 2 is parallel to the basal plane for the alumina grain at the bottom of the figure.

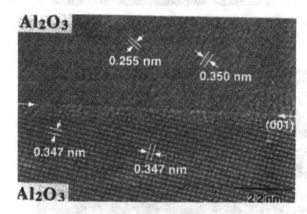

Fig. 2 Lattice fringe image of an alumina-alumina boundary (400 kV).
Crystallographic information is given for mating crystal

Similar observations were made for Al_2O_3-aluminate boundaries. Once again, no irrefutable experimental evidence could be found for the presence of a glassy phase at the boundary. Figures 3 show HREM picture for an Al_2O_3-aluminate boundary, where the basal plane of the aluminate runs parallel to the interface. A lower magnification view of the area depicted in Fig. 3 indicates that the boundary is strongly faceted, with an step depth of near 0.5 nm. It is in some triple junction boundaries where high resolution photographs offer evidence to the presence of a amorphous phase (not shown here). It should be noticed that the amorphous phase does not seem to wet the surrounding grain boundaries, indistinctly of being Al_2O_3-Al_2O_3 or Al_2O_3-aluminate interfaces. However, to reach such a conclusion more detailed and extensive observation of grain boundaries in this system is required.

Finally, an attempt to reveal the surface composition on grain boundaries that present an easy path for crack propagation was undertaken. Figure 4 shows a micrograph of the fracture surface of the composite. Intergranular fracture occurred leaving exposed grain facets which were subject to Auger electron analysis (probe 200 nm in diameter). A result, representative of the observed trend in elements distribution at fractured surfaces, is shown in Fig. 5. Besides the presence of Al and O in those surfaces, the coexistence of Si and La was always observed.

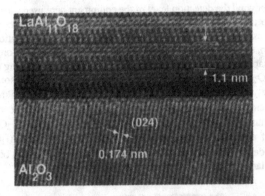

Fig. 3 Lattice fringe image of an alumina-aluminate boundary (200 kV).
Crystallographic information is given for mating crystal.

Fig. 4 Details of the fracture surface. Notice the predominantly intergranular fracture mode.

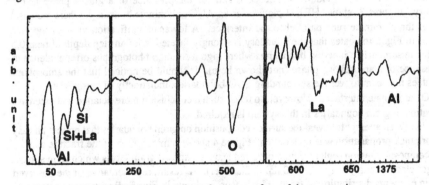

Fig. 5 AES Spectrum from fracture surface of the composite

A recent report by Chen and Chen[7] shows that alumina composites containing aluminate platelets ($LaAl_{11}O_{18}$, $LaMgAl_{11}O_{19}$, $SrAl_{12}O_{19}$, and $Mg_2NaAl_{15}O_{12}$) display fracture toughness of about 4.4 $MPa{\cdot}m^{1/2}$, which is 50 % higher than that of sintered high purity Al_2O_3, for composites containing 30 vol. % platelets. The authors indicate as a limitation to their materials, the intrinsic strength of aluminate for efficiently crack to occur. Considering our data on silica undoped materials[4] agrees well with those of ref. 7 and since crack path in these materials tends to follow interfaces alumina/aluminate, a serious consideration should be given to hetero-interfaces as a major contributor to the easy debonding. The further increase in fracture toughness we have observed in silica doped composites not only depends on inducing the growth of stronger elongated grains to bridge the cracks (i.e., plate-like alumina grains) but also on the presence of foreign elements such as Si and La at the grain boundaries, which it seems to weak.

CONCLUSIONS

It is obvious from the present experimental evidence that, grain boundaries in the system Al_2O_3/aluminate can not generally characterized by the presence of a thin glassy film at grain boundaries, claimed to weaken the boundary to promote bridging and pull-out of elongated grains in other systems (e.i., silicon nitride). However, the increase in fracture toughness achieved for this material associated to the elongated morphology of its components, alumina and aluminate, is notable. Besides the possible contribution to toughening of hetero-interfaces (i.e., alumina-aluminate boundaries), it is clear that bridging of cracks by strong plate-like alumina grains with weak bonded boundaries play an important role making the material tough. The reasons for grain boundary weakening in this material is still matter for examination.

ACKNOWLEDGMENTS

This work has been carried out as part of the Synergy Ceramics Project under the Industrial Science and Technology Frontier (ISTF) Program promoted by AIST, MITI Japan. The authors are members of the Joint Research Consortium of Synergy Ceramics.

REFERENCES

1) P. F. Becher, J. Am. Ceram. Soc. **74**, 255 (1991).
2) K. Hirao, M. Ohashi, M. E. Brito and S. Kanzaki, ibid **78**, 1687 (1995).
3) H-J. Kleebe, M.K. Cinibulk, R. M. Cannon and M. Rühle, ibid **76**, 1969(1993)
4) M. Yasuoka, K. Hirao, M. E. Brito and S. Kanzaki, ibid **78**, 1853(1995).
5) H. Song and R. L. Coble, ibid **73**, 2077 (1990).
6) H. Song and R. L. Coble, ibid **73** 2086 (1990).
7) P. F. Chen and I-W. Chen, ibid **75**, 2610 (1992).

EFFECT OF AGGREGATE CONTENT
ON FRACTURE BEHAVIOR OF CONCRETE

Y. XI*, F.E. AMPARANO*, AND ZONGJIN LI**
*Dept. of Civil and Architectural Engineering, Drexel University, Pennsylvania, PA 19104
**The Hong Kong University of Science and Technology, Kowloon, Hong Kong

ABSTRACT

Effect of aggregate content on fracture behaviors of concrete is studied by testing on geometrically similar three-point bend beams. The results are analyzed by using a size effect method in which the fracture behavior of concrete is characterized by two parameters, fracture energy G_f and effective fracture process zone c_f. Test results showed that with increasing volume fraction of aggregate in the range 45% - 75%: (1) the modulus of elasticity of concrete decreases slightly, (2) fracture energy G_f increases, but the rate is very small; (3) the size of the fracture process zone, c_f, decreases, which may be explained by changes in coarseness of grain structures defined in terms of mosaic patterns.

INTRODUCTION

The effect of rigid inclusions on fracture behaviors of composites has been a major research topic for many years. Concrete is a specific type of composite materials, which can be considered as a two phase composite with aggregates as inclusions and cement paste as the matrix. The fracture properties of concrete have been studied extensively. Buyukozturk and Lee [1] studied the behavior of mortar-aggregate interfaces. Kawakami [2] investigated the effect of aggregate type. Zollinger et al. [3] studied the effects of concrete age and aggregate size on fracture toughness of concrete. However, the effect of aggregate content on fracture properties of concrete has not been studied in detail, which is the main purpose of the present study.

It is generally found that both the critical stress intensity factor and the fracture energy increase with the addition of rigid inclusions, at least for low volume fractions. Also, the fracture energy is normally found to reach a maximum at a particular value of the volume fraction and then fall with the further addition of particles, implying that there may be a critical value of the volume fraction before which the composite is toughened by the added inclusions and beyond which the toughening mechanisms no longer apply. One of the objectives of the present study is to find if there exists such a critical volume fraction of aggregate.

One of the difficulties involved in experimental determination of fracture toughness of concrete is the size effect. The fracture toughness obtained from concrete specimens of same composition but different sizes are different. With increasing size of specimens, the obtained fracture toughness decreases. The difference is quite significant and cannot be simply explained by the Weibull statistical theory. In order to characterize the size effect, some theories have been developed. One of the size effect methods developed by Bazant [4] is adopted in the present study. The method will be briefly introduced first, and then the experiment of concrete beams will be described. The test results will be analyzed in two aspects. One is the effect of aggregate content on fracture toughness of concrete, the other is the effect on the size of the fracture process zone.

THE SIZE EFFECT METHOD

Geometrically similar specimens of brittle heterogeneous materials, such as concrete, rock, and ceramics, exhibit a pronounced size effect on their failure loads. The phenomenon, which is an important consequence of fracture mechanics, has been described by the size effect law proposed by Bazant [4]

$$\sigma_N = \frac{Bf_u}{\sqrt{1+\beta}} \qquad \beta = \frac{d}{d_0} \tag{1}$$

255

in which $\sigma_N = c_n P_u/bd$ = nominal stress at maximum load P_u; b = thickness of the specimen; d = size or characteristic dimension of the specimen (e.g. beam depth in bending test); c_n = coefficient introduced for convenience; B and d_0 = parameters determined experimentally; f_u is a measure of material strength, which can be taken as the tensile strength of the material. β is called the brittleness number. For geometrically similar specimens of different sizes, Eq. 1 can be algebraically transformed to a linear relationship

$$Y = AX + C \qquad (2)$$

in which $X = d$; $Y = (f_u/\sigma_N)^2$. The slope A and intersection C in the plot can be used to determine the two parameters B and d_0, $B = 1/\sqrt{C}$; and $d_0 = C/A$. And then, the two fracture parameters G_f and c_f can be determined based on B and d_0 [5]

$$G_f = \frac{(Bf_u)^2}{c_n^2 E} d_0 g(\alpha_0) \qquad (3)$$

$$c_f = \frac{d_0 g(\alpha_0)}{g'(\alpha_0)} \qquad (4)$$

in which G_f and c_f are defined as the energy required for crack growth and the effective length of the fracture process zone (FPZ), respectively, in an infinitely large specimen. E = Young's modulus of concrete; α_0 = initial relative crack length, a_0/d, a_0 is initial notch length; $g(\alpha_0)$ = a geometry-dependent function of relative crack length $\alpha = a/d$. The finite size of FPZ is a new fracture parameter which makes the fracture behavior of concrete differ from those of linear elastic fracture mechanics (LEFM). In LEFM, FPZ can be considered to be negligibly small.

An important consequence of the FPZ is the fracture resistance curve or R-curve. It was found that the R-curve for concrete strongly depends on the specimen geometry. A general expression of the R-curve equations from the size effect parameters is

$$R(c) = G_f \frac{g'(\gamma)}{g'(\alpha_0)} \frac{c}{c_f} \qquad \frac{c}{c_f} = \frac{g'(\alpha_0)}{g(\alpha_0)} \left[\frac{g(\gamma)}{g'(\gamma)} - \gamma + \alpha_0 \right] \qquad (5)$$

The R-curve given above is size-independent until the peak load but deviate afterwards into size-dependent horizontal branches. The size effect method reviewed was used in the present study to determine fracture properties G_f, c_f, and the effect of aggregate contents.

EXPERIMENTS

Three-point bend beam specimens were used in the present study. The geometry of the beam specimens is shown in Fig. 1. Thickness of the beams was 6.35 cm (2.5 in.). The ratio of the support span of the beam to the beam depth was 2.5 for all specimens. This ratio was chosen mainly due to the restriction of the support span of the loading system. The ratio of the notch length to the beam depth was 0.25 for all specimens. The ratio of total length to support span was 1.2. In this way, all beams are geometrically similar. Four different beam sizes were tested. The four beam depths were: 5.08, 7.62, 10.16, and 15.24 cm (2, 3, 4, and 6 in).

The aggregates used in the test were gravel and river sand. The maximum aggregate size was 3.81 cm (1.5 in.). The volume ratio of coarse to fine aggregate was 2, which was kept the same for all beams. So, the effect of aggregate size was excluded. The water-cement ratio was 0.5 for all beams. Along with the beam specimens, three 7.62 x 15.24 cm (3 x 6 in.) cylinders were cast for test of compressive strength. All specimens were tested after 14 days of curing in a fog room.

The three-point bend beam tests were performed on a Instron close-loop control testing system with crack mouth opening displacement (CMOD) control. The loading rate was controlled such that the beams reached their peak loads in about 10 min. To examine the effect of aggregate contents, four different volume fractions were used: 0.45, 0.55, 0.65, and 0.75. Four sizes of beams were tested for each volume fraction of aggregate, and three specimens for each size, and thus, totally 48 beams were tested. Typical load-CMOD curves are shown in Fig. 2.

ANALYSIS OF TEST RESULTS

To evaluate fracture energy G_f by Eq. 3, the modulus of elasticity of concrete, E, must be evaluated first, which is $E = 57,000\sqrt{f'_c}$ [10], where f'_c is compressive strength of concrete in psi. This formula has been used in general to convert the compressive strength of concrete to modulus of elasticity. Fig. 3 shows the results of E verses V_a. V_a is the volume fraction of aggregate. One can see that E decreases with increasing V_a, but the rate is not significant. However, this trend is completely different with the prediction of composite theories according to which the stiffness of composites should be enhanced by addition of rigid particles [9]. The possible explanation to this result is that the cement paste in concrete is different with the plain cement paste. The stiffness of the cement paste matrix is lower than that of plain cement paste due to the disturb of large amount of aggregate in the concrete mixing process.

G_f can be calculated based on B, d_0, and E of each group of specimens with a specific V_a. Fig. 3 shows the G_f verses V_a. As one can see, G_f can be practically considered to be independent of V_a, at least in the range of 45% - 75%. This means that the critical volume fraction corresponding to the peak value of G_f does not exist in the range of V_a higher than 45%.

Table 1 Results of size effect analysis

V_a	f'_c (psi)	E (10^7 psi)	G_f (lb/in.)	c_f (in.)	λ
0.45	5010	0.403	0.1233	1.024	1.212
0.55	4630	0.388	0.1028	0.45	1.481
0.65	4440	0.38	0.1075	0.6	1.905
0.75	4230	0.371	0.145	0.438	2.667

The size of FPZ (i.e. c_f) is determined by Eq. 4, which showed that FPZ decreases with increasing volume fraction of aggregate, as shown in Fig. 4. This can be explained by using a morphological parameter called coarseness λ. λ has been defined in the theory of mosaic pattern [6,7,8]. λ is a measure of the grain sizes of the internal structure of the material. Large λ corresponds to fine-grained structures and small λ means coarse-grained structures. In general, composites with fine grain structures (large λ) have small FPZ, while composites with coarse grain structures (small λ) have large FPZ. λ can be expressed in terms of average aggregate size E(D) and volume fraction V_a for a two phase composite [6]

$$\lambda = \frac{1}{E(D)(1 - V_a)} \tag{6}$$

From Eq. 6, with increasing volume fraction, λ increases, which means at a fixed size of aggregate, a high volume fraction of aggregate leads to a fine grain structure. Also, the effect of aggregate size can be evaluated by Eq. 6. For smaller size of aggregates, λ becomes higher, which means finer grain structure. Therefore, parameter λ is a comprehensive measure of coarseness of

Fig. 1 The geometry of the beam specimens

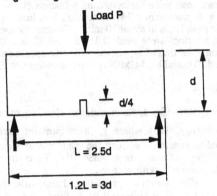

Fig. 2 Load-CMOD curves for beams with Va = 55%

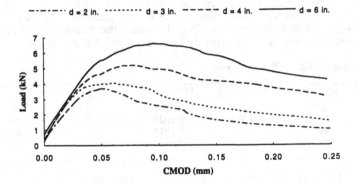

Fig. 3 Effects of aggregate content on E and Gf

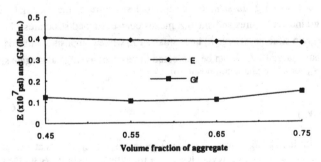

Fig. 4 Effect of aggregate content on fracture process zone

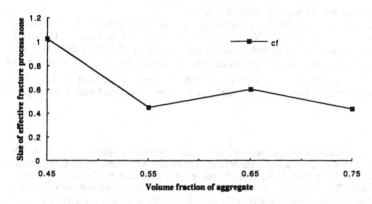

Fig. 5 Effects of coarseness of grain structure on Gf and cf

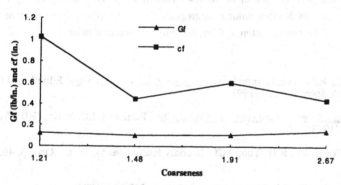

Fig. 6 R-curves of concrete beams with various aggregate contents

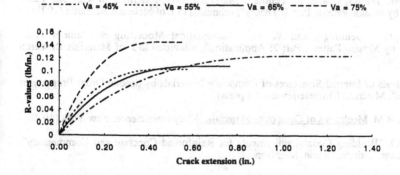

grain structures. The FPZs evaluated based on the present test results are plotted with coarseness λ in Fig. 5. All results of the analysis are listed in Table 1.

Based on G_f and c_f, the R-curves evaluated by Eq. 5 are plotted in Fig. 6. One can see from Fig. 6 that the R-curves for different V_a approach asymptotically to the values of G_f, and the crack extensions corresponding to the values of G_f are those of c_f. While c_f decreases with increasing aggregate content and constant aggregate size, G_f remains constant. This implies that aggregate enhances the fracture resistance within c_f. The general relationship of c_f and λ has been shown by the present results, but analytic model has not been developed and is an ongoing research topic.

CONCLUSIONS

The fracture properties of concrete with various aggregate contents were tested by geometrically similar three-point bend concrete beams. The results were analyzed by using the size effect method in which the fracture behaviors of concrete are characterized by two parameters, fracture energy G_f and effective fracture process zone c_f. Test results showed that with increasing volume fraction of aggregate in the range of 45% to 75%: (1) Modulus of elasticity of concrete decreases slightly, which is different from the prediction of conventional composite theories; (2) fracture energy G_f can be considered to be independent of aggregate contents in the range of 45% - 75%; (3) size of fracture process zone c_f decreases. This may be explained by changes in coarseness λ of grain structures due to addition of aggregates. With increasing volume fraction of aggregate, λ increases and the grain structure is finer, and thus c_f becomes smaller.

REFERENCES

1. O. Buyukozturk and K.M. Lee, in Interfaces in Cementitious Composites, Edited by J.C. Maso, (RILEM), E&FN Spon, London (1993).

2. H. Kawakami, in Interfaces in Cementitious Composites, Edited by J.C. Maso, (RILEM), E&FN Spon, London (1993).

3. D.G. Zollinger, T. Tang, and R.H. Yoo, ACI Materials Journal, 90(5), Sept. - Oct., p. 463-471 (1993).

4. Z.P. Bazant, J. of Engineering Mechanics, ASCE, 110(4), p. 518-535 (1984).

5. Z.P. Bazant, and M.T. Kazemi, International J. of Fracture, 44, p. 111-131 (1990).

6. Y. Xi, H.M. Jennings, and P. Tennis, "Mathematical Modelling of Cement Paste Microstructure by Mosaic Pattern, Part 1: Theory", submitted to J. of Materials Research (1995).

7. P. Tennis, H.M. Jennings, and Y. Xi, "Mathematical Modelling of Cement Paste Microstructure by Mosaic Pattern, Part 2: Application", submitted to J. of Materials Research (1995).

8. Y. Xi, "Analysis of Internal Structures of Composite Materials by Second Order Property of Mosaic Patterns", Materials Characterization, (in press).

9. Christensen, R.M. Mechanics of Composite Materials, Wiley-Interscience, New York (1979).

10. ACI 318-89, "Building Code Requirements for Reinforced Concrete and Commentary", American Concrete Institute, Detroit, Michigan.

INVESTIGATION OF CRACKING MECHANISMS OF PLASMA SPRAYED ALUMINA-13% TITANIA BY ACOUSTIC EMISSION

C.K. Lin, S.H. Leigh, R.V. Gansert, K. Murakami,[+] S. Sampath, H. Herman, and C.C. Berndt
The Thermal Spray Laboratory
Department of Materials Science and Engineering
SUNY at Stony Brook
Stony Brook, NY 11794-2275

[+]The Institute of Scientific and Industrial Research
Osaka University
Osaka 567, Japan

ABSTRACT

Free standing alumina-13% titania samples were manufactured using high power water stabilized plasma spraying. Heat treatment was performed at 1450°C for 24 hours and then at 1100°C for another 24 hours. Four point bend tests were performed on the as-sprayed and heat-treated samples in both cross section and in-plane orientations with *in situ* acoustic emission monitoring to monitor the cracking during the tests. Catastrophic failure with less evidence of microcracking was observed for as-sprayed samples. Energy and amplitude distributions were examined to discriminated micro- and macro-cracks. It was found that the high energy (> 100) and high amplitude (say > 60 dB) responses can be characterized as macro-cracks. Physical models are proposed to interpret the AE responses under different test conditions so that the cracking mechanisms can be better understood.

INTRODUCTION

Thermal spray deposits, where the lamellar structure consists of microcracks, pores, unmelted particles and metastable phases, are heterogeneous and complex. When deposits are examined using thermal and/or mechanical tests, both macro- and microcracking as well as other sources can release energy and be monitored using acoustic emission (AE) technology.[1-3] AE methods can be applied to understand coating degradation and failure mechanisms and may even be used to predict coating lifetime.[4-6]

The record of AE responses will be a combination of all possible noise origins which may have different AE characteristics. The study of AE using multiple transducers can locate the origins of AE within the material.[7,8] Acoustic emission events, amplitude and energy distributions as well as ringdown counts in either normal and/or cumulative forms are commonly used. However, in order to distinguish the differences in micro- and macrocracks, studies of AE characteristics from individual AE events are important. A "crack density function" (CDF)[9] which consolidates both the number of cracks and size of cracks has been proposed. For example, for plasma sprayed ceramic coatings under thermal cycling tests, it has been found that macrocracking tends to occur at low values of the CDF.

Acoustic emission technology was used to *in situ* monitor the cracking of samples, which were free standing alumina-13% titania prepared by water stabilized plasma (WSP) spraying. In this study, AE responses from different test conditions (i.e., as-sprayed and heat-treated samples in cross section and in-plane orientations) will be assessed and the energy and amplitude distributions will be examined to discriminate the macro-, and microcracks. Models will be proposed to interpret the AE responses.

261

EXPERIMENTAL PROCEDURES

Specimen Preparation

The water stabilized plasma spraying process was used to prepare alumina-13% titania specimens with average thicknesses of ~5 mm; the spray parameters are listed in Table 1. The mild steel substrates (0.25" x 2" x 9"), which were grit blasted and pre-sprayed with a thin layer of aluminum, were mounted on a rotating mandrel so that nine substrates were sprayed in one run.

Table 1 : Spray parameters for WSP.

Material	Alumina-13% titania
WSP	126 kW
Voltage	300-320 V
Ampere	400 A
Powder Injection	65° at 30 mm feed distance
Spray Distance	330 mm
Lathe Speed	90 rpm
Traverse Speed	~40 mm/second
Tube Outside Diameter	6.5 inches
Number of Passes	70
Feed Rate	34.4 kg/hr

The free standing plates were obtained by etching away the thin aluminum layer using hydrochloric acid and then they 5 mm wide cuts made to produce test coupons. The specimens were sintered at 1450°C for 24 hours, cooled to 1100°C (at 100°C/hr) and aged for a further 24 hours, then cooled at a rate of 100°C/hr to 700°C, and furnace cooled to room temperature. The porosity (as measured by the Archimedean method) of the as-sprayed and heat treated samples were ~10% and 3%, respectively. The major phase content (as determined by X-ray diffraction) was γ- and δ-alumina for as-sprayed samples, and α-alumina and titania after the heat treatment.

Four Point Bend Test with Acoustic Emission Monitoring

The experimental setup for four point bend tests is shown in Fig. 1, where an ATS universal test machine (model 1101) was connected to an IBM compatible PC with a data acquisition board (DAS 1602, Keithley®) to record the load-displacement curve, and an AET 5500 system (Hartford Steam Boiler Inspection Technologies) was used to monitor the cracking of the specimen during the four point bend tests. The lengths of the inner and outer spans are 20 and 40 mm respectively, and the crosshead speed was 1.0 mm/min. The sample codes are listed in Table 2 which indicates that from 10 to 13 tests were performed for each condition.

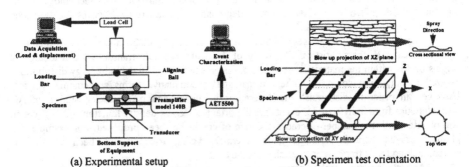

(a) Experimental setup   (b) Specimen test orientation

Figure 1 : Four point bend test with *in situ* acoustic emission monitoring. (a) experimental setup (not to scale) and (b) spray direction with respect to test direction. The specimen shown in the middle figure is tested in the in-plane direction. (The substrate is shown to only indicate the orientation of the test direction.) If the specimen is rotated 90° according to the X axis, then it will be tested in the cross section direction. The cross-sectional and top views of a single splat are shown on the right of each figure.

Table 2 : Sample identifications for four point bend tests with *in situ* AE.

As-Sprayed Samples		Heat-Treated Samples	
In-plane	Cross section	In-plane	Cross section
AS-B1 to *AS-B10*	AS-C1 to AS-C10	*AH-B1* to *AH-B12*	AH-C1 to AH-C13

Any AE signals were detected by a transducer (model AC175L) which was attached to the top middle of the bottom fixture, preamplified (model 140B) with a preamplifier filter (model FL12Y), and then output to the AET 5500 system. The AET 5500 system was also attached to an IBM compatible PC to record the AE responses. The threshold voltage was set to 0.4 V which was the voltage without observing any AE responses when the tensile test fixture was moved up or down. This arbitrary calibration accounts for any background noise that arises from the mechanical test device.

RESULTS AND DISCUSSION

Mechanical Properties and basic AE responses

Since this study will focus on AE responses, the results of four point bend tests are summarized below and discussed in detail elsewhere.[10] It is noted that there is an increase of MOR (*30* to *141* MPa when the test is performed in the in-plane orientation and 33 to 157 MPa in the cross section orientation) and elastic modulus (*11* to *129* GPa in the in-plane and 12 to 128 GPa in the cross section) after heat treatment. However, the differences in mechanical properties between tests performed in the two orientations are within one standard deviation range.

Various presentations of the AE response such as peak amplitude, events, ringdown counts and energy can be illustrated with respect to time in discrete and cumulative forms. Typical graphs for the peak amplitude and total energy versus time are shown in Fig. 2. It can be noted that catastrophic failure usually occurs when the test is performed in the in-plane direction (Fig. 2, solid curves) and microcracking occurs when tested in the cross section direction (Fig. 2, broken curves). The distinction of "catastrophic" and "microcracking" features is based on the AE response prior to the failure.

Since no significant differences can be found when tests were performed in two different orientations, the following discussion will be focused on the results from the cross section orientation for as-sprayed and heat treated samples.

AE responses vs. displacement

The AE responses versus displacement are illustrated in Fig. 3a and 3b for as-sprayed and heat-treated samples tested in both cross section directions, respectively. It is noticed that AE activity of the specimens which show microcracking before failure is observed at a relatively small displacement; say < 0.1 mm. However, for the specimens which

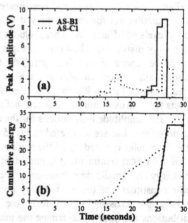

Figure 2 : (a) peak amplitude and (b) total energy versus time, where specimen AS-B1 shows catastrophic failure and specimen AS-C1 shows microcracking before failure.

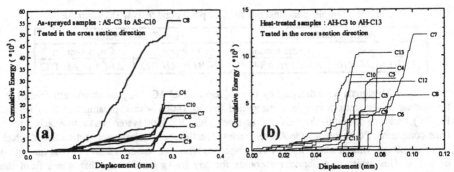

Figure 3 : Cumulative energy versus displacement for (a) as-sprayed and (b) heat-treated samples tested in the cross section orientation, respectively.

show catastrophic failure (indicated as bold letters, C3 and C9, in Fig. 3a), the AE activity only initiates after a displacement of ~0.2 mm. For the heat-treated samples (Fig. 3b), it is noticed that, although AE activity can be observed in the early period, most of the energy was released close to final failure when compared with the AE responses from the as-sprayed samples (note the large X axis scale differences for Figs. 3a & 3b). The average total displacement is 0.281 ± 0.012 and 0.073 ± 0.013 mm for as-sprayed and heat-treated samples tested in the cross section directions, respectively.

Energy and Amplitude Distributions

The distribution analysis of the overall AE response is important because it is postulated that micro- and macro-cracks have different characteristics. Figure 4 shows the energy distributions for as-sprayed and heat-treated samples tested in the cross section orientation. The energy distribution for the as-sprayed samples can be separated into three distinct regions by energy levels of 45 and 100. These regions are termed as micro-, "transitional", and macro-cracks. The energy distribution, however, changes after heat treatment, Fig. 4b. Though the distribution can still be separated into three parts, the threshold between microcracks and transitional cracks becomes ambiguous. In addition, the percentage of macrocracks for the overall AE events increases from below 8% to 12% after heat treatment. The change in the energy distribution reflects the microstructural differences before and after heat treatment.

The amplitude distributions for the as-sprayed and heat treated samples tested in the cross section orientation are illustrated in Fig. 5. It is noted that the sharp peak around 40 dB for as-sprayed samples is broader and the amplitudes of AE events are widely distributed between 40 to 60 dB after heat treatment. It is important to know whether there are any relationships between the energy and amplitude distributions. It can be determined that the high amplitude group is the major constituent of the high energy group. However, the groups of micro- and transitional cracking are cross-linked to form the low and medium amplitude groups. If amplitude distributions are used to determine the micro-, transitional and macro-cracks, the results will be different from the energy approach. The upshot of the somewhat ambiguous threshold of the low amplitude region infers that the amplitude approach may not be a suitable choice for discriminating differences among micro-, transitional, and macro-cracks.

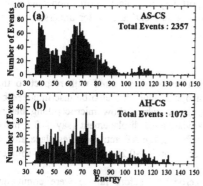

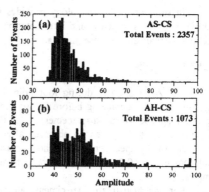

Figure 4 : Energy distributions for (a) as-sprayed and (b) heat treated samples tested in the cross section orientation.

Figure 5 : Amplitude distributions for (a) as-sprayed and (b) heat treated samples tested in the cross section orientation.

Cracking mechanisms and modeling

Physical models are illustrated in Fig. 6 to interpret, qualitatively, the AE results. For as-sprayed samples (Fig. 6a), cracks can propagate through the splat boundaries, unmelted particle-splat boundaries, porosity and from the pre-existing microcracks within the splats (see the enlarged figure of a single splat in Fig. 6a). In addition, since these sites are relatively weak when compared with heat-treated samples, the necessary stress to propagate the cracks is low. As a consequence, abundant AE activity can be detected for as-sprayed samples.

On the other hand, the microstructure of the deposits has changed after heat treatment. The cohesion strength increases and porosity decreases, but pore coalesce occurs and large-sized pores form after heat treatment (Fig. 6b). A significant decrease in crack propagation sites can be expected and these coalesced pores become the new sources for crack propagation and growth. The necessary stress level to propagate microcracks may be low because a large stress concentration will be induced around these pores. However, the elevated cohesion strength after heat treatment will reduce the susceptibility of these microcracks from growing into macrocracks. Once the stress level is large enough to force the microcracks to grow, macrocracks form and induce final failure. This physical model may explain why, for heat-treated samples, microcracks are commonly observed before failure and most of the AE activity is detected close to the final failure.

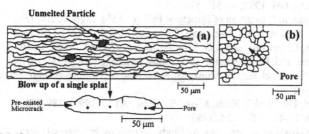

Figure 6 : Models for crack propagation sites within (a) as-sprayed and (b) heat-treated samples.

In summary, AE technology provides an opportunity to collect information concerning material responses to stress and interpretation of these results enable a better understanding of cracking mechanisms of these materials.

CONCLUSIONS

The AE responses during four point bending tests were evaluated with respect to the displacement. Microcracking before failure and catastrophic failure can also be discriminated by the AE response versus displacement data. For specimens which show microcracking before failure, the AE activity can be observed after a relatively small displacement (say < 0.1 mm). However, for the specimens which show catastrophic failure, the AE activity takes place after a displacement of ~0.2 mm. The average total displacement is ~0.28 mm for the as-sprayed samples.

For the heat-treated specimens, although microcracks before failure were commonly observed, most of AE activity was detected close to the final failure and the average total displacement is ~0.07 mm.

Although quantitative analysis of AE responses during four point bending tests is not yet viable; the qualitative analysis provides valuable information. The rich microstructure of as-sprayed samples provides enormous AE sources and the number of these sites decrease after heat treatment. The heat treatment coalesced large-sized pores and enabled microcracks to form in the early stage of bend tests, but the elevated cohesion strength inhibit these microcracks from propagating. When the applied stress is large enough, macrocracks form and lead to catastrophic failure of the materials.

ACKNOWLEDGMENT

This research program has been supported by the National Science Foundation under the STRATMAN Program DDM 9215846.

REFERENCES

1 M.M. Mayuram and R. Krishnamurphy, in Thermal Spray: International Advances in Coatings Technology, C.C. Berndt (Ed.), ASM International, Materials Park, OH, 1992, pp 711-715.

2 N. Iwamoto, M. Kamai and G. Ueno, in Thermal Spray: International Advances in Coatings Technology, C.C. Berndt (Ed.), ASM International, Materials Park, OH, 1992, pp 259-265.

3 H. Nakahira, Y. Harada, N. Mifune, T. Yogoro and H. Yamane, in Thermal Spray: International Advances in Coatings Technology, C.C. Berndt (Ed.), ASM International, Materials Park, OH, 1992, pp 519-524.

4 H. L. Dunegan, in Prevention of Structural Failure, ASM, Metals Park, OH, 1975 pp 86-113.

5 T. Tsuru, A. Sagara, and S. Haruyama, Corrosion-NACE, 43[11], (1987) 703-707.

6 F. Bordeaux, C. Moreau, and R.G. Saint Jacques, Surf. Coat. Technol., 54/55 (1992), 70-76.

7 S. Wakayama and H. Nishimura, in Fracture Mechanics of Ceramics, vol. 10, Eds. R.C. Bradt, D.P.H. Hasselman, D. Munz, M. Sakai, and V.Ya. Shevchenko, Plenum Press, NY, 1992, pp59-72.

8 M. Enoki, Y. Utoh and T. Kishi, Mater. Sci. Eng., A176 (1994), 289-293.

9 C.C. Berndt, J. Mater. Sci., 24 (1989), 3511-3520.

10 C.K. Lin, Statistical Approaches to Study Variations in Thermal Spray Coatings, Ph.D. Thesis, Dept. Mater. Sci. & Eng., SUNY at Stony Brook, NY, Dec. 1995.

STUDY OF FIBER COMPOSITE FAILURE CRITERION

S. J. Zhou*, R. Blumenfeld*, W.A. Curtin** and B. L. Holian*
* Theoretical Division and Center for Nonlinear Studies,
Los Alamos National Laboratory, Los Alamos, NM 87545
** Department of Engineering Science & Mechanics and Department of
Materials Science & Mechanics, Virginia Tech, Blacksburg, VA 24061

ABSTRACT

With a recently developed numerical technique, we have investigated failure processes in real fiber composites. This technique utilizes 3D lattice Green's functions to calculate load transfer from broken to unbroken fibers, and also includes the important effects of fiber/matrix sliding. It is found that the failure processes are more complex than previously thought: The fibers that break or slide are not necessarily located around the largest defect cluster. Rather, the breaking/sliding fibers accumulate, form several large clusters, coalesce and finally cause the fiber composite to fail. Only in some cases does one large defect dominate the damage evolution. Based on our observations of the irregular evolution of the damage, a simple model is developed to describe these different failure modes, and the failure criterion of the fiber composite is also discussed.

INTRODUCTION

It is generally assumed that a fiber composite begins to fail when one of the large defects (a cluster of broken fibers) reaches its critical size, and thus the composite failure strength can be determined in terms of this critical defect. This approach is true for a single crack in a homogeneous medium, where the critical crack length is given by the well known Griffith relation derived from continuum mechanics. When the crack length exceeds this critical value, the crack grows unstably. However, a failure criterion for a fiber composite may not be established in a similar way. The reason is that the fiber composite is *not* a homogeneous system because fibers have a strength distribution which follows the Weibull form, and load transfer function from a broken fiber to others is "local load sharing". Therefore two major factors control the failure of a fiber composite. One is the stress concentration around the critical defect (mechanical factor) and the other is the dispersion of the fiber strength distribution (statistical factor).

Previous fiber composite failure criteria [1-4] have made two main assumptions: (1) A critical defect is coplanar and consists of only compact broken fibers; (2) Load sharing is extremely localized. The probabilities of failure of the fibers are based on the chain-of-links scheme, where failure of the weakest link causes a complete failure of the fiber composite. Then the critical defect is identified and thus a failure strength is obtained. Based on these models, He, Evans and Curtin [5] developed a criterion to predict the transition from global load sharing to local load sharing by evaluating, around a broken fiber, the relative survival probabilities of the nearest-neighbor fibers (Φ_s^N) and the next-nearest-neighbor fibers (Φ_s^{NN}). When $\Phi_s^{NN} \leq \Phi_s^N$, i.e., the survival probability of the nearest neighbor exceeds that of the next-nearest neighbor, the development of a well-defined crack from an initial fiber failure cannot occur and thus a global load-sharing mechanism would apply. Fiber failure would be then more likely to occur far from an existing break, rather than in the nearest neighbors. However, these models overly simplify the failure process. As shown by Zhou and Curtin [6], the critical cluster configurations at failure mostly consist of sliding rather than broken fibers, and some strong intact fibers still survive within the cluster. The mechanism of failure, which is essential to establish a precise failure criterion, has not been well addressed.

Different load sharing rules of fibers in composites can also lead to quite different failure modes. In global load sharing, there is no stress concentration around defects and no particular region serving as a critical defect (cf. [6]). In fact, the global load sharing rule smears out some of the heterogeneous components of fiber composite systems due to the fiber statistical strength distribution and due to varying locations of defects. The failure criterion is therefore not directly related to a particular defect. The statistical factor plays a major role in evaluating the failure strength, but in this case analytic calculations are possible and a failure strength prediction based on statistics agrees quite well with experiment and simulation results [7,8].

Under local load sharing, however, there is a high stress concentration around breaking and sliding fibers. And, statistically there are regions containing weak fibers and others containing strong fibers. The dispersion of the fiber strength distribution decreases with increasing Weibull modulus. Weibull modulus generally varies from 2 to 15 in actual fiber composites. Fiber composites with higher Weibull modulus (e.g., $\rho = 10$) are closer to homogeneous systems and thus the mechanical factor dominates the failure process. A critical defect would act like a crack, and composite failure evolution is basically due to the unstable growth process of this defect. Consequently, a failure criterion could be derived from this critical cluster. However, this is not the case for composites with lower Weibull modulus (e.g., $\rho = 5$) where the dispersion of the fiber distribution is significant. Statistically the regions with strong fibers would block the spread of a "critical" defect, while weaker fiber regions would form new defects that grow under the load transferred from other defects. The early stage of composite failure evolution should then involve coalescence of several clusters, rather than the evolution of one "critical cluster". Once the newly formed cluster is sufficiently large that the stress concentration around it can overcome fairly strong fibers, then it would grow in a way similar to a propagating crack. Thus the statistical factor becomes dominant and the failure criterion is not solely determined by the "critical cluster".

For simplicity, we discuss here the tensile failure of fiber-reinforced composites, which is generally dominated by failure of a fiber bundle. Recently we have developed an innovative and powerful technique for studying the failure of fiber-reinforced composite with the flexibility of adjusting the load transfer from broken to unbroken fibers [6]. This new approach is also capable of exploring how the defects form, spread, and finally cause the composite to fail.

In this paper, we will investigate the evolution of composite failure to see what conditions lead to a failure criterion of a fiber composite based on the formation of a critical cluster.

SIMULATION RESULTS

The model composite studied here is identical to that in [6], where fibers are labelled by an index n ($1 < n < N_f$ for $N_f = N_x N_y$ fibers in a square array). Each fiber is divided into N_z small fiber elements of length $\bar{\delta}$ (total fiber length $N_z \bar{\delta}$) labelled by index m $(1 < m < N_z)$. The entire fibrous system thus consists of $N_x N_y N_z$ fiber elements labelled by (n,m), and the stress on an element is denoted $\sigma_{n,m}$. Each fiber element of size $\bar{\delta}$ is then represented by a tensile spring with spring constant k_t. Each spring is also assigned a strength $s_{n,m}$ chosen from the Weibull probability distribution

$$P_f(s) = 1 - e^{-(s/\bar{\delta})^\rho} \tag{1}$$

where $P_f(s)$ is the probability of fiber failure, and $\bar{\sigma} = (L_0/\bar{\delta})^{1/\rho}\sigma_0$ is the fiber scale strength at length $\bar{\delta}$ (L_0 and σ_0 are two scale parameters). The role of matrix and interface friction τ in transferring load among fiber elements is described by pure shear springs of spring constant k_s. The load transfer is then controlled by a parameter $\Omega^2 \equiv k_s/k_t$. In global load sharing ($\Omega \to \infty$), there exist characteristic fiber length δ_c and the relevant strength σ_c that control failure. They are

$$\delta_c = [\frac{\sigma_0 r L_0^{1/\rho}}{\tau}]^{\frac{\rho}{\rho+1}}, \qquad \sigma_c = \frac{\delta_c \tau}{r} \tag{2}$$

where r is the radius of a fiber. We maintain these quantities for scaling on local load sharing. (For more detailed discussion of this model, see [6])

To explore the failure mechanism of fiber composites, we choose a relatively small 400 fiber composite with $N_x = N_y = 20$. As shown in [6], a 400-fiber composite is sufficiently large compared to the critical cluster sizes which appear before failure under local load sharing. $N_z = 40$ is chosen with $\bar{\delta} = 0.05\delta_c$ to provide an accurate longitudinal discretization. In all the calculations, Ω is fixed at 0.001 (i.e. local load sharing); For $\rho = 5$ and 10, $\bar{\sigma} = 1.821\sigma_c$ and 1.349 σ_c, respectively. All stresses are referenced to σ_c, lengths to δ_c, and strains to σ_c/E_f. Periodic boundary conditions are used in the x and y directions.

At $\rho = 5$, we find that the failure stress increases with increasing dispersion of the spatial defect distribution. The more scattered the defects, the less the stress concentrates locally on the fibers, on average. This is not surprising since several small clusters certainly reduce the stress concentration in comparison to an equivalent large cluster. This can be seen by comparing the critical defects in the order of increasing failure stress. Similar results are also seen in 1600-fiber composites.

It is interesting to note that the evolution of failure is quite different for different Weibull moduli. For comparison, only the Weibull modulus is changed between Figures 1 and 2. The failure process, i.e. the evolution of fiber failure after loading up to the failure stress, for a low Weibull modulus is as follows (see Figure 1): First, the weaker fibers are gradually broken, which does *not* necessarily occur around the largest defect, then newly accumulated breaking and sliding fibers form several large defects. They coalesce and create a new giant defect, which causes the final failure of the fiber composite. We have also observed that a final failure doesn't necessarily take place in the layer containing the largest defects. As for the composite with high ρ (see Figure 2), the critical defects are smaller and there are few defects in the final failure layer. The whole failure process is caused by the evolution of a single large defect.

We attempted to correlate the composite failure stresses with the geometrical factors of a "critical cluster" or damage level in the final failure cross-section. Both the total number of the fibers (including breaking, sliding, and intact fibers) within the largest defect only and within the complete failure layer are used to characterize failure. Certainly this is not very accurate because the shapes and distributions of the defects are also crucial to influence the failure process. For example, stress concentration in a two dimensional rectangular cluster is quite different from a three dimensional spherical one. It is found that for $\rho = 5$ such a correlation does not exist in Figures (3a) and (3b) for both critical clusters and complete failure layers. This is strong evidence against the correlation of the failure strength and largest defects where the Weibull modulus is low (high dispersion). This finding also suggests that the composite failure is a many-body problem, where collective effects from all the defect clusters determine the failure process. Therefore, more parameters than size alone are needed to describe this process, such as the spatial distribution of the defect clusters and their shapes. But for a system with only one sufficiently large cluster, it is possible to define the upper limit of a critical cluster size, above which fiber composites fail in a definite crack-like mode, and the lower limit, below which the cluster has no immediate long-range effects on the failure process.

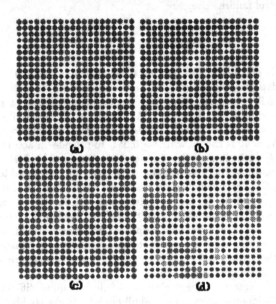

Figure 1. *Coalescence-growth-like failure evolution of a 400-fiber composite with Weibull modulus $\rho = 5$. Intact and sliding fibers are represented by the largest and the middle-size circles, respectively, with grey colors denoting their stresses. Broken fibers are indicated by the smallest black circles. (a) Critical state just before final failure; (b) New breaking and sliding fibers are created randomly; (c) New defects accumulate, link with two existing large clusters, and form a giant cluster; (d) The giant cluster spreads and the composite fails.*

Comparing Figures (3a) and (3b), we note that total numbers of sliding/breaking fibers in the failure layers are about 2.5 times higher in 1600-fiber composites than in 400-fiber composites, on the average, but the sizes of the largest defect are slightly smaller in 1600-fiber composites than those in 400-fiber composites. These results indicate that under local load sharing, the failure of fiber composites with low ρ are controlled locally by defect clusters, whose sizes decrease with increasing system size. This is understandable because smaller critical clusters can exist in larger systems.

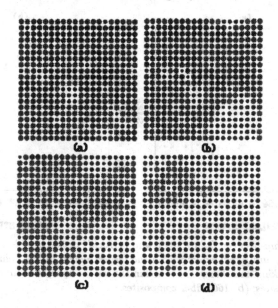

Figure 2. Crack-growth-like failure evolution of a 400-fiber composite with $\rho = 10$. (a) Critical state just before final failure; (b) New breaking and sliding fibers are continuously created around one large cluster at the right-bottom corner; (c) The newly enlarged cluster merges an existing large cluster, and (d) causes the composite to fail.

Even though the composite failure strength with low ρ and local loading sharing is difficult to predict, some failure features can still be obtained. For example, we have observed that the strength of a composite with 1600 fibers on average is lower than that with 400 fibers (see Figure 4). This is because as the system size increases, there are more potential locations for a critical defect to occur and to grow unstably, and hence a higher probability for such a defect to occur at a lower applied stress (For more detailed discussion, see [9]). With the probalility distribution of failure, we can obtain detailed information on the strength of larger fiber composites.

DISCUSSION AND CONCLUSIONS

The failure processes of fiber composites are quite complicated. In this paper, we have investigated the failure evolution of composites with 20×20 fibers. The influence of system size needs to be taken into account. Increasing system size could make a fiber composite fail in a more crack-like mode because the probability of having a dominant defect increases with system size. Furthermore, statistically the failure strength of a fiber composite with a low value of ρ is lower on the average than that with high ρ. Therefore the combined effects of different loading levels and statistical strength ditribution on the evolution of the critical clusters need to be considered.

An analytical model that addresses this issue has recently been introduced [10], which takes into account these results. In this model one calculates the probability of failure of

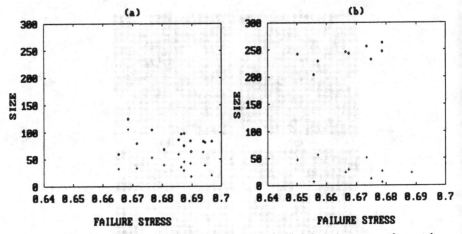

Figure 3. Relationship between failure stress and "critical" defect size; + are the numbers of breaking, sliding, and intact fibers within the largest defect in the failure layer; ◇ are the numbers of breaking and sliding fibers within the failure layer; Weibull modulus $\rho = 5$; (a) 400-fiber composites; (b) 1600-fiber composites.

a fiber away from a compact defect cluster. For example, for a defect cluster of just one fiber at $\mathbf{r}_0$ the model predicts that the failure probability of a fiber at $\mathbf{R} + \mathbf{r}_0$ is

$$P(R) = \exp\left[((\sigma(\mathbf{r}_0) - \sigma_1)\exp(-\rho a/\xi))^\rho\right]\exp\left\{-\left[\sigma(R) - \sigma_1\right]^\rho\right\}\left\{1 - \exp\left[\sigma_1^\rho - \sigma^\rho(R)\right]\right\} . \tag{4}$$

In this relation all stresses are measured in units of $\bar{\sigma}$, $\sigma(\mathbf{r}_0)$ is the stress at the center of the defect, σ_1 is the external tensile stress, a the distance between nearest neighbor fibers, ξ the typical distance over which the lattice Green's function decays, and $\sigma(R)$ is the stress at position R away from $\mathbf{r}_0$, which is approximated well by $(\sigma(\mathbf{r}_0) - \sigma_1)\exp(-R/\xi)$. The rather forbidding form (4) cannot be obtained by simple intuitive considerations, yet it explains nicely the above observations. Additionally, it yields a surprising prediction: The highest probability of the occurence of a local crack growth mode is when $\rho = 2$ for low values of the external applied stress, but for high values it is when $\rho = 10$! Furthermore, the calculation predicts the existence of an interesting crossover regime where neither $\rho = 2$ nor $\rho = 10$ are the most susceptible to failure through local growth but rather intermmediate values of ρ take this role.

In summary, we have studied the failure processes of fiber-reinforced composites with computer simulation. For composites with low Weibull moduli, the fibers that break or slide are not necessarily located around the largest defect cluster. Rather, the breaking/sliding fibers accumulate, form several large clusters, coalesce and finally cause the fiber composite to fail. Consequently, a failure criterion cannot be derived solely from this largest cluster. In contrast, in composites with high Weibull moduli, fibers fail around the largest defect cluster, and it is the growth of this largest defect that causes the final failure. In this case, the failure criterion is indeed determined by the largest cluster. These simulation results can be qualitively understood with a simple analytic model.

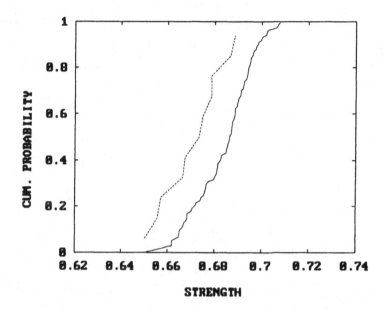

Figure 4. Cumulative probability of failure versus applied stress, for both 400-fiber (solid line) and 1600-fiber (dashed line) composites with $\rho = 5$.

REFERENCES

1. B.W. Rosen, J. Am. Inst. Aeronautics Astronautics, **2** , 1985 (1964).
2. P.M. Scop and A.S. Argon, J. Compos. Mater. **1**, 92 (1967).
3. R.L. Smith and S.L. Phoenix, J. Appl. Mechan. Trans. ASME, **48**, 75(1981).
4. S.B. Batdorf, J. reinf. Plastics and Comp. **1**, 153 (1982).
5. M.Y. He, A.G. Evans and W.A. Curtin, Acta Metall. **41**, 871 (1993).
6. S.J. Zhou and W.A. Curtin, Acta Metall. **43**, 3093 (1995).
7. W.A. Curtin, J. Am. Ceram. Soc. **74**, 2837 (1991).
8. W.A. Curtin, J. Mech. Phys. Solids **41**, 35 (1993).
9. M. Ibnabdeljalil and W.A. Curtin, to be published.
10. R. Blumenfeld and S.J. Zhou, in preparation.

MICROMECHANICAL BEHAVIOR OF POLYMER INTERFACE REINFORCED WITH COPOLYMERS

Q. Wang*, F. P. Chiang*, L. Guo**, M. Rafailovich** and J. Sokolov**
*Dept. of Mechanical Engineering, SUNY at SB, Stony Brook, NY11794-2300
** Dept. of Materials Science and Engineering, SUNY at SB, Stony Brook, NY11794-2275

ABSTRACT

The fracture toughness of interface reinforced with dps-b-dpmma copolymer between immiscible polymers of PS and PMMA is tested by asymmetric double cantilever beam. The local deformation field at the interfacial crack tip is determined by the technique of SIEM. Normal and tangential crack opening displacements are calculated. A weak singularity is shown to exist near crack tip. Direct observation on the fracture process inside an environmental scanning electron microscope shows the large effect of mode mixity.

INTRODUCTION

The interface between two immiscible polymers are usually very weak. However, when a properly chosen block copolymer is added as an adhesive between them, the interfacial bond strength can be greatly improved[1-4]. If the copolymer is chosen such that each of its blocks is miscible with one or the other of the homopolymers, segregation will tend to occur at the interface for thermodynamics reasons. This causes both the decrease of the interfacial tension and the improvement of the adhesion between the homopolymers.

Interface adhesion can be quantified using the concept of interface fracture toughness. It is found that an effective test method for the fracture toughness of a polymer/polymer interface is the asymmetric double cantilever beam (ADCB) specimen and a convenient way to generate the asymmetry is to bond the sample to a substrate that is stiff with respect to the cantilever beams, as shown in figure 1(a). Specimen is wedged open to generate an interfacial crack. From the

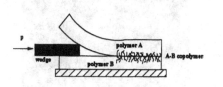
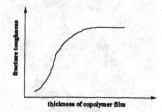

Figure1(a)ADCB (b) toughness increases with the copolymer film thickness

crack length, fracture toughness can be obtained from the model of a single cantilever beam on an elastic foundation [5]

Mat. Res. Soc. Symp. Proc. Vol. 409 ® 1996 Materials Research Society

$$G = \frac{3ED^3u^2}{8a^4[1 + (0.64D / a)]^4} \tag{1}$$

where a is the crack length, u the thickness of wedge, D the beam depth and E the Young's modulus. Experiments indicates that the toughness with copolymer can be one order of magnitude higher than that without copolymer in PS/PMMA system, as schematically shown in figure 1(b).

Although the basic principle for the reinforcement from copolymer is known, the detailed micromechanical processes at the crack tip is not well understood. In this work an environmental scanning electron microscope is employed to perform a direct observation on fracture process and an experimental micromechanics technique SIEM is applied to give a quantitative measurement on deformation at crack tip.

EXPERIMENTAL

The materials system chosen in this work consists of polystyrene (PS) and polymethylmethacrylate (PMMA) homopolymers. Sheets (40 mm x 10 mm x 2 mm) of PS and PMMA were compression molded at 270°F. They were joined by a thin layer of dPS-b-dPMMA diblock copolymer with molecule weight of 285400. The copolymer was dissolved in Toluene and spun onto the PS from the solution. The film thickness was measured by an ellipsometry (with an error of ±3 angstrom) on the copolymer film spun on a silicon wafer under the same conditions of solution concentration and spin speed. Then the sandwich of PS, PMMA homopolymers and PS-PMMA copolymer was heated to joining temperature 270°F and annealed for 30 minutes. The PS side of the sample was adhered to an rigid substrate. The interfacial crack was initiated and propagated by inserting a razor blade into the interface.

Figure 2 (a) speckle pattern before and (b) after applying the load

Tests were also performed inside an environmental scanning electron microscope Hitachi S-2460N, which allows an in situ direct observation on fracture process along or near the interface at very small scale. It is not necessary for the specimen to be coated with any conductive layer. The deformation field at the crack tip was also measured quantitatively by technique SIEM (speckle interferometry with electron microscopy). This technique has been successfully applied to determine the mechanical properties of an fiber/matrix interphase layer in a metal matrix composite material[6]. The speckle patterns at crack tip before and after opening crack is shown in figure 2. The image was digitized and then processed to obtain displacement fields.

RESULTS AND DISCUSSIONS

Bimaterial interfaces are susceptible to both debonding and sliding. The mode mixity is its inherent feature. The ratio between K_1 and K_2 (phase angle of a complex K) plays a significant role in interface fracture behavior[7]. In order to investigate the phase angle effect, tests on beams under three point bending were performed. As shown in figure 3, under certain loading condition the phase angle has such a sign and value that interfacial crack is preferred to propagate into PMMA. It is clear that although the crack propagates straight along the boundary between two homopolymers, the crazes in PS and inclined microcracks in PMMA are always present at crack tip.

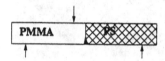

Figure 3. Crack tip features and loading condition

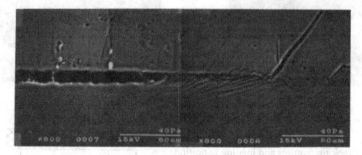

Figure 3. Crack grows through the formation of craze and microcracks

When loading condition changes as shown in figure 4, both sign and value of phase angle change. In this case the crack becomes preferred to grow into PS. Since PS has less craze resistance than PMMA, more crazes and microcracks are formed. Crack shows a kink path of propagation. More energy is dissipated in this case and interface appears tougher.

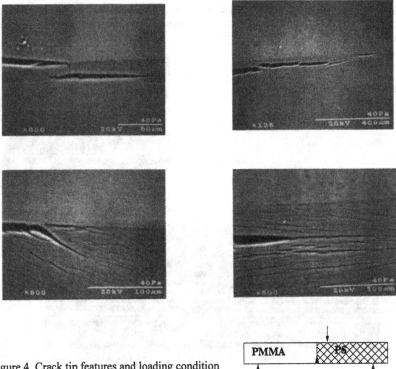

Figure 4. Crack tip features and loading condition

Tests on asymmetric cantilever beam show that the interface without copolymer gives toughness of about 5 (J/m²) while the interface reinforced with copolymer gives about 150 (J/m²). The deformation field at crack tip is determined by SIEM. Both U (parallel to the crack plane) and V (normal to the crack plane) displacement fields, as shown in figure 5, are calculated from the speckle patterns. Strong interface shows larger normal crack tip opening displacement (NCOD) than weak interface but approximately the same tangential crack tip opening displacement (TCOD). The phase angle is defined as

$$\hat{\psi} = \tan^{-1}(\frac{\sigma_{xy}}{\sigma_{yy}})_{r=\hat{L}} \tag{2}$$

Another way to evaluate the mode mixity is the ratio of NCOD/TCOD. If ϕ is defined as

$$(\tan \phi)_r = \frac{TCOD}{NCOD} \tag{3}$$

where r is in the K-annulus. The ϕ is easy to obtain from figure 5. The ψ can be calculated from a relation [8]

$$(\phi)_r = \hat{\psi} + \varepsilon \ln(\frac{r}{\hat{L}}) - \tan(2\varepsilon) \tag{4}$$

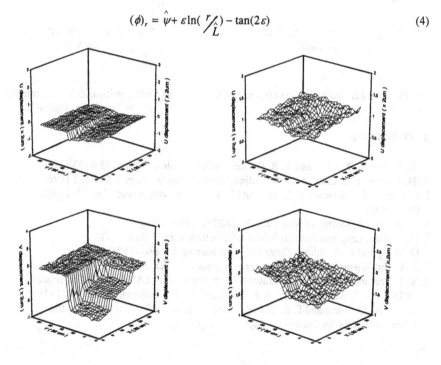

Figure 5. Displacements field at crack tip of (a) strong interface   (b) weak interface

The stress singularity is obtained as shown in figure 6. A weak singularity is found at crack tip, which is consist with the results of [8]. This may attribute to the fact that many crazes and microcracks exist at crack tip.

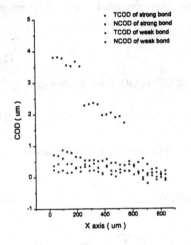

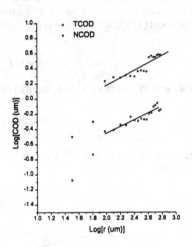

Figure 6. (a) Crack opening displacements

(b) plotted in logarithmic coordinates straight lines have slope of 0.5

REFERENCES

1. H. R. Brown, K. Char and V. R. Deline, Maromolecules, 26, pp4155-4163 (1993)
2. H. R. Brown, K. Char and V. R. Deline, Maromolecules, 26, pp4164-4171 (1993)
3. C. Creton, E. J. Kramer, C. Y. Hui, and H. R. Brown, Macromolecules, 25, pp3075-3088 (1992)
4. H. R. Brown, Journal of Materials, 25, pp2791-2794 (1990)
5. M. F. Kanninen, International Journal of Fracture, 9, pp83-92 (1973)
6. Q. Wang and F. P. Chiang, Composite Engineering, in press (1995)
7. J. W. Hutchinson and Z. Suo, Advances in Applied Mechanics, 29, pp63-191 (1992)
8. X. T. Yan and F. P. Chiang, Ultrasonic Charaterization and Mechanics of Interfaces, AMD-Vol.177, (ed S. I. Roklin, S. K. Datta and Y. D. S. Rajapakse), pp97-111 (1993)
9. X. T. Yan, Q. Wang and F. P. Chiang, Proceedings of the 9th Conference of the American Society for Composites, pp214-219 (1994)

THE COMPOSITION OF POLYMER COMPOSITE FRACTURE SURFACES AS STUDIED BY XPS

Donald A. Wiegand and James J. Pinto
ARDEC, Picatinny Arsenal, NJ

ABSTRACT

The composition of the fracture surfaces of a composite made up of a polycrystalline organic nonpolymeric filler and a binder composed of a copolymer was studied by XPS. Because the binder and the filler of the composite each have at least one element not in common it is possible to easily distinguish between the binder and filler by XPS. A measure of the relative amounts of binder and filler on the fracture surfaces, therefore, could be made as a function of the sample temperature, T, and the strain rate during fracture. The ratio of filler to binder, F/B, increases with decreasing T at constant strain rate and is least sensitive to strain rate at T's below T_G, the quasi static glass transition T. At higher T, F/B increases with strain rate at constant T. These results indicate that as the binder becomes stronger and stiffer due to a decrease in T or an increase in strain rate more of the fracture processes take place in the filler whose properties are expected to be less sensitive to T and strain rate. These results are related to the fracture properties as observed by uniaxial compression.

INTRODUCTION

By x-ray photoelectron Spectroscopy (XPS) it is possible to obtain a quantitative measure of the relative numbers of atoms or ions which are present in the surface layers of materials. For composite materials it is then possible to obtain a measure of the relative amounts of each component of the composite in the surface layers if each component has a detectable element which the others do not. If components have detectable elements in common they can still be distinguished and the relative amounts determined if the chemical shifts are sufficiently different so that unique separations can be made.

In this work the relative amounts of composite components in the surface layers of fracture surfaces were investigated as a function of conditions during fracture, i.e. temperature and applied strain rate. The material studied is a polymer composite and the temperature and applied strain rate during fracture are of interest relative to its viscoelastic properties. While other composites were investigated only one is discussed in detail here because of space limitations. This composite is made up of a filler and a copolymer binder. The filler contains nitrogen and the binder does not, while the binder contains fluorine and chlorine which the filler does not. The filler can thus be uniquely identified by nitrogen and the binder can be uniquely identified by the other elements so that the relative amounts of filler and binder can be easily studied.

EXPERIMENTAL

Samples were prepared by pressing powders of composite 9502 into large billets and machining to size.[1,2] The filler is 1,3,5-triamino-2,4,6-trinitrobenzene (TATB) (95%) while the copolymer is chlorotrifluoroethylene/vinylidine fluoride (KEL F 800). The filler particle size is in the micrometer range.

The samples were in the form of right circular cylinders one inch in length and one half inch in diameter and were temperature conditioned at -45°, -10°, 25° and 75° C in steel compression fixtures in a separate conditioning chamber for at least two hours before compression. The samples were moved from the conditioning chamber to the loading platen in the steel compression fixtures for compression. All temperatures given are the temperatures of the conditioning chamber. It is estimated that the sample temperature at the time of compression did not deviate from the conditioning temperature by more than a few degrees centigrade. Samples were compressed along the cylinderical axis using an MTS servo-hydraulic system operated at constant strain rates of 0.001, 0.01, 0.1 and 1.0 per second.[3]

One to five samples were compressed at each of the four temperatures and the four strain rates. After fracture the samples were stored in air for up to several months before XPS measurements were made. The XPS spectra were taken with a Kratos ES300 spectrometer and the pressure during measurements was in the range of 10^{-8} to 10^{-9} torr.

RESULTS AND DISCUSSION

The number of photoelectrons, $M_i(B)$, collected in a given time period from the energy level np of atoms/ions B of the ith component of the composite can be expressed as

$$M_i(B) = I \ A_i \ C_i(B) \ \sigma_{np} \ \lambda_{np} \ S_{np} \ T_{np} \ K_i \qquad (1)$$

where I is the number of photons/area in the incident beam in the same time period, A_i is the area of the ith component projected onto to a plane normal to the photon beam, $C_i(B)$ is the concentration of B in the ith component, σ_{np} is the cross section for the photoelectric process from the np energy level, λ_{np} is the inelastic mean free path, S_{np} is the asymmetry parameter and T_{np} is the analyzer transmission coefficient for electrons having the kinetic energy associated with the np energy level and the incident photon energy.[4,5] K_i is a parameter for the ith component to account for the attenuation of the number of photoelectrons counted due to such factors as surface contamination and surface roughness (shadowing). In all cases gaussians were fitted to the curves of counts versus binding energy and $M_i(B)$ was taken as the area of the pertinent gaussian.

In equation (1) all quantities are either known or can be determined except A_i and K_i (see below). However, in this paper the concerned is with ratios. Therefore, solving equation (1) for $A_i K_i$ and taking the ratio for two components i and j

$$A_i/A_j \ K_{ij} = [M_i(B)/M_j(D)] \ [C_j(D) \ \sigma_{mq} \ \lambda_{mq} \ S_{mq} \ T_{mq}/C_i(B) \ \sigma_{np} \ \lambda_{np} \ S_{np}T_{np}]$$
$$(2)$$

is obtained where $K_i/K_j = K_{ij}$ and the photoelectrons are from energy levels np and mq of atoms/ions B and D of the ith and jth components respectively. Equation (2) contains the ratio $\lambda_{mq}/\lambda_{np}$. While λ is sensitive to the particular element(s) in the electron is path, the ratio of λ's is relatively insensitive to the type of elements.[4] Therefore, the ratio of λ's in equation (2) is insensitive to surface contamination.

Equation (2) also contains the ratio of transmission coefficients, T_{mq}/T_{np}. The Kratos spectrometer was used in the retarding potential mode and the transmission coefficients were measured for photoelectrons from the Cu $2p_{1/2}$, $2p_{3/2}$ and 3s and the Au $4p_{3/2}$ and $4d_{5/2}$ energy levels relative to the transmission coefficients for the Cu 3p (77 ev) and the Au 4f(85 ev) coefficients respectively. The Cu and Au samples were cleaned by Argon bombardment before measurement. Equation (1) is solved for the transmission coefficient and ratios taken to give

$$T_{mq}/T_{np} = [M_{mq}/M_{np}] \ [\sigma_{np} \ \lambda_{np} \ S_{np}/\sigma_{mq} \ \lambda_{mq} \ S_{mq}] \qquad (3)$$

where M_{mq} and M_{np} now refer to different energy levels of the same element. The relative transmission coefficients obtained using equation (3) were plotted versus binding energy and a smooth curve fitted to the points. Relative coefficients were then obtained from this curve for use with equation (2).

In Figure 1 $A_F/A_B \ K_{FB}$ is given versus temperature with strain rate as a parameter. This quantity, which is referred to here as an effective area ratio, increases with decreasing temperature and

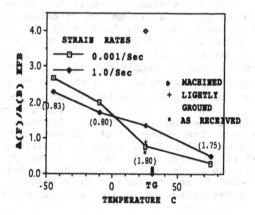

Figure 1. Effective surface area ratio, $A_F/A_B \ K_{FB}$, versus temperature for varying conditions.

increases with increasing strain rate in the high temperature range but decreases somewhat with increasing strain rate in the low temperature range. However, the strain rate dependence is much greater for the high temperatures than for the low temperatures as evidenced by the ratios of A_F/A_B K_{FB} at high to low strain rates given in parentheses at each temperature of Figure 1. The effective area ratio is also given in Figure 1 for an external machined surface from one of the deformed samples and for two powder samples, one measured as received and the other after light grinding. The deformation from grinding appears to increase this ratio somewhat while machining increases it very significantly. The strain rate during grinding is larger than the lower strain rate used here and the strain rate during machining is much larger than the higher strain rate used. While the surface temperature during machining is unknown, a lubricant was used so that this temperature may not have been too much above 25 C. These results are all consistent with the conclusion that this effective area ratio increases with strain rate at 25° C. Although the glass transition temperature, T_G, of the composite is unknown, the T_G of the polymer binder is indicated in the Figure.[6] Therefore, it can be concluded that the ratio of Figure 1 is insensitive to strain rate at temperatures significantly below this T_G, while it is strain rate dependent close to and above this temperature.

A_F/A_B K_{FB} increases by over a factor of ten between the high and low temperatures of Figure 1. This result indicates a change in the ratio of filler to binder on the surface by a factor of ten with these changes in temperature and strain rate during fracture if the relative surface roughnesses do not change. It is also important to note that this ratio is small compared to the ratio of the bulk concentrations of filler to binder of approximately 20. Therefore, if K_{FB} is of the order of unity, the fracture surfaces contain much higher percentages of binder than the bulk. This is supported in part by the much larger effective area ratio for the external machined surface since it indicates that it is possible to decrease the relative amount of binder on the surface. Partial failure by dewetting, i.e. failure at the binder-filler interface could account for the small value of A_F/A_B K_{FB} since equal areas of binder and filler are produced for this type of failure. However, if K_{FB} is significantly less than unity, then A_F/A_B must be significantly larger than the values of Figure 1. A small K_{FB} could be due to greater attenuation of photoelectrons from the filler than the binder because of surface roughness and/or surface contamination differences. Fracture must always take place in part in the binder if the binder completely coates the filler particles. Therefore, there is a maximum possible value for the effective area ratio determined by the filler particle size and the binder coating thickness. However, this ratio would be zero if the fracture process takes place completely in the binder. A determination of K_{FB} is required for further interpretation.

The results of Figure 1 were obtained using the fluorine in the binder as a measure of relative binder effective area, and similar results were obtained using the chlorine in the binder. The ratio $M(F)/M(Cl)$ for photoelectrons from the fluorine and the chlorine of the binder was also calculated by the use of equation (1) and found to be 3.26. The experimental values are independent of temperature and strain rate as expected and the average for thirteen measurements of eleven samples is 3.26 +/- 0.24. This is in excellent agreement with the calculated value and lends support to the general approach used here.

The results for another composite, 9404, are similar to those of Figure 1 for 9502 except that the transition from a strain rate dependent to strain rate insensitive effective area ratio takes place at a lower temperature. This shift to a lower temperature is consistent with the lower T_G of 9404.[7]

The compressive stress for composite 9502 at the higher two temperatures of the figure initially increases linearly with increasing strain, then curves over and passes through a maximum, and decreases gradually with further increases of strain. At even larger strains the stress decreases more rapidly with increasing strain to values near zero.[1,2] At the lower two temperatures of the figure the behavior with increasing strain is similar up to the maximum. However, at strains somewhat beyond the maximum, the stress abruptly decreases to zero or near zero.[1,2] The transition between these two types of behavior is not precise and the abrupt decrease of the stress was observed in at least one case at 25° but never at 75° C. After the abrupt decrease of the stress at -45° and -10° C the sample is fractured into several pieces. The sample is also fractured into several pieces at 25° and 75° C after strains much larger than the strain at the maximum stress although an abrupt decrease of the stress is not observed for these conditions. Thus, the fracture surfaces used for the effective surface area measurements at the two low and two high temperatures were produced under different conditions of stress and strain.

From the initial linear portion of the stress versus strain curves at all temperatures a modulus, E, is obtained which is independent of strain rate and temperature for the lower two temperatures of Figure 1 within a scatter of data having standard deviations of about 20%. E appears to decreases with increasing temperature above -10° C and has a clear strain rate dependence at 75° C but not at 25° C. In addition, the maximum compressive stress, the compressive strength, which may be taken as a measure of the failure stress,[2] is strain rate insensitive at -45° C, has a strain rate sensitivity which increases with increasing temperature for the other three temperatures, and decreases with increasing temperature at all strain rates.

All of the observations are consistent with the following interpretation: At the highest temperature the deformation process takes place primarily in the polymer binder which is above its T_G. Therefore, the modulus and the compressive strength are strain rate and temperature dependent, the effective fracture surface area is rate dependent and has the largest observed polymer component (smallest A(F)/A(B) K_{FB}), and fracture takes place by slow crack growth initiated probably at or by craze formation. At 25° C the polymer is stiffer and stronger relative to the filler than at 75° C so that more of the deformation and failure processes take place in the filler. Thus, the strain rate and temperature sensitivity of the modulus and the strain rate sensitivity of the compressive strength are decreased, and the effective surface area has a larger filler component relative to the polymer, all compared to 75° C. The polymer apparently has rubber-like properties at 25° C because of the strain rate dependence of the effective surface area ratio, the modulus and the compressive strength. It is assumed here that the filler properties are not strain rate sensitive. At the lower two temperatures the polymer is clearly below its T_G of about 30° C and so is in a glass state and brittle fracture takes place by rapid crack growth through the (assumed brittle) filler and the polymer binder. Therefore, the effective surface area ratio, the modulus and the compressive strength are either strain rate independent or strain rate insensitive, and more of the fracture takes place in the filler as evidenced by the larger effective surface area ratios.

SUMMARY AND CONCLUSIONS

The temperature and strain rate dependence of the effective surface area ratio indicate that the fracture process shifts from the polymer binder to the filler with decreasing temperature and increasing strain rate. These changes in the effective surface area ratio are also consistent with the glass transition temperature of the polymer and the observed temperature and strain rate dependencies of the mechanical properties of the composite.

ACKNOWLEDGEMENT

The authors wish to thank J. Sharma for many very helpful discussions regarding the XPS part of this work and to A. Mansour for a discussion of the asymmetry parameter. This work was supported in part by Sandia National Laboratories, Albuquerque.

REFERENCES

1. D. A. Wiegand, C. Hu, A. Rupel and J. Pinto, 9th International Conference on Deformation, Yield and Fracture of Polymers (Churchill College, Cambridge, U.K.), The Institute of Materials, London (1994) p64.

2. D. A. Wiegand, 3rd International Conference on Deformation and Fracture of Composites, University of Surrey, Guildford, U.K. (1995) p558.

3. D. A. Wiegand, J. Pinto and S. Nicolaides, J. Energetic Materials 9, 19 (1991).

4. N. Ikeo et al, Handbook of X-ray Photoelectron Spectroscopy, JEOL Ltd, Akishima, Tokyo, Japan (1991) p12.

5. R. F. Reilman, A. Msezane and S. T. Manson, J. Electron Spectroscopy and Related Phenomena 8, 389 (1976).

6. B. M. Dobratz and P. C. Crawford, LLNL Explosive Handbook, Properties of Explosives and Explosive Simulants, UCRL 52297 Change 2, p6-6 (1985).

7. B. M. Dobratz and P. C. Crawford, Ibid. p6-8.

Part V

Dynamic Instabilities in Fracture

SMALL-SCALE SIMULATIONS OF LATTICE FRACTURE

M. MARDER
Department of Physics and Center for Nonlinear Dynamics
The University of Texas at Austin, Austin TX 78712
marder@chaos.ph.utexas.edu

ABSTRACT

Many properties of rapid fracture may profitably be studied in atomic scale computer simulations involving relatively small numbers of atoms. A first result of such a study is that qualitative properties of Mode III fracture change little when one explores various shapes of the interparticle potential, introduction of randomness, and elevated temperatures. A second result is that Mode I fracture is considerably more susceptible to instability than had previously been understood, and that to obtain stable Mode I fracture may require non-central forces between atoms.

INTRODUCTION

The fracture of brittle solids is a physical process which naturally connects large and small scales[1,2]. Stresses and strains which cause the fracture are applied on macroscopic scales, while the end result is the severing of bonds on an atomic scale. Therefore, it seems reasonable to assume that computer simulations of the fracture process that account for phenomena at the atomic level must be very large. Several such simulations are now being carried out, in systems involving as many as hundreds of millions of atoms[3-5].

The first goal of this brief article is to show that certain properties of brittle fracture may profitably be studied in simulations that are much smaller, involving only a few thousand atoms. The second goal is to use these simulations to address various questions about the fracture process. It appears that brittle fracture is not strongly affected by shapes of interparticle potentials, small amounts of randomness, or small increases of temperature above zero. However, brittle fracture in Mode I is found to be much less stable than had previously been understood, and the problem is traced to the use of central forces between atoms.

THEORY

The starting point for construction of small-scale atomic simulations of fracture is a collection of analytical results for fracture in lattices. These results are only available for a very specific models, but allow one to check the magnitudes of errors introduced by working with very small systems.

The models are slight generalizations of those solved by Slepyan[6] and co-workers[7], with solutions related to earlier work by Atkinson and Cabrera[8], Celli and Flytzanis[9], and Thomson[10,11]. For example, a two-dimensional model of Mode III fracture in a triangular lattice $2(N + 1)$ atoms high is described by the equations[12,13,14]

$$\ddot{u}\left(\vec{r}\right) = \frac{1}{2} \sum_{\substack{\text{nearest} \\ \text{neighbors } \vec{r}'}} \mathcal{F}\left[u\left(\vec{r}'\right) - u\left(\vec{r}\right)\right], \tag{1}$$

where the location of each mass point is described by a single spatial coordinate $u(\vec{r})$, that should be interpreted as the height of mass point $\vec{r}$ into or out of the page. The coordinates $\vec{r}$ range over sites in a triangular lattice. The function

$$\mathcal{F}\left(u\right) = u\theta\left(2 - |u|\right) \tag{2}$$

represents the brittle nature of the springs, and θ is the step function. The boundary condition which drives the motion of the crack is that for $[x, y]$ the coordinates of any lattice point on the top row of the strip,

$$u\left(\left[x, \pm y\right]\right) = \pm\Delta\sqrt{2N + 1} \tag{3}$$

The three main conclusions from analytical study of this model are [14]

- Steady crack motion is forbidden in a range of velocities beginning at 0, and ranging up to around 20% of the wave speed.
- Thereafter, steady cracks moving at a velocity $v(\Delta)$, which can be calculated, are perfectly stable until
- at an upper critical velocity, v_c, they fall prey to a micro-cracking instability which break bonds off the crack line parallel to the direction of motion.

One can find analytically the speed v of a crack moving in steady state as a function of the applied strain Δ. In addition, one can calculate a velocity v_c at which steady crack motion becomes impossible due to a microbranching instability.

Suppose one sets the goal of finding out the velocity at which this microbranching instability first occurs as a function of the height of the system, $2(N + 1)$. Using the analytical methods of Ref. 14, one finds that for a strip 20 lattice points high, the instability occurs for a crack of velocity $v = .576$, for a strip 200 lattice points high it occurs when $v = .550$, and for a strip 2000 lattice points high it occurs when $v = .552$. Thus, in a strip 20 lattice points high the instability occurs only 5% away from where it occurs in a strip of 2000. Similarly, the velocity of a crack v as a function of loading, Δ, computed for a strip 20 atoms high reproduces within less than 1% the velocity of a crack in a strip 200 atoms high. Thus, I conclude that in order to understand steady crack motion, or the onset of instability of steady crack motion, a strip 20 atoms high gives results that are within a few percent of the answers that would be obtained for an infinite system.

The rigid upper and lower boundaries described by Eq. (3) are essential in arriving at this conclusion. Rather than describing a very large system in which elastic waves travel away from the crack tip and never interfere with it, these boundary conditions require the crack tip to adopt a steady state in interaction with the waves that it emits, and which bounce off upper and lower boundaries. The thinner the strip in which the crack moves (the smaller is N), the more rapidly this equilibrium comes about. Once the crack is in steady state, its behavior is essentially independent of the height of the strip. These two facts provide thin strips with computational advantages over large systems.

SIMULATIONS

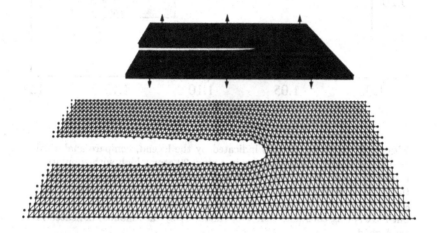

Figure 1: Positions of atoms during an ideally brittle Mode III fracture, in a strip of height $2(N+1) = 20$, and width 80. The upper portion of the figure shows the geometry of the loading, and the lower shows a snapshot of positions of atoms during steady motion.

The analytical results for strips of height $2(N+1)$ still involve systems with infinite numbers of particles in the horizontal direction. This fact turns out not to provide a difficulty for computation. The strategy I have adopted is to consider a system 20 atoms high, and on the order of 80 atoms in width, as depicted in Fig. 1. The location of the crack tip is monitored, and every time the crack moves right by one lattice spacing, all particles are picked up and moved by one latice spacing to the left, so that the simulation tracks the moving crack. The problem of wave reflections at left and right boundaries is handled by creating short regions near those boundaries with a large Stokes damping, so that all elastic waves arriving there are almost completely absorbed. This procedure is accurate to within around 1%, as illustrated in Fig. 2.

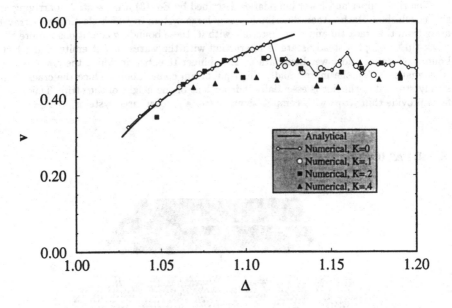

Figure 2: The two lines, as indicated by the legend, compare analytical results for steady crack velocities in a strip 20 atoms high with simulation results in a strip 20 high and 60 wide. The velocities are plotted as a function of the driving strain Δ. The symbols display the effect of random bond strengths on crack velocities. The springs all snap at the same extension, but have spring constants that vary randomly by the amounts indicated in the legend.

Application to Mode III – With these investigations as background, one can begin to conduct numerical studies of systems which cannot easily be solved analytically. I have so far checked three different effects; the effect of changing the interparticle potential from the simple form given in Eq. (2), making the bond strengths in the lattice vary randomly, and having the crack propagate in a lattice maintained at nonzero temperature. A brief summary of these numerical searches would be that small changes in the basic model Eq. (1) do not alter the qualitative conclusions listed following Eq. (3). For example, in Fig. 2, one can see the effect of making bond strengths random by varying amounts. The randomness is implemented by setting the spring constants between neighbors in Eq. (1) to $(1 + k_{\vec{r}_1 \vec{r}_2})1/2$, where $\vec{r}_1$ and $\vec{r}_2$ are nearest neighbors, and $k_{\vec{r}_1 \vec{r}_2}$ is a random variable ranging with uniform probabliity from $-K$ to K. The springs still snap at the extension given by Eq. (3). The qualitative changes introduced by the randomness are that it becomes possible for the crack to jump up or down by a lattice spacing, as shown in Fig. 3, and it also becomes possible for the crack tip to encounter a particularly tough bond, and arrest prematurely. Fig. 2 was produced by obtaining a moving crack at $\Delta = 1.2$, and then very slowly ramping Δ down to 1, with velocities calculated by measuring the time needed for the crack to progress 20 lattice spacings. The fluctuations visible in Fig. 2 are to be understood as resulting from this procedure; if enough averaging were carried out, no fluctuations in velocity would be visible despite the presence of randomness.

Figure 3: In the presence of randomness, cracks no longer travel along straight-line paths. This simulation is carried out with $K = 0.2$, for a strip 20 atoms high and 80 long

As a second example, one can introduce the effect of temperature. I carried this out by making use of the energy-absorbing regions at the left and right hand sides of the sample. If $b_{\vec{r}}$ is the coefficient of Stokes dissipation at site $\vec{r}$, then in accord with the fluctuation dissipation theorem, particles receive a random kick at each time interval of magnitude $f\sqrt{6b_{\vec{r}}Tdt}$, where dt is the time step, T is the temperature with units of energy, and f varies randomly from -1 to 1. Since $b_{\vec{r}}$ is only nonzero near the left and right boundaries of the system, the effect is that of a purely Hamiltonian system in contact with a heat bath. The effect of temperature on crack motion is depicted in Fig. 4. As before, crack velocities were measured by determining the time needed for the crack tip to advance by 20 lattice spacings, and the fluctuations observed in the crack velocity have to be understood in this context. Low to moderate temperatures also do not much affect the qualitative features of crack motion. Like static disorder, nonzero temperature pushes the crack to wander from a straight line, but unlike static disorder it does not lead to early crack arrest.

Application to Mode I –In proceeding from Mode III to Mode I models of fracture, the numerical studies show that my previous analysis of Mode I fracture in Ref. 14 was incorrect. Mode I fracture is subject to an additional instability that was overlooked.

In Mode I fracture, it is important to specify the ratio between the distance particles move to cause bonds to snap, and the interparticle spacing. For a lattice model to correspond as closely as possible to linear elasticity, this ratio should be very small, so that bonds sever when particles move almost imperceptibly from their equilibrium locations. This limit is not realistic, but it allows the forces between particles to be taken as purely linear functions of their displacements, up to the point of bond rupture, and therefore permits analytical solutions of Mode I problems.

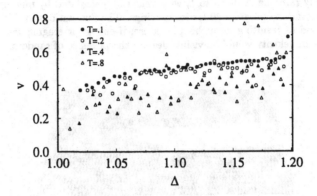

Figure 4: Crack velocity versus driving strain Δ for various values of temperature. Crack velocity is measured by monitoring the time needed for the crack to move twenty lattice spacings.

The Mode I problems that are amenable to analytic treatment are described by the equations

$$\ddot{\vec{u}}(\vec{r}) = \frac{1}{2} \sum_{\substack{\text{nearest} \\ \text{neighbors} \; \vec{r}'}} \vec{\mathcal{F}}\left[\vec{u}\left(\vec{r}'\right), \vec{u}\left(\vec{r}\right)\right], \tag{4}$$

where now $\vec{u}$ has both x, and y components describing the deviation of each particle from its equilibrium location, and $\vec{r}$ ranges over a triangular lattice. In the limit where particle displacements are small when bonds snap, one can take

$$\vec{\mathcal{F}}\left[u\left(\vec{r}'\right), u\left(\vec{r}\right)\right] = \{k_{\parallel} \hat{d}_{\vec{r}\vec{r}'}^{\parallel} \hat{d}_{\vec{r}\vec{r}'}^{\parallel} \cdot \left(\vec{u}\left(\vec{r}'\right) - \vec{u}\left(\vec{r}\right)\right)$$
$$+ k_{\perp} \hat{d}_{\vec{r}\vec{r}'}^{\perp} \hat{d}_{\vec{r}\vec{r}'}^{\perp} \cdot \left(\vec{u}\left(\vec{r}'\right) - \vec{u}\left(\vec{r}\right)\right)\} \tag{5}$$
$$\theta\left(|\vec{r}' - \vec{r}| + \delta - |\vec{r}' - \vec{r} + \vec{u}\left(\vec{r}'\right) - \vec{u}\left(\vec{r}\right)|\right).$$

Here $k_{\parallel}$ and $k_{\perp}$ are spring constants parallel and perpendicular to the interparticle separation, $\hat{d}_{\vec{r}\vec{r}'}^{\parallel}$, and $\hat{d}_{\vec{r}\vec{r}'}^{\perp}$ are unit vectors that are parallel and perpendicular, respectively, to the vector $\vec{r} - \vec{r}'$, and δ gives the small extension from equilibrium at which bonds snap. Boundary condition Eq. (3) still holds, now for the u_y component of displacements, with u_x fixed at zero at upper and lower boundaries. An analytical solution of this model is contained in Ref. 14.

If one assumes that particles interact by central forces alone, so that $k_{\perp}$ vanishes, then the simplest steady fracture propagation is unstable. The instability has a simple geometrical origin, and can be understood from Fig. 5. In order for the crack to propagate steadily, the next bond to snap needs to be the one between particles a and b. In the case of Mode III fracture, the bonds $c - a$ and $f - a$ would be pulling up on particle a, but in Mode I fracture they exert no vertical force, so the attempt to fracture becomes a tug-of-war between $b-a$ $g-a$ on one side, and $d-a$ $e-a$ on the other. In fact, as the crack

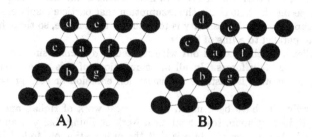

Figure 5: When particle displacements are very small compared to interparticle spacing, there is no preference for snapping bond $a - b$ as opposed to $a - d$, as shown in A). However, when displacements are no longer small, bond $a - b$ will be the first to break, as indicated in B).

tries to reach steady state, bond $d-a$ snaps. The end result is that the crack executes extremely irregular motion in the manner depicted in Fig. 6. This behavior is unexpected in a model which was designed to be ideally brittle.

Figure 6: An attempt to propagate a Mode I crack with purely central forces, $\delta = .1$, and driving strain $\Delta = 1.6$ results in the extremely irregular crack path depicted here.

CONCLUSIONS

The interplay of numerical and analytical work on discrete models of fracture is bringing suprises. It has been clear for some time that Mode III steady-state cracks can become unstable to branching. However, it now appears that all the Mode I steady-state crack solutions analytically studied to date are unstable at all velocities.

I am now searching for ways to control these instabilities, and am investigating three interrelated mechanisms. The first is to adopt a nonzero $k_\perp$, so that particles no longer interact by purely central forces, and bonds $c-a$ and $f-a$ exert a vertical force. The best way to implement this idea practically is to have particles interact through a three-body potential. The second is to abandon the assumption that particle displacements are small when bonds snap. The third possibility is to re-orient the sample, so that the crack is not driven to travel parallel to a line of bonds.

These studies are being carried out with molecular dynamics simulations using both two-body and three-body forces. In all my simulations of this type carried out so far steady-state crack motion is unstable to dislocation emission or void growth away from the tip.

Similar instabilities have been noted in numerical work of Holian and Ravelo[5], and more suprisingly in continuum-scale analytical work of Ching, Langer, and Nakanishi[16]. The only fact which now seems certain is that the propagation of Mode I brittle fracture is quite sensitive to details of lattice structure and bonding, and it will take some time to unravel all the possibilities.

ACKNOWLEDGEMENTS

I am grateful for financial support from the Alcoa Foundation, and for travel funds from the U. S., Israel Binational Science Foundation, Grant 920-00148/1.

REFERENCES

1. L. B. Freund, *Dynamic Fracture Mechanics* (Cambridge University Press, New York, 1990).

2. M. F. Kanninen and C. Popelar, *Advanced Fracture Mechanics* (Oxford New York 1985).

3. F. F. Abraham, D. Brodbeck, R. A. Rafey, and W. E. Rudge, *Physical Review Letters* **73** 272 (1994).

4. A. Nakano, R. K. Kalia, and P. Vashishta, *Physical Review Letters* **73** 2336(1994).

5. B. L. Holian and R. Ravelo, *Physical Review* **B51** 11275-11288 (1995).

6. L. I. Slepyan, *Soviet Physics Doklady* **26** 538 (1981).

7. Sh. A. Kulakhmetova, V. A. Saraikin, and L. I. Slepyan, *Mechanics of Solids* **19** 101-108 (1984).

8. W. Atkinson and N. Cabrera, *Phys. Rev.* 138, A764 (1965).

9. V. Celli and N. Flytzanis, *J. Appl. Phys.* **41** 4443 (1970).

10. R. Thomson, C. Hsieh, and V. Rana, *J. Appl. Phys.* **42** 3154 (1971).

11. R. Thomson, *Solid State Physics* **39** 1 (1986).

12. M. Marder and X. Liu, *Phys. Rev. Lett.* **71** 2417 (1993).

13. X. Liu *Dynamics of fracture propagation* (Dissertation, University of Texas, 1993).

14. M. Marder and S. Gross, *Journal of the Mechanics and Physics of Solids* **43** 1-48 (1995);.

15. S. J. Zhou *et. al.*, *Physical Review Letters* **72** 852 (1994).

16. E. Ching, J. S. Langer, and H. Nakanishi, unpublished..

ON THE RESPONSE OF DYNAMIC CRACKS TO INCREASING OVERLOAD

P. GUMBSCH
Max-Planck-Institut für Metallforschung, Seestr. 92, 70174 Stuttgart, Germany

ABSTRACT

One of the most interesting questions in the dynamics of brittle fracture is how a running brittle crack responds to an overload, i.e. to a mechanical energy release rate larger than that due to the increase in surface energy of the two cleavage surfaces. To address this question, dynamically running cracks in different crystal lattices are modelled atomistically under the condition of constant energy release rate. Stable crack propagation as well as the onset of crack tip instabilities are studied.

It will be shown that small overloads lead to stable crack propagation with steady state velocities which quickly reach the terminal velocity of about 0.4 of the Rayleigh wave speed upon increasing the overload. Further increasing the overload does not change the steady state velocity but significantly changes the energy dissipation process towards shock wave emission at the breaking of every single atomic bond. Eventually the perfectly brittle crack becomes unstable, which then leads to dislocation generation and to the production of cleavage steps. The onset of the instability as well as the terminal velocity are related to the non-linearity of the interatomic interaction.

INTRODUCTION

During brittle failure of structural materials as well as in most laboratory fracture experiments a brittle crack travels in regions of large overload. With increasing overload the initially smooth and mirrorlike fracture surface commonly evolves into a rough, hackled region. Recent experiments show that the transition occurs once a critical crack velocity v_c, of the order of 1/3 to 1/2 of the Rayleigh wave speed v_R, is reached [1, 2, 3]. These experiments further reveal that crack velocity violently oscillates after the transition but hardly ever exceeds six tenth of v_R. In contrast, continuum fracture theory predicts that cracks accelerate up to the Rayleigh wave speed [4]. Continuum analysis can also give reasons for the appearance of dynamical instabilities [5]. However, those are only expected beyond about 2/3 of the Rayleigh wave speed. The problem with our current understanding of the dynamics of brittle fracture processes therefore is that linear elastic continuum fracture mechanics predicts terminal velocities and critical velocities which are almost twice as high as experimentally observed.

The purpose of this paper is to analyze the energetics of this problem atomistically using molecular dynamics (MD) techniques. Treating the crack and its surrounding as a thermodynamical ensemble it can be shown [6] that the mechanical energy release rate G per unit crack advance must exceed the surface energy γ of the two crack surfaces to allow for crack advance; $G \geq 2\gamma$. On the atomic length scale, the same considerations apply. However, the discreteness of the lattice causes the crack to remain stable and not to advance/heal until loads somewhat larger/smaller than $G = 2\gamma$ are reached [7, 8]. This finite stability range is known as *lattice trapping*. In the context of the dynamics of a running crack one can simply ask the question: How does a crack respond to an overload, i.e. how can a crack release stored elastic energy which exceeds the surface energy of the two crack surfaces? For this question it is principally not relevant to know the origin of the overload. However, amongst many loading situations in which a crack becomes mechanically unstable the longer it grows, lattice trapping can also be seen as a possible origin of an overload.

Mat. Res. Soc. Symp. Proc. Vol. 409 ° 1996 Materials Research Society

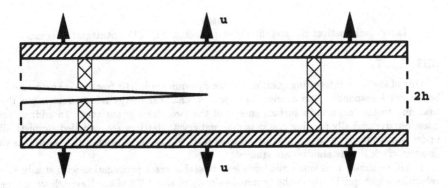

Figure 1: Schematic outline of the geometry used for the atomistic modelling of dynamically running cracks.

In principle, a crack has several possibilities to release a surplus of energy:
1.) the crack can accelerate, if it has some inertia,
2.) it can create heat (and noise), which can be detected experimentally [3],
3.) it can create more surface than minimally needed, which leads to cleavage steps or bifurcations (see for example [9] and references therein) and
4.) it can either emit or move dislocations.
Marder and coworkers [10, 11] recently investigated these possibilities by subjecting a 2D triangular lattice with snapping spring force laws to various overloads. They found that there appears to be a lower range of crack velocities $0 < v_{tip} < 0.3 v_R$, which can not be reached as steady state velocities. Furthermore, they detected a dynamical instability at $v_{tip} \approx 2 v_R/3$, where the bonds to the side of the mean fracture plane break before the bonds crossing the fracture plane. This instability can therefore be interpreted as an instability to crack branching.

In this paper dynamically running cracks will be analyzed atomistically in real 3D crystal structures and with the use of "realistic" interatomic interaction models. It will be shown that the non-linearity of the interatomic interaction causes drastic changes in the velocity distribution around the crack tip at increasing overloads. Shock waves are emitted at the breaking of every single atomic bond at intermediate overloads, which limits the critical velocity to $v_{tip} \approx 0.4 v_R$. Beyond the instability the crack creates cleavage steps or emits dislocations.

MODEL GEOMETRY AND TECHNICAL REMARKS

The model geometry used for the atomistic crack propagation simulations should provide the crack with well defined and constant loading conditions to allow for precise studies of the response to an overload. The geometry used in this study is schematically displayed in Figure 1. It consists of a thin strip of material with a central crack which is loaded by displacement boundary conditions on the upper and lower boundary regions. Comparing the stored (elastic) energy in the homogeneously strained region far in front of the crack tip with the energy of the crack surfaces in the fully relaxed region far behind the crack

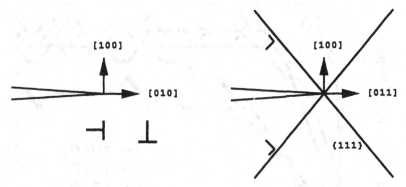

Figure 2: Crystallographic orientation of the crack systems, specified by the crack plane and the crack front direction, with respect to the glide systems. The (100)[001] crack system is shown on the left, the (100)[0$\bar{1}$1] crack system on the right. A detailed description of the dislocations, which can be generated in these geometries is given in the text.

tip, one can easily show that crack propagation in this model can be studied under the condition of constant energy release rate. This condition holds as long as the crack tip is not interacting with the free surfaces to the right and left of the thin strip. The dimensions of the model are about 50 lattice parameters in height, $2h$, and 200 lattice parameters in length, L. Initially, the crack tip is placed at $L/4$.

Periodic boundary conditions of minimal periodic length are applied along the crack front to model plane strain boundary conditions. The restrictions imposed through the application of periodic boundary conditions are probably of less importance for the analysis of brittle crack propagation but will severely curtail the possibility of dislocation generation. Whereas dislocations could be generated in the form of dislocation loops in a real crystal, here they must be infinitely long and straight. Additional restraints to dislocation emission may result from the geometry of the crack system in relation to the slip systems: Depending on the orientation of the crack front, the natural Burgers vectors can be completely suppressed. Using these features of the periodic boundary conditions, one can therefore study brittle crack propagation in an otherwise ductile material.

In this paper, crack propagation is studied on the {100} plane in a fcc crystal. The interatomic interaction in the atomistic region is modelled by an EAM potential which is fitted to the cohesive energy, the lattice parameter and the elastic constants of Ni [12]. The crack front is oriented either along a <001> direction or along <011>. The crack systems, specified by the crack plane and the crack front direction, are therefore denoted as (100)[001] and (100)[0$\bar{1}$1]. In the (100)[001] crack system, the natural glide planes of the fcc crystal, the {111} planes, are all inclined against the crack front and dislocation emission on these planes is therefore suppressed. Dislocations which could possibly be generated are either of the Lomer type with a <001> Burgers vector as depicted in Figure 2(left) or must have larger Burgers vectors. The second crack system provides two sets of {111} glide planes at 54° to the crack plane (Figure 2(right)). The orientation of the 1/6<211> partial dislocations, which could be emitted on these glide planes is such that the edge partial is the

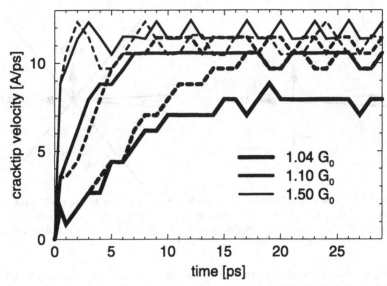

Figure 3: Crack tip velocity versus time is plotted for 3 different load levels and the two different damping coefficients. The damping coefficient for Ni (full lines) as well as a damping coefficient reduced by a factor of 30 (dotted lines) are used.

leading partial in backward orientation and the 60° partial is leading in forward direction.

Both these crack systems have already been studied quite extensively with the same EAM potential in static crack tip simulations of brittle fracture processes and dislocation nucleation [13, 14, 15]. The static calculations showed that brittle crack propagation on the original (100) crack plane is expected for both crack systems under opening (mode I) loading conditions [13]. Increasing shear (mode II) loading did not change the failure mode of the (100)[001] crack system (Figure 2(left)) up to a phase angle $\Psi = arctan\,(K_{II}/K_I) = 21°$, where the crack deviated onto an inclined {110} plane. Further increasing the shear component to a phase angle of $\Psi = 28°$ resulted in the nucleation of a Lomer dislocation on the (100) crack plane. The (100)[0$\bar{1}$1] crack system (Figure 2(right)) first showed a deflection of the crack from the original (100) plane onto an inclined {111} plane at a phase angle $\Psi = 10°$. Upon further increasing the shear component, the backward oriented edge partial dislocation was nucleated at the crack tip [13, 14] at a phase angle of $\Psi = 14°$.

The molecular dynamics technique employed here is a classical leap frog integration scheme for Newton's equation of motion. However, a local temperature control [16] which resembles an electronic heat bath for the ions is employed since the dynamically running crack is expected to act as a local heat source. The equation of motion is therefore given by:

$$\ddot{x}_i = \frac{f_i}{m_i} - \mu_i\,\dot{x}_i$$

with

$$\mu_i = \alpha \left(\frac{T_i - T_0}{(T_i^2 + \epsilon^2)^{1/2}} \right),$$

T_0 being the temperature of the electronic heat bath and T_i is taken as the kinetic energy of particle i. ϵ is a small number ($\frac{1}{10}$ K) to prevent μ_i from going singular for atoms at rest. This atomistic thermostat is essentially equal to the method proposed by Berendsen et al. [17] except for the local measure of the temperature. The viscous damping coefficient α is usually chosen to mimic the electron phonon coupling of Ni [16]. To study the influence of the local damping on the behaviour of the crack, the damping coefficient is sometimes also decreased to a 30th of this value. At the upper and lower borders of the model as well as at the free surfaces to the right and to the left (Figure 1), the damping is gradually increased to prevent the reflection of incoming phonons (see also [18]).

The models are first relaxed at boundary displacements which correspond to the Griffith load, G_0, i.e. at boundary displacements which (in front of the crack tip) lead to a homogeneous strain energy per unit length large enough to balance the surface energy of the two crack surfaces. At this load the model is then equilibrated for 10000 time steps equivalent to 20 ps at a temperature of $T_0 = 10$ K. This temperature is also kept as the temperature of the electronic heat bath during subsequent crack propagation. The models are then loaded to a higher strain level by scaling all atomic displacements relative to the ideal crystal positions. This scaling is applied instantaneously between two MD steps. Such scaling of the atomic displacements does not change the overall shape of the elastic strain field. The scaling therefore instantaneously changes the load level without creating any shock waves or otherwise disturbing the crack field. Such rescaling of the displacements can in principle be used any time during a MD run to either increase or decrease the load level.

RESULTS AND DISCUSSION

Dynamic Brittle Fracture: The (100)[001] crack system, displayed in Figure 2 (left), is exclusively used for the study of the dynamics of perfectly brittle, atomically sharp cracks. The dependence of the crack tip velocity on the load level is displayed in Figure 3 for different load levels and the two different damping coefficients. Crack tip velocity is evaluated from the momentary crack tip positions which are determined in 0.5 ps time intervals. The plot of the crack tip velocity versus time (Figure 3) shows that the crack tip accelerates in all cases to a terminal velocity. The acceleration infers a small but finite inertia of the crack in the given geometry. The inertia appears to depend only very slightly on the damping coefficient.

The terminal velocity reached at the different load levels shows a few interesting aspects. At small overloads, between $1.00 \, G_0$ and $1.02 \, G_0$ the crack is stationary due to lattice trapping. At $1.04 \, G_0$ the crack starts to move and accelerates up to a terminal velocity which already is a sizeable fraction of the Rayleigh wave speed. This suggests that the steady state propagation of a perfectly brittle crack is limited to velocities above a certain lower critical velocity. Such a lower critical velocity has previously been found [11] in analytical studies of simple 2D crack models and is hereby confirmed in a full MD study. The precise physical interpretation of the origin of this forbidden band of low velocities, however, remains still unclear and should be studied further with analytical models.

The terminal steady state velocity reached at small overloads (see Figure 3) strongly depends on the damping coefficient. The damping coefficient acts locally and most strongly on the atoms with the highest velocities. In this sense the local damping could be interpreted

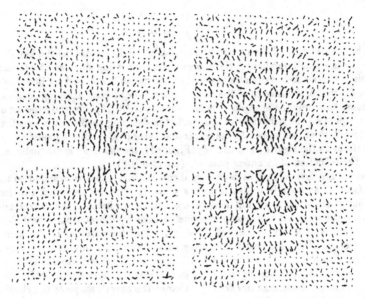

Figure 4: Velocity distribution of individual atoms around two (100)[001] cracks moving at the same steady state velocity but subjected to different overloads: (left) $\Delta G = 0.1\,G_0$, (right) $\Delta G = 0.5\,G_0$.

as reducing the amplitude of the shear waves which carry the information that the crack propagates. The damping would thereby essentially reduce the apparent overload and therefore reduce the steady state crack tip velocity.

Upon increasing the load to a value of $1.10\,G_0$, the crack reaches a terminal velocity of about $0.4\,v_R$. Further increasing the load up to $1.50\,G_0$ does not significantly change the terminal velocity. Furthermore, the crack tip velocity is almost independent of the damping coefficient in this loading regime. Averaged over longer times the steady state crack tip velocity is even somewhat lower for the lower damping coefficient at $1.50\,G_0$.

At this point one should ask the question how the crack manages to travel at almost identical velocity but to dissipate such different amounts of energy. An answer to this question may be found by analyzing the velocity fields around the crack tips of cracks moving at the same steady state velocity at overloads of $0.10\,G_0$ and $0.50\,G_0$. These velocity fields are displayed in Figure 4. At small overloads the velocity field around the crack is smooth and continuous as expected from continuum theory. At increasing overloads, however, the crack creates a shock wave at the breaking of every single atomic bond. The shock waves are mainly of longitudinal compressive character in the direction perpendicular to the crack plane but are characterized by a significant transverse component at an angle of 45° to the crack propagation direction. No shock waves are visible in the angular section between about -45° and 45° around the crack propagation direction.

To better understand the origin of the shock waves one can investigate the instantaneous crack tip configurations one of which is shown in Figure 5, which displays the crack tip configuration halftime between the breaking of the last and the next bond at the crack tip.

Figure 5: Snapshot of the atomic configuration at the tip of a shock wave emitting crack halftime between the breaking of the last atomic bond and the next.

It is clearly seen that the atoms above and below the actual crack tip have been pushed far away from the two crack tip atoms. Such behaviour can be rationalized as a consequence of the non-linearity of the interatomic interaction, which strengthens in compression and weakens in tension. The last broken bond obviously caused the atoms, which are now just behind the crack tip, to impact into the surface, thereby pushing their neighbours with strong compressive forces into the crystal. The interatomic interaction of the impacting atoms and their neighbours with the crack tip atoms thereby weakens due to the non-linearity of the interatomic potential. The reduced coupling strength (interaction) in turn reduces the velocity at which information is transferred along the crack surface and the non-linearity in the interatomic interaction could therefore be made responsible for the rather low terminal steady state velocity.

The hypothesis that the non-linearity of the interatomic interaction is responsible for the rather low terminal crack tip velocity of the moving brittle crack can be tested by repeating the dynamic calculations with a simple harmonic pairwise snapping spring force law, which only extends to a distance halfway between the first and second nearest neighbours of the fcc lattice. These calculations are a bit tricky since the short range snapping spring force law results in a rather large lattice trapping of $\Delta G > 1.0\,G_0$. Furthermore second neighbours may come within interaction range and make the effective interaction non-linear. To circumvent these difficulties, the neighbour list of the atoms is frozen in the initial configuration and not updated during crack propagation. To start the crack, it was loaded beyond the lattice trapping range for only 1 ps and then instantaneously unloaded to much smaller overloads whereby it could then reach the steady state. One then finds that it is possible to drive the perfectly brittle crack up to 0.85 of the appropriate Rayleigh wave speed at very moderate overloads of only $0.40\,G_0$. The resulting velocity distribution around the running crack tip is shown in Figure 6. It clearly reveals the perfectly continuous velocity distribution expected from continuum theory and shows no signs of any shock waves (although one might have guessed that the extremely non-linear snapping of the bonds may cause such behaviour).

In summary, the study of dynamically moving perfectly brittle cracks showed very close similarities to experimental observations [2], in that the maximum steady state velocity that can be attained by such cracks is found to be limited to about 40% of the Rayleigh wave speed. After reaching the terminal velocity, cracks release the surplus of mechanical energy by emitting shock waves, caused by the non-linearity in the interatomic interaction,

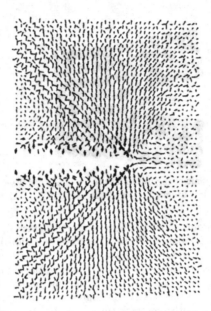

Figure 6: Velocity distribution around a (100)[001] crack modelled with a harmonic snapping spring force law. The crack is loaded at $G = 1.4\,G_0$, where it reaches a steady state velocity of $0.85\,v_R$.

as seen by comparing different interaction models. The non-linearity of the interatomic interaction is responsible not only for the generation of shock waves but also for the 'low' terminal velocity which can be reached by the moving brittle crack.

Dynamic Instability and Dissipation at Large Overloads: Driving the crack at higher loads results in crack tip instabilities. The loads necessary to create an instability depend quite strongly on the actual crystallographic orientation and on the damping. For the crack system studied above the critical load is approximately $1.9\,G_0$ if the damping coefficient for Ni is used and only about $1.7\,G_0$ for the reduced damping. The (100)[0$\bar{1}$1] crack becomes unstable already at a load of $1.4\,G_0$.

Evaluating the atomic positions around the tip of the shockwave-emitting, steady-state (100)[001] crack of Figure 5, already shows that instability in this crack system is reached when the bonds between the impacting atom and the next crack tip atom and/or the bonds between them and the atoms just above or below break before the next crack tip bond. The instability is however mainly of transverse character trying to shear the {110} planes at 45° into <$\bar{1}$10> directions. Despite the large shear on these planes, dislocations of such large <$\bar{1}$10> Burgers vectors at first can not successfully be created. Similarly, dislocations with [100] Burgers vectors normal to the crack plane are not observed and the overload is not large enough to allow for bifurcations. As a consequence, the crack tip which is just going unstable at a load of $2.0\,G_0$ has only two possibilities: it can either create a Lomer dislocation with a Burgers vector parallel to the crack propagation direction, which is occasionally observed, or it can attempt branching. Such branching attempts are seen

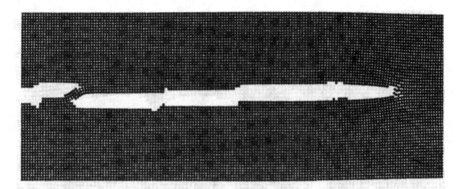

Figure 7: Fracture surface of a (100)[001] crack at a load of $G = 2.0\,G_0$. The crack is loaded just beyond the regime where it can reach a constant steady state velocity. It periodically develops crack tip instabilities, which lead to the cleavage steps observed on the fracture surface.

most often. They eventually lead to monoatomic or diatomic steps on the fracture surface.

After each instability event, the crack accelerates again up to terminal velocity, builds up the shock waves, slows down a little and is then stopped by the next instability event. This leads to a periodic repetition of instability events and violent crack tip velocity oscillations. Those crack tip velocity oscillations bear some resemblance with those found experimentally [2], however, they occur on a much shorter time scale and can therefore not be compared directly. The fracture surface created by such periodic instability events is displayed in Figure 7 for a crack propagating at $G = 2.0\,G_0$.

Loading the crack at a constant strain rate of $5\,10^{-4}$/ps the crack first moves in a perfectly brittle way. It then becomes critical and the same instability events as discussed above occur with increasingly higher frequency until the emission of a <110> dislocation becomes possible at a load level of approximately $G = 3.0\,G_0$. Such an event is shown in Figure 8. The dislocation emission occurred at the V-shaped contour on the fracture surface and then moved away from the crack tip. The crack also generated dislocations with Burgers vectors parallel to the crack propagation direction which it is pushing in front of it. Although the <110> dislocation is apparently very difficult to move, the high stress field of the crack is at least strong enough to push the dislocation far enough away from the crack surface so that it is not drawn back by the image stress. Similarly, Lomer dislocations are relatively difficult to move, but in the near tip field they can move even if their Burgers vector is parallel to the crack propagation direction and therefore sees only relatively little driving force.

At such high load levels dislocations can even climb in the near tip field. An example of such a climb process is shown in Figure 9, where shortly after the instantaneous increase of the load level to $G = 4.0\,G_0$ and after only a few steps of crack propagation, a Lomer dislocation climbed up three lattice spacings from the crack tip and was then carried with the crack tip. The crack tip velocity is not oscillating very strongly in this case. It appears as if the sharp brittle crack under such high overload is not even stable for very short periods of time. The crack is rather constantly made up of a heavily deformed region which appears as if it is almost molten.

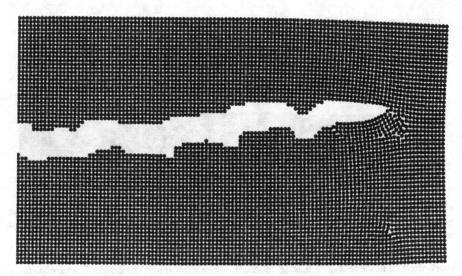

Figure 8: (100)[001] crack at increasing loading showing the emission of a <110> dislocation at a load level of approximately $G = 3.0\,G_0$. The dislocation is located in the lower right corner of the displayed section of the model.

Although these examples show a very rich variety of energy dissipating processes which can be associated with the propagation of a crack in a brittle material, no definite conclusions can be drawn from them yet. Much more systematic studies and also studies of the long time evolution of the cracks will be needed to understand which of these processes are really important. However, it is already clear from these few examples that unexpected processes like the climb of dislocations or the coupled motion of cracks and dislocations with Burgers vectors in crack propagation direction can occur at very high overloads.

As a last example, the crack tip configuration of the (100)[0$\bar{1}$1] crack is shown in Figure 10 at a load of $1.4\,G_0$. The crack first propagated, reached the instability, and was then stopped by the emission of a full $1/2$<110> dislocation to one side and a micro-twin to the other side. The behaviour of this crack on a macroscopic length and time scale would be interpreted as ductile response, although the crack initially (or under smaller overload) is able to propagate in a perfectly brittle manner. Understanding whether the crack will be stopped by dislocation emission or whether it will further grow as in the case displayed in Figure 8 is of course one of the major goals of the MD modelling of fracture processes. However, until now this difference is not fully understood. One may speculate that the velocity at which dislocations are generated at the crack tip and/or the velocity at which they can move away from the crack could be the decisive factor but further studies will be needed to clarify this issue.

At last, it seems worth noting that dislocation nucleation from a dynamically running crack is markedly different from the nucleation from a static crack. Whereas the static crack tip field in both crack systems studied here would favour at first a deflection of the crack from the original cleavage plane and give dislocations on the backward oriented glide plane

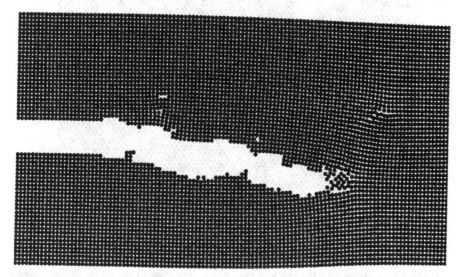

Figure 9: (100)[001] crack at $G = 4.0\,G_0$ showing dislocation climb processes in the near tip field.

first [13], the dynamic crack apparently favours dislocation nucleation on forward oriented glide planes at an angle around 45°, which of course are much more effective in shielding the crack.

CONCLUSIONS

The MD simulation of the dynamic fracture process can yield insight into new and unexpected phenomena at crack tips and may lead to a better understanding of the fundamental physical processes determining the brittle or ductile response of a material. Furthermore MD modelling of the dynamics of a brittle crack can sometimes be directly related and compared to experimental observations. As an example of the latter I showed in the first part of the paper that the maximum steady state velocity which can be attained by a dynamically moving, perfectly brittle crack is limited to a value well below the critical velocity obtained from continuum mechanical analysis. In the case of the (100)[001] crack in fcc Ni, it is limited to only 40% of the Rayleigh wave velocity, which incidentially compares very well with the maximum velocities found in experiments on PMMA and glass [2, 3]. Cracks at speeds close to the terminal velocity and at high overloads create shock waves at the breaking of every single atomic bond, whereas cracks at small overloads lead to a continuous velocity distribution around the tip. Comparing a perfectly harmonic force law with a more realistic EAM potential, it is shown that the harmonic potential always leads to a continuous velocity distribution around the crack tip and does not emit shock waves. It therefore dissipates less energy and the crack consequently reaches a higher velocity, close to the Rayleigh wave speed.

At even higher overload, the crack eventually becomes unstable and creates defects beyond just creating the perfect cleavage surface. These additional defects can be either cleavage steps, dislocations with Burgers vectors parallel to the crack propagation direc-

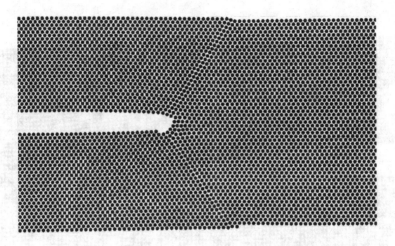

Figure 10: $(100)[0\bar{1}1]$ crack at $G = 1.4\,G_0$ being stopped by the emission of a dislocation an the creation of a microtwin.

tion, which then move with the crack or even climb very close to the crack, or dislocations with blunting Burgers vectors. The details of these instability processes depend on the crystallographic orientation of the glide systems with respect to the crack system and also depend on the level of overload.

ACKNOWLEDGMENTS

It is a pleasure to thank Brad L. Holian for introducing me to the modelling of dynamic fracture processes, inviting me to Los Alamos National Laboratories, discussing the subject of this paper with me in detail and then carefully reading the manuscript. Financial support from KSB-Stiftung, Stuttgart is gratefully acknowledged.

REFERENCES

[1] J. Fineberg, S. P. Gross, M. Marder, and H. L. Swinney, Phys. Rev. Letters **67**, 457 (1991).

[2] J. Fineberg, S. P. Gross, M. Marder, and H. L. Swinney, Phys. Rev. B **45**, 5146 (1992).

[3] S. P. Gross *et al.*, Phys. Rev. Letters **71**, 3162 (1993).

[4] L. B. Freund, *Dynamical Fracture Mechanics* (Cambridge University Press, New York, 1990).

[5] E. H. Yoffe, Philos. Mag. **42**, 739 (1951).

[6] A. A. Griffith, Philos. Trans. R. Soc. **221A**, 163 (1921).

[7] R. Thomson, C. Hsieh, and V. Rana, J. Appl. Phys. **42**, 3154 (1971).

[8] J. E. Sinclair, Phil. Mag. **31**, 647 (1975).

[9] B. Lawn, *Fracture of Brittle Solids - Second Edition* (University Press, Cambridge, UK, 1993).

[10] M. Marder and X. Liu, Phys. Rev. Letters **71**, 2417 (1993).

[11] M. Marder and S. Gross, J. Mech. Phys. Solids **43**, 1 (1995).

[12] S. M. Foiles, M. I. Baskes, and M. S. Daw, Phys. Rev. B **33**, 7983 (1986).

[13] P. Gumbsch, J. Mat. Res. **10**, 2897 (1995).

[14] P. Gumbsch and G. E. Beltz, Modelling Simul. Mater. Sci. Eng. **3**, 597 (1995).

[15] R. G. Hoagland, M. S. Daw, and J. P. Hirth, J. Mater. Res. **6**, 2565 (1991).

[16] M. W. Finnis, P. Agnew, and A. J. E. Foreman, Phys. Rev. B **44**, 567 (1991).

[17] H. J. C. Berendsen *et al.*, J. Chem. Phys. **81**, 3684 (1984).

[18] B. L. Holian and R. Ravelo, Phys. Rev. B **51**, 11275 (1995).

PARALLEL SIMULATIONS OF RAPID FRACTURE

Farid F. Abraham

IBM Research Division, Almaden Research Center, K18/D2, 650 Harry Road, San Jose, CA 95120-6099

ABSTRACT

Implementing molecular dynamics on the IBM SP2 parallel computer, we have studied the fracture of two-dimensional notched solids under tension using million atom systems. Brittle materials are modelled through the choice of interatomic potential function and the speed of the failure process, and our interest is to learn about the dynamics of crack propagation in ideal materials. Recent laboratory findings occur in our simulation experiments, one of the most intriguing is the dynamic instability of the crack tip as it approaches a fraction of the sound speed. A detailed comparison between laboratory and computer experiments is presented, and microscopic processes are identified. In particular, an explanation for the limiting velocity of the crack being significantly less than the theoretical limit is provided.

INTRODUCTION: A Computational Approach to Materials Failure

Continuum fracture theory typically assumes that cracks are smooth and predicts that they accelerate to a limiting velocity equal to the Rayleigh speed of the material.[1,2] In contrast, experiment tells us that, in a common fracture sequence, an initially smooth and mirrorlike fracture surface begins to appear misty and then evolves into a rough, hackled region with a limiting velocity of about six-tenths the Rayleigh speed. In some brittle materials, the crack pattern can also exhibit a wiggle of a characteristic wavelength. Recent experiments have clearly shown that violent crack velocity oscillations occur beyond a speed of about one-third the Rayleigh speed and are correlated with the roughness of the crack surface.[3] The authors concluded that the fracture dynamics may be universal, or structure independent, and that a dynamical instability of the crack tip governs the crack velocity behavior and the morphology sequence of 'mirror, mist and hackle.' All of these features are unexplained using continuum theory, though recent theoretical advances (e.g., by Langer[4] and Marder,[5]) are providing very important insights into this difficult problem. This suggests that a fundamental understanding may require a microscopic picture of the fracturing process. Pioneering atomistic simulations of crack propagation by Ashurst and Hoover[6] and the brittle to ductile transition by Cheung and Yip[7] were too small in size to study the dynamical crack stability issue.

With the advent of scalable parallel computers, computational molecular dynamics can be a very powerful tool for providing immediate insights into the nature of fracture dynamics. We have studied the rapid fracture of two dimensional triangular solids with up to 1500 atoms on a side, or about one half of a micron in length.[8] If we were to do three dimensions for an equivalent number of atoms, a cube would be only 130 atoms on a side! But like experiment,[3] our interest was to study two-dimensional 'mode one' loading. We were able to follow the crack propagation over sufficient time and distance intervals so that a comparison with experiment became feasible.

The molecular dynamics simulation technique is based on the motion of a given number of atoms governed by their mutual interatomic interactions described by continuous interatomic potentials and requires the numerical integration of Hamilton's equations of motion.[9] The interatomic forces are treated as central forces, modeled as

311

a combination of a Lennard-Jones (LJ) 12:6 with a spline cutoff.[10] We express quantities in terms of reduced units. Lengths are scaled by the parameter σ, the value of the interatomic separation for which the LJ potential is zero, and energies are scaled by the parameter ϵ, the depth of the minimum of the LJ potential. Reduced temperature is therefore kT/ϵ. Our choice of these simple interatomic force laws is dictated by our desire to investigate the qualitative features of a particular many-body problem common to a large class of real physical systems and not governed by the particular complexities of a unique molecular interaction. Richard Feynman summarizes this viewpoint well: "If, in some cataclysm, all scientific knowledge were to be destroyed, and only one sentence passed on to the next generation of creatures, what statement would contain the most information in the fewest words? I believe it is the atomic hypothesis that all things are made of atoms - little particles that move around in perpetual motion, attracting each other when they are a little distance apart, but repelling upon being squeezed into one another. In that one sentence, you will see, there is an enormous amount of information about the world, if just a little imagination and thinking are applied."

For an nice description of fast parallel algorithms for short-range molecular dynamics, we refer the reader to Plimpton's JCP article.[11] Our parallel molecular dynamics program is implemented on the IBM SP2 using 64 nodes. For 2,027,776 atoms, the update time per time step is 0.4 seconds with 96 percent efficiency.

METHOD AND RESULTS: THE FRACTURE MODEL AND COMPUTER EXPERIMENTS

Our system is a 2D rectangular slab of atoms with L atoms on a side, where $L = 712$ for the half million atom system and 1424 for the two million atom system. The slab is initialized at a reduced temperature of 0.00001. In our preliminary simulations, triangular notches of 10 lattice spacings are cut midway along the lower horizontal slab boundary, and an outward strain rate $\dot{\epsilon}_z$ is imposed on the outer most columns of atoms defining the opposing vertical faces of the slab. A linear velocity gradient is established across the slab, and an increasing lateral strain occurs in the solid slab. We found that a strain rate of $\dot{\epsilon}_z = 0.0001$ is sufficiently small for our size systems to prevent multiple fracture accompanying fracture at the notch. With this choice, the solid fails at the notch tip when the solid has been stretched by ~ 3 percent. We have also used an order of magnitude less strain rate of $\dot{\epsilon}_z = 0.00001$. At the onset of crack motion, the imposed strain rate remains constant (experiment 1) or is set to zero (experiment 2), and the simulation is continued until the growing crack has traversed the total length of the slab.

Figures 1 and 2 graphically summarize our nonzero strain rate simulation (experiment 1) for $L = 712$. Figure 1 shows (a) the crack tip position (in units of reduced length) and (b) the crack tip speed (in units of the Rayleigh sound speed), both as a function of reduced time. Figure 1(c) is an expanded view of the crack tip speed for early time. From Figure 1a, we find that the crack tip achieves a limiting speed equal to 0.57 of the Rayleigh speed c_R. However, the "instantaneous" tip velocity is very erratic (Figure 1b) after reaching a speed of $0.32c_R$. Before a time of about 70 and a speed less than $0.32c_R$, the acceleration of the crack tip is quite smooth (Figure 1c); but with the onset of the erratic fluctuations of the tip speed, there is significant deceleration of the propagating crack. Each of these features in Figs. 1a, b and c are obtained for experiment 2 and for the larger system ($L = 1424$), and they are in agreement with Fineberg's et al. experiments.[3] Like in the laboratory experiment, the influence of physical boundaries are a concern when sound and dynamical defects

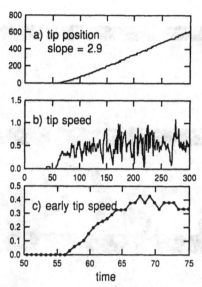

Fig. 1. (a) The crack tip position (in units of reduced length) as a function of reduced time. The slope is the limiting speed in reduced units which corresponds to 0.57 of the Rayleigh speed c_R; (b) The crack tip speed (in units of the Rayleigh sound speed) as a function of reduced time; (c) An expanded view of the crack tip speed for early time.

reflect from them. It should be noted that the onset of the instability relative to tip motion ($\delta t = 15$) occurs significantly earlier than it takes sound to travel from the tip to a lateral boundary and return ($\delta t = 80$). Hence, the transition seems to be an intrinsic instability.

Figure 2 shows the time evolution of the propagating crack using a gray-scale rendering of the instantaneous local transverse velocity v_z, or transverse strain rate. The scale goes from dark grey for the most negative v_z to light grey for the most positive v_z. Initially, the brittle crack propagates in a straight line and leaves "mirror" cleaved surfaces. Periodic stress waves immediately appear with motion. Corresponding to the onset of the erratic oscillations of the tip speed, at a tip speed of $\sim 0.32c_R$, the crack first begins to roughen and then to oscillate back and forth at approximate angles of ≤ 30 degrees from the vertical (the original direction of crack motion) and along symmetry lines of the crystal. Accompanying the oscillating 'zigzag' excursions of the growing crack is significant relaxation in the regions immediately next to the newly created surfaces. Prominent 'zigs' or 'zags' in the crack direction were accompanied by a propagating atomic displacement along two adjacent rows of atoms (a slip plane) that are ± 30 degrees to the vertical; it is initiated at the vertex of the change of direction and travels at about the longitudinal sound speed c_l. These 'dislocations' appear as slanted, inverted V being emitted from the moving crack tip, first to the right and then to the left. They may be traced back to their origin by constructing an imaginary line 30 degrees from the vertical and passing through the V. The vertical separation between neighboring dislocations equals the wavelength of the oscillating crack and is ~ 115. The V is simply an acoustical wake created by the moving dislocation. When

Fig. 2. The time evolution of the propagating L-J crack at constant strain rate of 0.0001 using a gray-scale rendering of the instantaneous local transverse velocity v_x, or strain rate, which goes from dark grey for the most negative v_x to light grey for the most positive v_x. The time sequence goes from left to right and top to bottom. The frames are for reduced times 100, 150, 175, 200, 225, 250, 275, 300 and 325.

the dislocation hits the top free boundary, an atomic step is formed. In experiment 1, the solid is being stretched at a nonzero strain rate, and it is "keeping up" with the lateral boundary expansion beyond its elastic limit through the creation and growth of a single crack seeded by the notch, as well as by necking through the slippage of adjacent atomic rows of atoms. For experiment 2, where the initially imposed strain rate is an order of magnitude smaller and set zero after the onset of crack propagation, we observe less dislocation emission accompanying the zigzag motion of the crack tip except at the final failure of the slab. This is in contrast to experiment 1 where increasing the strain during tip propagation forces the slab to relieve the additional strain by the sliding of atomic rows at each turning point of the crack direction at 30 degrees for forward. These off-angle slips provide an excellent signature for crack oscillation.

Another class of dislocations exists which are not apparent in the gray-scale pictures of the transverse velocity fields as presented in Figure 2. In Figure 3, the dislocation trajectories are shown for experiment 1 at various times during fracture tip motion. The off-angle slips are quite evident as lines emitted from the crack tip at plus and minus 30 degrees from the direction of motion. What is surprising is the appearance of an abundance of dislocations being emitted near the crack tip at right angles

Fig. 3. Snapshot pictures of the dislocation trajectories for the L-J fracture simulation in Figure 8. The time sequence goes from left to right and for reduced times 130, 235 and 360.

with respect to the forward motion! By examination, we observe that the transverse dislocations go out some distance, then return to the fracture surface where they disappear. The spacing between these dislocations is quite regular. We can understand their origin as a transverse slippage between two neighboring rows of atoms arising from a growing shear stress at the crack tip as the ever-increasing cascade of broken bonds in the forward direction allows equalling increasing parted rows of atoms to want to relax laterally. This buildup of severed atomic rows wanting to relax will eventually be sufficiently large enough to overcome the barrier to slippage. The front to slippage will manifest itself as a dislocation. Of course, this is a repeating process, hence a continual creating of transverse dislocations. Also, the return of dislocations to the crack surface and the healing of the surface is a consequence neighboring bands of matter bounded by slip planes relaxing to equilibrium.

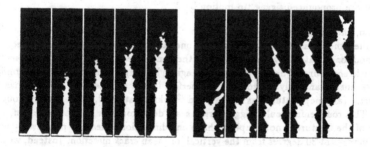

Fig. 4. a) The onset of crack instability, in reduced time intervals of 7 and beginning at reduced time 85. b) Late zigzag crack propagation, in reduced time intervals of 7 and beginning at reduced time 220.

To highlight the microscopic features of the failure dynamics, we present Figure 4 which is a short-time interval sequence of close-up views of the crack tip at an early time and at a late time. We see the onset of the crack instability beginning as a roughening of the created surfaces which eventually results in the pronounced zigzag tip motion; i.e., "smooth → rough → zigzag" corresponds to "mirror → mist → hackle." The times of onset for these various regions is shown in Figure 5. Note that roughening occurs before transverse slippage.

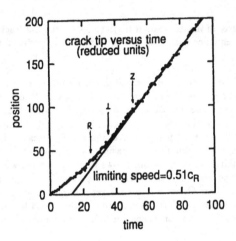

Fig. 5. The crack tip position (in units of reduced length) as a function of reduced time for experiment 2. The onset of the crack instability beginning as a roughening of the created surfaces is denoted by R; the first transverse dislocation by D; and the onset of the pronounced zigzag tip motion by Z.

Also, the onset of roughening corresponds to a point in the crack tip dynamics where the time it takes the tip to traverse one lattice constant approximately equals the period for one atomic vibration. Hence, the bond-breaking process no longer sees a symmetric environment due to thermal averaging, but begins to see local atom configurations "instantaneously" distorted from perfect lattice symmetry. This gives rise to small-scale (atomic) fluctuations in the bond-breaking path and, hence, atomic roughening. This roughening could trigger larger scale deviations. In the hackle region, the growth of the fracture is not simply a sharp one-dimensional cleavage progressing in a zigzag manner at 30 degrees from the vertical, or mean crack direction. Instead, we see a stair-step growth of connected "ideal 30 degree segments," resulting in a net forward angle of less than 30 degrees from the horizontal before changing the local direction by $\sim$ 60 degrees. The "ideal 30 degree crack segments" open at a velocity approximately equal to the Rayleigh speed c_R ! The origin of the erratic velocity oscillations is associated with the stair-step branching and connecting of regions of failure at and preceding the crack tip. The oscillating zigzag motion of the crack tip and the segmented stair-step growth contribute to the effective "forward" crack speed being less than theoretical prediction. The local fracturing is "brittle."

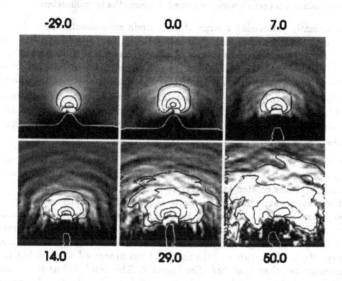

Fig. 6. Contour plot of the dynamic crack-tip stress $\sigma_{\theta\theta}$ in a region spanning a length $60\ \sigma$ and as a function of reduced time (see Figure 5). The stress contours have been multiplied by a factor of 100, the outer most contour being 50 followed by 100, 200, 300, etc.

Figure 6 shows a contour plot of the dynamic crack-tip stress $\sigma_{\theta\theta}$ in a region spanning a length $60\ \sigma$ and as a function of reduced time. The white line at the bottom of the figure gives the silhouette of the of the crack profile. We note that as the crack tip accelerates, the stress field flattens in the forward direction and expands laterally. Crack roughening occurs at approximately time 24 (Figure 5). In this time region, the stress distribution around the forward direction is flat, albeit noisy. With the disappearance of the prominent forward peak in the stress field, the crack can fluctuate off center, and the perfect forward motion is destroyed. By the time the crack tip has achieved its prominent zig-zag wandering (time equal to 50) the stress field is very broad and distorted, consistent with the erratic motion.

This microscopic branching has associated with it a larger-scale characteristic wavelength and a growing coarse-grain roughness The wavelength of our rough surface is about two orders of magnitude smaller than the millimeter wavelength measured by experiment. From the profiles, the roughness, or width, as a function of this distance are calculated.[12] We find that the width w scales with length L, according to the relation $w \sim L^{\zeta}$, where $\zeta = 0.81$. This result is in agreement with recent conjectures of a "universal" roughness exponent for the crack surface of real brittle materials.[12] In a continuum simulation of brittle fracture for PMMA, Xu and Needleman argue that the cohesive force between fracturing surfaces is approximately one-half micron.[13] This may be a proper "effective" force law for this complex material where surfaces do not separated by breaking short-range atomic bonds on the scale of angstroms. Hence, a new size scale of microns enters the macroscopic picture of fracture where the surface roughness, or wavelength, is millimeters. Simply scaling our results from atomic dimensions to microns transforms our observed wavelengths to millimeters.

We are presently performing a series of fixed strain experiments with a slit geometry for the crack and different initial crack lengths of 10, 20, 40 and 60.[14] The critical tension for crack propagation is given by $\sqrt{l_{tip}}T_c \sim constant$. We have learn several scaling rules. For example, the dynamics of crack position L versus time t is found to be homogeneous first-order; i.e., $\lambda L = \lambda function(t) = function(\lambda t)$ and collapses to a single curve if L and t are scaled by $l_{tip}^{-3/4}$ or $T_c^{3/2}$. For all experiments with different initial crack lengths, we find that the crack tip begins to roughen at a speed around 0.33 of the Rayleigh speed c_R. Also, the crack length at the onset of roughening scales as $\sqrt{l_{rough}}T_c \sim constant$.

We received a real surprise when we did our same fracture simulations on the two-dimensional triangular L-J crystal notched at the bottom horizontal boundary, but now the crystal was rotated by 90 degrees from our original orientation! That is to say, in the previous experiments, the notch is pointed perpendicular to horizontal atomic rows spaced by $\sqrt{3}/2a$ intervals, a being the lattice constant. Now, the notch is pointed perpendicular to horizontal atomic rows separated by $1/2a$ interval spacings, this being termed the cleavage direction since the created surface by fracture has the lowest energy. Much to our surprise, the crack did not proceeded upward, but turned toward the horizontal, then branched. See Figure 7. This result led us to investigate the elastic behavior of our simple crystal under large strains. We found a profound anisotropy in the dependence of Young's modulus and Poisson ratio on Mode I strain for the 2D L-J crystal for the original lattice orientation and the 90 degree rotation of the original lattice orientation. Such an anisotropy in the elasticity for large strains leads us to suspect that this was the origin for the different fracture behavior for the different notch orientations and loadings.

SUMMARY

A detailed comparison between laboratory and computer experiments has been presented. Many of the recent laboratory findings occur in our simulation experiments, one of the most intriguing being the dynamic instability of the crack tip and its associated properties. Microscopic processes have been identified, and explanations of certain features have been suggested; in particular, the reason for the limiting velocity being significantly less than the theoretical limit. Like Gross et al., we conclude that the fracture behavior appears to be universal, or independent of materials structure, simply because we observe the laboratory phenomena for our two-dimensional simple atomic system. Of course, as is often the case, the time and length scales in an atomistic simulation are significantly different from typical laboratory experiment. In contrast to millimeters and microseconds, we observed the phenomena to occur on scales of tens of nanometers and pico- to nano- seconds. However, our "simulation microscope" is validated by comparison and agreement with experiment. As mentioned earlier, the influence of physical boundaries is a concern when sound and dynamical defects reflect from them; e.g., the appearance of Wallner lines on fractured surfaces may occur in brittle materials and are known to be caused by stress waves reflected from nearby surfaces. Comparison with the larger system suggests that the immediate effects concerning the overall features of the slab fracture were minor. The "real bottleneck" is time; we need to simulate to microsecond time-scales!

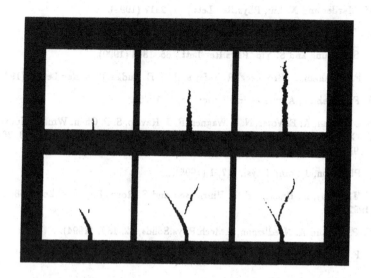

Fig. 7. The time evolution of the propagating L-J crack at constant strain rate of 0.0001. The time sequence goes from left to right. The top row is for the original lattice orientation and the bottom row is for a 90 degree rotation of the original lattice orientation.

ACKNOWLEDGEMENTS

This research was conducted in part, using the resources of the Cornell Theory Center, which receives major funding from the National Science Foundation and New York State with additional support from the Advanced Research Projects Agency, the National Center for Research Resources at the National Institutes of Health, IBM Corporation and members of the Corporate Research Institute.

REFERENCES

1. L. B. Freund, Dynamical Fracture Mechanics (Cambridge Univ. Press, New York, 1990).

2. H. J. Herrmann and S. Roux, editors, Statistical Models for the Fracture of Disordered Media (North-Holland, Amsterdam, 1990).

3. J. Fineberg, S. P. Gross, M. Marder and H. L. Swinney, Phys.Rev.Lett., 67, 457 (1991); J. Fineberg, S. P. Gross, M. Marder and H. L. Swinney, Phys.Rev., B45, 5146 (1992). S. P. Gross, J. Fineberg, M. Marder, W. D. McCormick and H. L. Swinney, Phys.Rev.Lett., 71, 3162 (1993).

4. J. S. Langer, Phys.Rev.Lett., 70, 3592 (1993); J. S. Langer and H. Nakanishi, Phys.Rev.E, 48, 439 (1993).

5. M. Marder and X. Liu, Phys.Rev.Lett., 71, 2417 (1993).

6. W. T. Ashurst and W. G. Hoover, Phys.Rev., B14, 1465 (1976).

7. K. S. Cheung and S. Yip, Phys.Rev.Lett., 65, 1804 (1990).

8. F. F. Abraham, D. Brodbeck, R. Rafey and W. E. Rudge, Phys.Rev.Lett., , (1994).

9. F. F. Abraham, Advances In Physics, 35, 1 (1986).

10. B. L. Holian, A. F. Voter, N. J. Wagner, R. J. Ravelo, S. P. Chen, Wm. G. Hoover, C. G. Hoover, J. E. Hammerberg and T. D. Dontjie, Physical Review A, 43, 2655 (1991).

11. S. Plimpton, J.Comp.Phys., 117, 1 (1995).

12. K. J. Måløy, A. Hansen, E. L. Hinrichsen and S. Roux, Phys.Rev.Lett., 68, 213 (1992).

13. X.-P. Xu and A. Needleman, J.Mech.Phys.Solids, 42, 1397 (1994).

14. F. F. Abraham, unpublished.

STABILITY ANALYSIS OF CRACKS PROPAGATING IN THREE DIMENSIONAL SOLIDS

H. Larralde, A. A. Al-Falou and R. C. Ball,
Cavendish Laboratory, Madingley Rd., Cambridge CB3 0HE, UK.

Abstract

We present a theory for the morphology of the fracture surface left behind by slowly propagating cracks in linear, isotropic and homogeneous three dimensional solids. Our results are based on first order perturbation theory of the equations of elasticity for cracks whose shape is slightly perturbed from planar. For cracks propagating under pure type I loading we find that all perturbation modes are linearly stable, from which we can predict the roughness of the fracture surface induced by fluctuations in the material. We compare our results with the classical results for cracks propagating in two dimensional systems, and discuss the effects in the three dimensional analysis which result from taking into account contributions from non-singular terms of the stress field, as well as the effects arising from finite speeds of crack propagation.

The morphology of surfaces created by fracture of brittle materials has surprisingly universal features at various length scales [6, 8]. At large length scales the surface presents a structure which is characterized by a very smooth initial region associated with the early stages of crack propagation at which the speed of propagation is still relatively low. This region is referred to as the "mirror" region. The surface then evolves into the increasingly rough "mist" and finally "hackle" regions, which correspond to high speeds of crack propagation [6]. In this paper we present results concerning the mesoscopic structure of the fracture surface created by slowly propagating cracks driven by mode I loading in three dimensional systems.

Specifically we consider a semi-infinite crack propagating quasistatically in an infinite three dimensional linear and isotropic material, subject to nominal type I loading at infinity. We assume that the shape of the crack is almost planar and treat the deviation from planarity as a perturbation parameter. Evidently, these "out-of plane" perturbations are responsible for the shape of the fracture surface. "In-plane" perturbations of the crack edge may affect the speed of propagation, but, to first order in perturbation theory, the in- and out-of-plane modes cannot mix [1, 2], so we focus on the out-of-plane case for a crack whose edge moves at a small constant velocity v_o.

The assumption that the crack propagates quasistatically allows us to perform the stress analysis as for a static crack, and, perhaps even more valuable, it allows us to circunvent the problem of the criterion of propagation. We assume that the unperturbed crack propagates under type I loading, the perturbations of the crack shape give rise to shear fields in the vicinity of the crack edge, of these, from symmetry considerations, the type III fields cannot affect the direction of propagation to first order. Thus, it is only the type II fields arising from the perturbed crack shape that can affect the posterior direction of propagation. It is straight forward to show that if $\bar{\Sigma}_I(r,\theta) = \frac{K}{r^{1/2}}F_I(\theta)$ is the stress field corresponding to type I loading and $\bar{\Sigma}_{II}(r,\theta) = \frac{K}{r^{1/2}}F_{II}(\theta)$ that for type II loading then $\bar{\Sigma}_I(r,\theta) + \epsilon\bar{\Sigma}_{II}(r,\theta) = \bar{\Sigma}_I(r,\theta+2\epsilon)$, correct to linear order in ϵ [1]. Thus, as long as we keep to linear order, the presence of a small amount of type II loading in addition to the type I loading field is indistinguishable to the original type I field slightly rotated from its original direction. This fact, together with the initial assumption that the crack propagates under type I loadings uniquely determines the direction of posterior propagation. It should be noted that the direction of propagation predicted by this argument coincides with that expected from the most common criteria of crack propagation[6].

While a complete description of the calculation does not belong here, it is worthwhile mentioning some of the salient points. First of all, it should be noted that the singular stress fields of a straight semi-infinite planar crack subject to type I loading satisfy the boundary conditions very strongly. So much so, that the surface of a slightly perturbed crack in the presence of those fields remains unloaded to first order in the perturbation [1]. Nevertheless, the shape perturbations displace the crack edge out of the plane, so the line at which the fields of the unperturbed crack become singular does not coincide with the locus of the edge of the perturbed crack. Thus, the fields must be locally displaced so the singularity coincides with the perturbed crack edge. The displacement of the edge manifests itself in the outer expansion of the stress fields as an unphysically divergent term, namely a term that diverges as $r^{-3/2}$ as $r \to 0$, where r is the distance to the position of the unperturbed crack edge. The actual stress fields arising from the perturbed crack shape are obtained from the next term of the solution, which indeed diverges as $r^{-1/2}$. This situation is in contrast to what happens in purely two dimensional systems. In these, the unperturbed crack can be chosen so its tip coincides with that of the perturbed crack, so there are no fields arising from the displacement. Further, as it continues to be true that the perturbed crack surfaces remain unloaded to first order in the presence of the unperturbed singular fields, in order to obtain a non zero answer it is necesary to consider the contributions arising from the nonsingular terms of the unperturbed fields (generically referred to as transverse or T stresses) [5]. In three dimensional systems these contributions are also present and are calculated using conventional weight function methods[4], yet, unless transverse stresses are applied to the system, the contributions of the nonsingular fields are negligible compared to those arising from the displacement of the edge for large enough cracks [2].

Since the complete calculation is done solely to first order in the perturbation the problem remains a linear one. This, in addition to translational invariance along the direction of the crack edge, justifies considering separate Fourier modes of the perturbation. Thus, without loss of generality, we considered perturbations varying as $\eta(x,k)\cos kz$, where we consider the unperturbed crack to lie on the semi-infinite $x-z$ plane (with $x < 0$), so its edge coincides with the z axis, and $\eta(x,k)$ is the perturbation parameter. Writing $\eta(x=0,k) \equiv \eta_0(k)$, the edge of the perturbed crack lies on the locus $(0,\eta_0\cos kz,z)$.

The central result of our calculation is that the stress intensity factor (SIF) characterizing the type II field in the vicinity of the crack edge as a result of the perturbed crack shape, neglecting the contributions of the T stresses, is given by [1, 2]:

$$K_{II}^{(1)} = \frac{1}{2} |k| \frac{2 - 3\nu}{2 - \nu} \eta_0(k) K_I,$$

(1)

where K_I is the SIF induced by the type I loading on the unperturbed crack and ν is the Poisson's ratio of the material. It should be noted that this SIF correspond to the type II field as measured in the frame of reference of the unperturbed crack, as opposed to being measured in the proper frame of the perturbed crack, the difference being an extra "tilt" term.

The above result has several features, the most striking is that it only depends on the position of the perturbed crack edge, and not on the previous shape of the crack. Also, as expected, it is symmetric in k and vanishes as $k \to 0$, which corresponds to the two dimensional limit. With this result, and given the previous discussion on the direction of propagation as a function of the local stress field around the crack edge, we are led to the following equation for the slope of the propagating crack edge [2],

$$\frac{\partial \eta(x, k)}{\partial x} = -\frac{2 - 3\nu}{2 - \nu} |k| \eta(x, k).$$

(2)

Since the factor $A(\nu) = \frac{2-3\nu}{2-\nu}$ is positive for all real values of ν, it is readily apparent that all modes of the perturbation are linearly stable. Indeed, from eq.(2) we see that sinusoidal corrugation decays as

$$\eta(x, k) \propto \exp[-A(\nu)|k|x].$$

(3)

This prediction was tested with a simple experiment on a brittle foam marketed for (dry) floral arrangements called *Oasis*. Samples of Oasis were cut with a corrugated knife 40% of the way through the sample, and then carefully prized apart propagating the initial crack across the sample. The peak to peak amplitude of the corrugation was measured at fixed intervals along the direction of propagation. These results are shown in figure (1a). The value of Poisson's ratio determined from the slope of the best line fit, $\nu = 0.40 \pm 0.01$ agrees well with the value obtained by mechanical methods on the same material, $\nu = 0.41 \pm 0.06$ [2].

The fracture surface left behind by initially flat cracks in Oasis was also studied. The features of these surfaces are presumably driven by local fluctuations of the material, and we assume that the effect of these fluctuations takes the form of an additive uncorrelated white noise in eq.(2). Under these assumptions, the power spectra for cuts both parallel and perpendicular to the crack edge are given by [2]

$$< |\eta(k)_\perp|^2 > = < |\eta(k)_{||}|^2 > \propto |k|^{-1}.$$

(4)

This prediction is supported by the plots of the power spectra obtained from fractured samples of Oasis shown in figure (1b).

As emphasised earlier, these results apply to cracks propagating slowly due to type I loading in three dimensional materials under conditions in which the nonsingular transverse T stresses can be neglected. The extension to the more realistic case in which the speed of propagation of the crack is high enough to require a full elastodynamic treatment poses several problems. Not the least of these is the actual determination of the perturbation fields. Our efforts thus far [3], while incomplete, have shown that the type II stress intensity

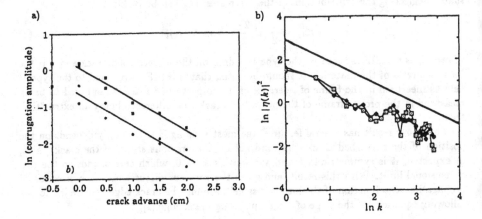

Figure 1: a) Semi-log plot of the peak to peak corrugation amplitude as a function of crack advance for sinusoidal cracks propagated through *oasis*. Both faces of two samples were measured, the straight lines have slopes 0.871 ± 0.004 corresponding to a Poisson's ratio of 0.4. b) Power spectra of the fracture surface propagated from an initially planar crack. The light squares correspond to sections perpendicular to the crack edge, the diamonds to sections parallel to the edge. The data was averaged over 4 samples. The straight line has slope -1.

factor induced by slowly varying perturbations in shape of fast cracks shows very little variation from the static value up to the speed set by the Yoffe criterion [6, 7]. This speed being already higher than the terminal speed of crack propagation observed experimentally [6]. Nevertheless, the ubiquitous morphological structure of fracture surfaces is suggestive of the existence of a dynamical instability at which the increase of work of fracture associated with the increase of fracture surface would hinder further acceleration of the crack [2]. Indeed, dynamical instabilities have been observed recently in essentially two dimensional systems [9]. While inconclusive, our results concerning fast cracks would suggest that the usual criterion for crack propagation, namely that cracks propagate in the direction of maximum hoop stress, while appropiate for slow cracks as discussed above, might not be correct for fast cracks. If so, a closer look at the physics of propagation will be required to elucidate the correct criterion.

If we again restrict ourselves to slowly propagating cracks, for which the direction of propagation is well established, and consider the effect of applied T stresses on the propagation of perturbation modes, a surprising variety of phenomena occurs [4]. The contribution arising from the T stresses can be calculated using weight function methods [7] and, thanks to the linearity of the problem, added to the contribution arising from the displacement of the crack edge. If we consider solely the contribution of T_{xx} (i.e. the tensile stress parallel to the direction of propagation), then the equation of evolution for each perturbation mode

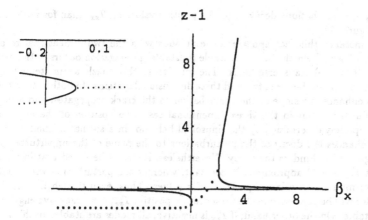

Figure 2: Plot of the roots of eq.(6). The connected lines denote real roots while the dotted lines denote the real part of the complex roots. For each value of β_x the root with largest real part corresponds to the dominant growth or decay mode depending on whether it lies above or below the origin. The inset shows the detail of the plot near $\beta_x = 0$, note that near the origin the rate of decay is faster for positive values of β_x than for negative ones.

$\eta(x, k)$ becomes [4]

$$\frac{\partial \eta}{\partial x} = -\frac{2-3\nu}{2-\nu} |k| \, \eta + \frac{2T_{xx}}{K_I} \sqrt{\frac{2}{\pi}} \int_{-\infty}^{x} dx' \frac{e^{-|k|(x-x')}}{\sqrt{x-x'}} \left[\frac{\partial \eta}{\partial x'} \left(1 + \frac{2\nu}{2-\nu} |k|(x-x') \right) \right] \quad (5)$$

In the limit $k \to 0$, this equation essentially reduces to that of Cotterel and Rice [5] describing the shape of slightly curved or kinked cracks in two dimensional systems. In their work, they showed that the propagation was stable for positive T_{xx} and unstable for negative T_{xx}. In the three dimensional case described by equation (5) the situation is far more complex. Assuming that the form of the crack is a known Laplace transformable function from $-\infty$ to the origin, say, it can be shown that the leading behavior of $\eta(x, k)$ is of the form $e^{(\zeta-1)|k|x}$ where ζ is the root with largest real part of

$$z^{3/2}\left(z - \frac{2\nu}{2-\nu}\right) - \beta_x(z-1)\left(z + \frac{\nu}{2-\nu}\right) = 0 \quad (6)$$

and $\beta_x \equiv \frac{2T_{xx}}{K_I}\sqrt{\frac{2}{|k|}}$ [4]. The behavior of the roots of eq.(2) is shown in figure (2). From this plot we infer that, for a fixed value of k, $\zeta - 1$ remains negative, so perturbations are exponentially supressed, for values of T_{xx} from $-\infty$ up to a specific positive value. Beyond that value $\zeta - 1$ becomes complex (the real part being shown by dots in the figure), indicating that the system presents an oscillatory stable behavior. As T_{xx} increases, the real part of $\zeta - 1$ becomes positive, corresponding to an oscillatory unstable growth of the perturbations. And finally, for large enough T_{xx}, $\zeta - 1$ becomes real and positive which corresponds to pure exponential growth of the perturbations. Further, and rather surprisingly, for small value of $|T_{xx}|$, perturbations decay faster for positive values of T_{xx} than for negative ones (see inset figure (2)).

In first instance this fact appears to be at odds with the two dimensional results of Cotterell and Rice [5] which state that stable (unstable) propagation occurs in the presence of negative (positive) transverse fields. The solution to this puzzle again stems from the fundamental difference between two and three dimensional systems. In both cases a positive T_{xx} acts to enhance the slope of the perturbation as the crack propagates, and a negative T_{xx} to diminish it. But in the three dimensional case, the position of the unperturbed crack is completely determined by the sinusoidal behavior in z so that a small increase in the slope enhances the decay of the perturbations to the plane of the unperturbed crack, while a negative T_{xx} hinders the decay. Nevertheless, it should be noted that in an actual experiment T_{xx} is held approximately constant, whereas the perturbation will usually be composed of modes over a large spectrum of k values. Thus, even though the small wave length modes will be supressed in the presence of a positive T_{xx}, the large wavelength modes will be unstable. On the other hand, if T_{xx} is negative, all modes are stable. In this respect, the qualitative behavior of cracks propagating three dimensional is not unlike that of cracks propagating in two dimensional systems.

To conclude, we find that truly three dimensional out-of-plane perturbations of a quasistatic crack relax stably under the dominant influence of the singular stress fields. Only when the wavelength λ of the perturbations becomes significant compared to sample dimensions, so that $T_{xx}\sqrt{\lambda}$ compares with K_I, do we find the T stresses which dominate the two dimensional results becoming important.

A. A. Al-Falou gratefully acknowledges financial support from the DAAD HSPII. Dr. H. Larralde was funded by EPSRC (formerly SERC).

References

[1] R. C. Ball and H. Larralde; International Journal of Fracture 71, pp. 365-377, (1995).

[2] H. Larralde and R. C. Ball; Europhys. Lett., (30), pp. 87-92, (1995).

[3] H. Larralde and R. C. Ball; Dynamic stress intensity factors and stability of three dimensional planar cracks; submitted to J. Mech. Phys. Solids (1995).

[4] A. A. Al-Falou, H. Larralde and R. C. Ball; Effect of T-stresses on the path of a three dimensional crack propagating quasistatically under type I loading; submitted to J. Mech. Phys. Solids (1995).

[5] B. Cotterell and J. R. Rice; International Journal of Fracture, Vol. 16, No. 2, pp. 155-169, (1980).

[6] Lawn, B.; Fracture of Brittle Solids; 2nd ed. (Cambridge University Press, Cambridge, 1993).

[7] Freund, L. B.; Dynamic Fracture Mechanics; (Cambridge University Press, Cambridge, 1990).

[8] J. P. Bouchaud, E. Bouchaud, G. Lapasset and J. Planes; Phys. Rev. Lett. 71, pp. 2240, (1993).

[9] S. P. Gross, J. Fineberg, M. Marder, W. D. McCormick and H. L. Swinney; Phys. Rev. Lett. 71, pp. 3162, (1993).

THE ROLES OF ATOMIC-SCALE DYNAMICS AND STRUCTURE IN THE BRITTLE FRACTURE OF SILICA

THOMAS P. SWILER*, TANSEN VARGHESE**, JOSEPH H. SIMMONS**
*Sandia National Laboratories, Albuquerque NM 87185
**Department of Materials Science and Engineering, University of Florida, Gainesville FL 32611

ABSTRACT

We modeled the initiation of fracture in vitreous silica at various strain rates using molecular dynamics simulations. We avoided biasing the location for fracture initiation within the sample so that we could study the effects of dynamics and structure on determining the path to fracture, defined as the particular bonds that break during the course of fracture. We sought to show that the path to fracture would be primarily determined by the local variations in the structure of the vitreous phase at low strain rates, with diminished sensitivity on structural variations at higher strain rates. However, the results of our model indicate that the path to fracture is dependent not only on the initial structure of the system and the applied strain rate, but also on the initial phase of the thermal vibrations. This underscores the importance of atomic dynamics in determining the path to fracture in brittle materials and provides a justification for extending the analysis of fracture surfaces to the near-atomic scale.

INTRODUCTION

Although brittle fracture is a seemingly familiar process, there is no consensus on the atomistic processes involved. This stems from the fact that brittle fracture is a rapid process, and current analytical techniques are not capable of imaging materials on both the appropriate size and time scales involved. The fractoemission analysis technique of Dickinson et al.[1] is one of the only currently available techniques that can provide dynamic fracture information on the activities of individual atoms and breaking bonds. Although this technique provides important insight into the dynamics of brittle fracture, it cannot provide a comprehensive understanding of the atomic mechanisms involved.

We chose to perform molecular dynamics simulations to elucidate the roles of atomic-scale dynamics and structure during brittle fracture. Molecular dynamics (MD) gives us the ability to see with detail all aspects of a system undergoing fracture. The tricky part of using MD to model fracture is not necessarily performing the simulations, but analyzing the results of the simulations. For example, there is a tendency for people to draw a direct comparison between the stress-strain curves derived from MD simulations with those obtained from experiment. We will explain what the limitations to direct comparison are, and we will show that the simulation of a small system having simple potential functions, with proper analysis, can provide insight into brittle fracture that the simulation of much larger systems, lacking proper analysis, cannot.

METHOD

We performed molecular dynamics simulations on silica-like systems using a two-body interatomic potential function proposed by Soules[2]. The characteristics of the simulated vitreous silica systems resulting from this potential are summarized by Ochoa et al.[3] For this study, we used systems containing 1152 atoms with full periodic boundary conditions.

We obtained simulated 300 K vitreous silica structures by quenching a 12,000 K liquid derived from melting a high-cristobalite structure. Both the quench and temperature ramp-up

327

were performed at a rate of 1 K•fs⁻¹. Because of limitations in the potential, the absence of free surfaces, and the short simulation time, the melting temperature of this system, 10,000 K, is greatly elevated over the accepted melting temperature of cristobalite, 2090 K.

We simulated fracture by uniaxially straining systems to failure. Because of the small size of our system, we essentially model the small volume of material near a flaw or in front of an advancing crack tip that undergoes significantly greater strain than the macroscopic sample as a whole. Strain was applied by scaling the size of the periodic cell and uniformly scaling the positions of particles within the cell at periodic intervals during the simulation. This method ensures that there is no bias to fracture at any particular location. The range of applicability of this method of strain application extends to strain rates as high as about 0.2 ps⁻¹ for samples of the size used here when using arguments based on crack propagation velocities, as shown in Fig. 1. Our simulations find that the fundamental mechanical and fracture behavior of the system changes at this same strain rate, indicating that it is near this strain rate that the processes that limit the rate of brittle crack propagation occur.

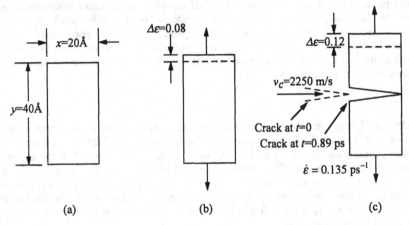

Figure 1. Geometric interpretation of maximum strain rates, based on crack propagation velocities, system size, and strain at failure. (a) Initial dimensions. (b) Average 0.08 strain in material outside of crack region, as observed in typical fracture experiments. (c) Additional 0.12 strain added in strain region ahead of crack, equal to strain in controlled-atmosphere experiments. We assume the width of the high strain region to be on the order of our 20 Å sample width. A crack propagating through this region at a velocity is 2250 m/s (0.6•v_S for v-silica) gives a high strain period of 0.889 ps for the final 0.12 strain, a strain rate of 0.135 ps⁻¹. Uncertainties in predicting the size of this high-strain region in a real system suggest the actual value should lie within an order of magnitude of this value. Since the elastic modulus of our system is larger that of vitreous silica, a maximum strain rate of about 0.2 ps⁻¹ is suggested for our system.

We simulated fracture at a range of strain rates and under various conditions to probe the intricacies of brittle fracture. We applied strain rates ranging from 0.005 ps⁻¹ to 50 ps⁻¹ to examine the gross system response at wide range of strain rates. We then concentrated on strain rates near 0.2 ps⁻¹, ranging from 0.05 ps⁻¹ to 0.5 ps⁻¹, to see how the fracture behavior changes in this strain rate region, and examined the effects of structural perturbations on the ultimate path to fracture and the strain rate sensitivity of these effects. The results presented in this study are mainly stress-strain curves and real-space images of the path to fracture in simulated systems, and our analysis link these results to results from real fracture experiments.

RESULTS

We will focus on three key results of MD fracture simulations. The first is that brittle fracture on the atomic scale looks different than fracture on the macroscopic scale. The second is that the nature of the fracture process is dependent on the strain rate. The third is that the path to fracture is influenced by individual atomic motions.

A distinguishing feature of brittle fracture is that it results in little macroscopic deformation beyond the creation of fracture surfaces. The simplest model of this process is the "unzipping" of atomic bonds between a pair of atomic planes with no associated permanent deformation beyond those atoms. This model may be appropriate for describing the cleavage fracture of single crystals such as ionic salts, but not the fracture of amorphous solids where there are no neatly aligned planes on which this kind of orderly bond breakage can occur. Additionally, it has been observed that even cleavage fracture in single crystal samples will eventually come to an end as cracks advance across large samples, giving way to mist and hackle as seen in materials such as glass[4]. Clearly, we need to rethink this model that is applicable only to very limited conditions.

Our simulations suggest that brittle fracture models of amorphous materials need not constrain bond breakage to only those bonds that would otherwise span the crack, the amorphous equivalent of the bonds between cleavage planes. Macroscopic measures of brittle fracture in these materials do not require that a reconstructing strain be limited to a single layer of atoms in a brittle material, only that it be limited to a small volume as compared to the entire system. We find that structural rearrangements occur throughout the simulated near-crack-tip region sample by the breakage of isolated bonds resulting from inhomogeneous bond environments under high strain. Since there is no long-range ordering of these rearrangements to relieve stress in an extended volume, fracture would still appear to be brittle from a macroscopic standpoint. The energy released as this reconstructed high-strain layer falls back into an unstrained state after fracture may result in the latent fractoemission observed by Dickinson et al.

A comparison of the stress strain curves obtained by simulation to those generated from experimental results is shown in Fig. 2. The simulated systems fracture at about 15-20% strain because they are not large enough to allow for stress-intensity effects. This is comparable to the maximum strains observed before fracture of flaw-free pristine silica fibers in vacuum which are also free of stress-intensity effects. The simulated elastic modulus is higher than that of real silica, resulting in a significantly higher fracture stress in the simulated systems.

We find that brittle fracture in amorphous materials is a strain rate dependent process, in agreement with previous MD simulations by Ochoa and Simmons[5]. Both the maximum stress and the strain at maximum stress increase with strain rate, as shown in Fig. 3. Note that the structure becomes significantly stiffer at strain rates above 0.5 ps^{-1}, approaching a maximum value at 20 ps^{-1}. The increase in both the system stiffness and the overall fracture energy at the strain rate corresponding to the maximum crack propagation velocity indicates that this behavior may be responsible for limiting crack propagation velocities. Marder et al.[6] point out that although continuum models predict that crack propagation velocities should approach the Rayleigh wave velocities, they in fact rarely exceed 0.4 to 0.6 of this speed. The failure of these continuum models to accurately predict crack propagation velocities may reflect that they only account for the limits in the ability of the propagating crack to make its presence known to the material in its path and do not account for the lattice effects that make fracture more difficult as the local strain rate approaches a limiting value.

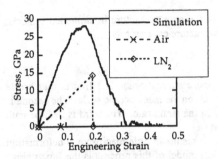

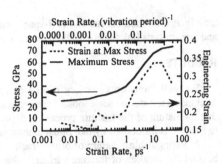

Figure 2. Stress-strain curves from simulation and experiment are distinctly different. Simulation curve is for a system strained at 0.05 ps^{-1}, "air" is for a system fractured in air, after Kurkjian[7], "LN$_2$" is for a sample fractured in liquid nitrogen, after Proctor[8]. Linear elastic modulus of 72.5 GPa is assumed for fracture strains in experimental systems.

Figure 3. Both the rigidity of the structure (maximum stress) and the strain at maximum stress increase with strain rate. The strain rate here is plotted both in units of ps^{-1} and in terms of a thermal vibrational period.

The initial aim of our fracture simulations was to show that there are structural variations in amorphous structures that determine the path to fracture. We developed a "free volume sphere" method to find largest interstices in these structures, with the expectation that the fracture path would include these voids. However, we found that fracture did not always occur at the largest voids present. We also found that the location of fracture in a structure could be completely different at different strain rates, as is shown in Fig. 4 . Our simulations also show that the probabilistic path to fracture becomes more diffuse, and may bifurcate, as the strain rate increases towards its upper limit, as is evidenced by the images of the fracturing systems shown in Figs. 4 & 5. Thus we find that the path to fracture can be influenced by the dynamics driven by the application of strain as well as the structure of the system.

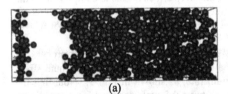

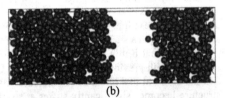

(a)   (b)

Figure 4. Fracture locations in same system at different strain rates. (a) 0.005 ps^{-1}, (b) 0.1 ps^{-1}.

Our simulations also show that the particular path to fracture is influenced by thermal vibrations and other small perturbations in structure. In order to investigate these effects, we performed a series of simulations at strain rates of 0.05 ps^{-1} and 0.5 ps^{-1} where we fractured the sample after different hold times to probe the effects of the vibrational state of the system, after stopping and re-starting the simulation to perturb all atomic positions by reducing the position with which they were specified, and after slightly perturbing the position of one atom in the system. This is an expansion of previously reported work by these authors[9].

Of the three types of perturbations examined, annealing the system for different times before fracture had the largest effect. In this case, the structure is still essentially the same because the system is well below its glass transition temperature. However, thermal vibrations perturb each atom in the structure by about 0.1 Å, whereas the other perturbation methods perturb atomic positions by at most 10^{-5} Å. The effects of thermal vibrations on the path to fracture of a sample strained at 0.5 ps^{-1} is illustrated in Fig. 5. Note that the location of fracture occurs in completely different regions of the sample at different times. In general, we find that the effects of this type of perturbation on the particular path to fracture are enhanced at the higher strain rates. The variations of the paths to fracture for samples strained at 0.05 ps^{-1} are significantly less. The effects of perturbations are evident in the stress-strain curves at both strain rates as shown in Fig. 6.

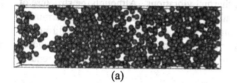
(a)
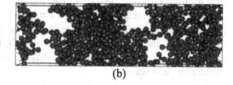
(b)

Figure 5. Fracture locations in same system after different anneal times at 0.5 ps^{-1} strain rate. (a) 1 ps, (b) 2 ps.

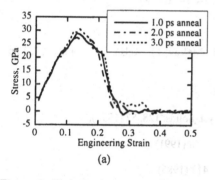
(a)
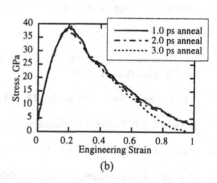
(b)

Figure 6. Variation in stress strain curves for the same systems at different anneal times and strain rates. (a) 0.05 ps^{-1}, (b) 0.5 ps^{-1}.

The other methods of perturbing the structures, round off and single atom perturbation, had a lesser effect on fracture. Both methods resulted in real space images of fracture that were scarcely different, but effects of both could be discerned in stress-strain curves. The effects of round off were greater than the effects of single atom perturbation. Since the effects of both grew over time, these effects were more pronounced at the lower strain rate than at the higher strain rate, because fracture took longer and system behaviors had more time to diverge.

CONCLUSIONS

Computer simulations of brittle fracture currently provide the best means of studying this process on the atomic scale because currently available analytical techniques cannot image fracture at this scale during the event. With some care in the interpretation of results, one can gain important insights into the brittle fracture process, even with simple potentials and systems of limited size.

We find that the fracture behavior of simulated systems changes markedly as strain rates approach those corresponding to maximum crack propagation velocities. As the system is pushed to higher strain rates, the structure becomes more rigid, the fracture path becomes more diffuse, and bifurcation of the fracture path may occur. This provides an explanation for dynamic crack branching during rapid fracture of a brittle material.

We also find that the path to fracture shows extreme sensitivity to all aspects of a dynamic system, from the applied strain rate to the state of thermal vibrations. Although the structure of an amorphous system clearly plays a role in its path to fracture, it is not the sole determining factor of the path to fracture in an otherwise unbiased system. This behavior should not be limited to the fracture of amorphous materials, but may occur in all materials where individual atomic motions can cause a crack to change its path, even those that would otherwise tend to fracture by cleavage.

ACKNOWLEDGMENTS

This work performed in part at Sandia National Laboratories, supported by the U.S. Department of Energy under contract number DE-AC04-94AL85000.

REFERENCES

1. J.T. Dickinson, S.C. Langford, L.C. Jensen, G.L. McVay, J.F. Kelso, C.G. Pantano, J. Vac. Sci. Technol. A **6** (3), 1084 (1988).

2. T.F. Soules, J. Non-Cryst Solids **123**, 129 (1990).

3. R. Ochoa, T.P. Swiler, J.H. Simmons, J. Non-Cryst. Solids. **128**, 57 (1991).

4. Y.L. Tsai, J.J. Mecholsky, Jr., J. Mater. Res. **6**, 1248 (1991).

5. R. Ochoa, J.H. Simmons, J Non-Cryst Solids **75**, 413 (1985).

6. M. Marder, S Gross, J. Mech. Phys. Solids **43**, 1 (1995).

7. C.R. Kurkjian, U.C. Paek, Appl. Phys. Lett. **42** (3), 251 (1983).

8. B.A. Proctor, Physics and Chemistry of Glasses **31** (2) 78 (1990).

9. T.P. Swiler, T. Varghese, J.H. Simmons, J. Non-Cryst. Solids **181**, 238 (1995).

TEMPORAL INSTABILITIES (DISSIPATIVE STRUCTURES) IN CYCLICALLY DEFORMED METALLIC ALLOYS

MICHAEL V.GLAZOV *, DAVID R.WILLIAMS ** AND CAMPBELL LAIRD *

* Department of Materials Science and Engineering, LRSM, The University of Pennsylvania, Philadelphia, PA 19104; **Department of Mechanical Engineering, Cornell University, Ithaca, NY 14850.

ABSTRACT

The existing models for the "classical" Portevin-Le-Chatelier effect have been analyzed, and the non-linear dynamical model has been proposed in order to quantify the nature of temporal instabilities in fatigued metallic alloys. The model employs the concept of a positive feed-back among the populations of mobile, immobile and Cottrell-type dislocations with atmospheres of point defects. Three major types of loading have been numerically simulated: pure sinusoidal, creep fatigue ("the Lorenzo-Laird bursts") and ramp loading ("the Neumann bursts", when the amplitude of otherwise cyclic loading grows linearly with time). Computer movies of the temporal evolution of stress and dislocation densities have been prepared as an aide for analysis and illustration. The model successfully reproduces stress serrations in terms of the underlying dislocation mechanisms and thus for the first time establishes a fundamental link between the micro- and macromechanics of cyclic deformation.

INTRODUCTION

Strongly non-equilibrium systems of different physico-chemical nature may exhibit a tendency to self-organization either in the form of emerging temporal rhythms, or spatial scales, or both, [1], and we demonstrate that fatigued metallic alloys fall into this behavior as well. In our previous work we have introduced the concept of Self-Organized Dislocation Structures ("SODS") in fatigued metals, and we have shown how different types of dislocation patterning (i.e. persistent slip bands, mazes, dislocation grids overlying grain boundaries etc.) can be understood using nonlinear dynamics and self-organization theory, [2,3].

In this work we make an attempt to show that spatial instabilities (i.e. dislocation rearrangements and patterning) are intimately connected with temporal instabilties - stress serrations and strain bursts accompanying such processes. In monotonically tested metals this phenomenon is widely known as the Potevin-Le-Chatelier (PLC) effect. The same is true with respect to cyclically tested metals: the analog of the PLC-effect in fatigue is known as the "Yan-Hong-Laird bursts", [4].

Our main goal was to establish, via nonlinear dynamic modeling, the fundamental connection between the microscopic dislocation mechanisms and the macroscopic mechanical response of cyclically deformed metals.

Mat. Res. Soc. Symp. Proc. Vol. 409 © 1996 Materials Research Society

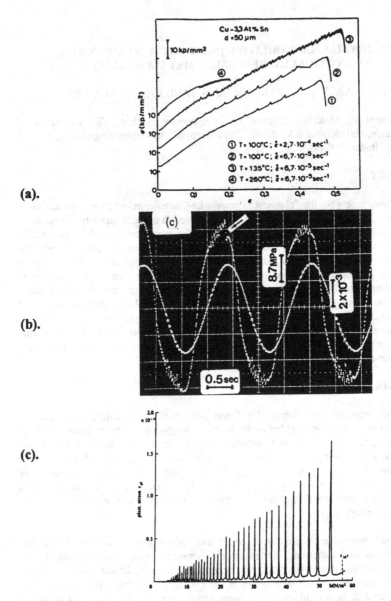

(a).

(b).

(c).

Figure 1. Several examples of temporal instabilities ("temporal dissipative structures", using Prigogine's terminology, [1]) in monotonically and cyclically tested metals. **(a).** The "classical" Portevin - Le-Chatelier effect in a Cu-3.3at.%Sn binary alloy at 255⁰C and different values of applied strain rate, [5]; **(b).** The "Yan-Hong-Laird bursts" - stress serrations on stress vs time curve - in a cyclically deformed binary alloy Cu-16at.%Al under strain control, [4]; **(c).** Strain bursts in a ramp-loaded Al single crystal, [6].

EXPERIMENTAL DATA AND THE MODEL

Several examples of temporal instabilities are given in Fig.1a,b,c. The "classical" PLC-effect manifests itself in the form of small stress spikes on the stress vs time curves, Fig.1a, [5]. A successful attempt has been made to reproduce these temporal instabilities in a strain-controlled fatigue tests, i.e. under cyclic loading, - Fig.1b illustrates stress serrations in a strain-controlled fatigue test for Cu-16at.%Al single crystals, [4]. Fig. 1c gives an example of the "Neumann bursts"- plastic strain bursts in Al single crystals which occur during gradual increase of the stress amplitude from zero while the frequency of loading remains constant, [6].

Although several models have been proposed in order to explain the "classical" PLC-effect, to the best of our knowledge no attempt has been made so far to model temporal instabilities in cyclically deformed metals. However, without proper understanding of the PLC-effect in monotonically deformed metallic alloys it is impossible to understand even more complex "Yan-Hong-Laird bursts" or "Neumann bursts". We have thus been motivated to look for a nonlinear dynamic model that could be used, after the necessary corrections and modifications, to reproduce the stress serrations in a "classical" PLC-effect. After a careful analysis which is described elsewhere, [7], we have chosen the model of Ananthakrishna and Valsakumar that successfully reproduces stress serrations in monotonically deformed metals, [8]. The model introduces three dislocation populations - g , mobile (glissile) dislocations, s - immobile (sessile) and i - dislocations with atmospheres of point defects that mimic Cottrell's idea of dynamic strain aging, [8]. Specifically, the following quasi-chemical reactions among dislocations of these three types were considered:

$$
\begin{aligned}
g &\xrightarrow{\theta V_g} g + g \\
g + g &\xrightarrow{\kappa \mu'/2} s + s \\
g + g &\xrightarrow{(1-\kappa)\mu'/2} 0 \\
g + s &\xrightarrow{\mu} 0 \\
s &\xrightarrow{\lambda} g \\
g &\xrightarrow{\alpha} i \\
i &\xrightarrow{\alpha'} s
\end{aligned}
\tag{1}
$$

The 1st reaction describes the process of dislocation generation by the multiple cross-glide mechanism (or any other suitable process such as a Frank-Read, or a Bardeen-Herring source), θ is the breeding constant and V_g - the average velocity of the glissile dislocations. This process makes it possible to connect mesoscale dislocation dynamics to macroscopic stress-strain response of the material (see below). The second reaction in the set describes the conversion of two glissile dislocations into two Cottrell-type ones. Annihilation of dislocations is taken care of in the next two equations. The last three reactions describe the mutual transformations of dislocations (mobilization due to the applied stress or thermal activation, transformation of mobile dislocations into dislocations with atmospheres and their subsequent trapping). These three equations form the feedback loop, [8]

$$s \rightarrow g; \quad g \rightarrow i; \quad i \rightarrow s \qquad (2)$$

which under certain well established conditions can serve as a physical reason for the occurrence of stress serrations in jerky flow.

After all the dislocation reactions are written down, one can obtain a set of three nonlinear differential equations for the process rates and to couple these equations to the machine equation representing the load sensed by the load cell. The final set of four differential equations looks as follows, [8]:

$$N'_g = \theta V_g(\sigma^*) N_g - \mu N_g - \alpha N_g + \lambda N_s - \mu' N_g N_s$$
$$N'_s = k \mu N_g^2 - \mu' N_g N_s - \lambda N_s + \alpha' N_i$$
$$N'_i = \alpha N_g - \alpha' N_i \qquad (3)$$
$$\sigma'_a = K[\varepsilon' - b_0 N_g V_g(\sigma^*)]$$

In the last equation of the system the coefficient K stands for some generalized effective compliance of the "sample+machine" system, ε' is the imposed strain rate (i.e. the sum of the elastic and plastic strain rates) and b_0 - the Burgers vector. The second term in this equation is the plastic strain rate, ε'_p which is proportional to the Burgers vector b_0, the number of glissile dislocations N_g, and the velocity of gliding $V_g(\sigma^*)$, [8].

This model satisfactorily reproduces stress serrations in monotonically tested metallic alloys, but is inapplicable to cyclic deformation. In order to take account of that phenomenon, we modified the machine equation of the model in the following way: 1). for the process of pure cyclic deformation we put $\varepsilon' = f \sin(wt)$; 2). for creep fatigue it is necessary to put $\varepsilon' = e + f \sin(wt)$, where e is a small constant rate superimposed on the otherwise cyclic process; 3). for ramp loading one gets: $\varepsilon' = f t \sin(wt) - q \cos(wt)$, since in this case the total strain will change as $\varepsilon(t) = -q t \cos(wt)$, i.e. describes the process of ramp loading.

THE RESULTS OF COMPUTER MODELING

We began by applying pure sinusoidal loading with various forcing frequencies and amplitudes to find a set of parameters that qualitatively reproduced the phenomena observed in experiment. However, in all cases all indications of activity on the stress vs time curve died out by the second period of oscillations, (for details and the values of parameters see [7]). After tedious numerical experimentation it has been found that adding a relatively small constant strain rate to the sinusoidal forcing function caused stable and reproducible stress serrations that last practically indefinitely in time (this type of loading corresponds to the conditions of creep fatigue). The first three periods of creep fatigue solution are given in Fig. 2. A comparison of the stress curve to the mobile dislocation density curve shows a very clear correspondence of peaks in dislocation density and peaks in stress: when stress goes down, the immobilized dislocations break free from the atmospheres of point defects (or any other pinning centers) and the mobile

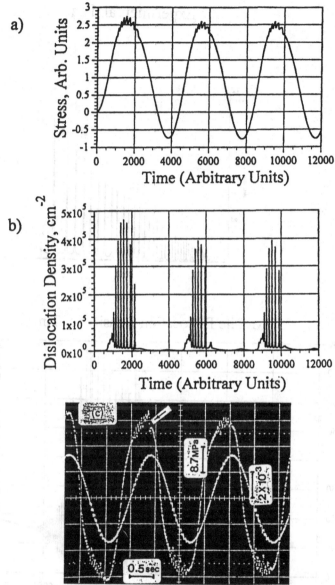

Figure 2. The results of modeling of stress serrations in a cyclically deformed metallic alloy (corresponds to the conditions of "creep fatigue", see text). **(a)**. Stress serrations vs time in a strain-controlled fatigue test; **(b)**. Mobile dislocation density bursts accompanying stress serrations; **(c)**. Experimental results of Hong and Laird for binary alloy Cu-16at.%Al, [4].

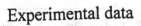

Experimental data

(a)

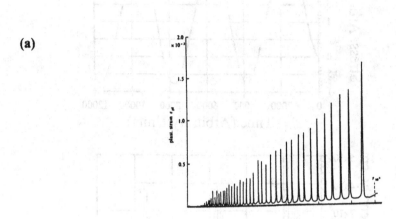

Mobile Dislocation Density vs. Time

(b)

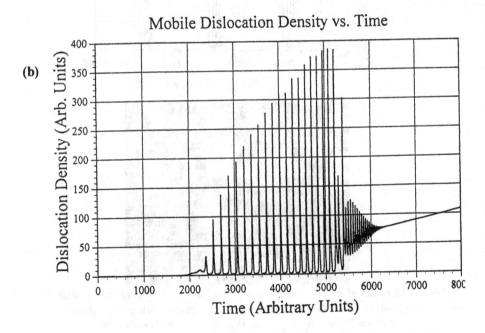

Figure 3. Ramp loading: **(a).** Neumann bursts in a ramp-loaded Al single crystal, [6]; **(b).** The results of our computer modeling - mobile dislocations density vs time.

dislocation density undergoes a dramatic increase. This physically consistent result indicates that the model is capable of reproducing even the finest effects of dislocation dynamics in fatigued metals. The observation that a constant strain rate term is necessary in order to maintain the stress serration effect is consistent with the experimental work of Hong and Laird, [4]. These researchers observed that when stress serrations occur under purely sinusoidal loading, the effect is not stable and dies out fairly rapidly.

We attribute this constant strain rate requirement to the difference in the frequency spectra of the two different forcing functions, *fsin(wt)* and *e + fsin(wt)*. For a system of differential equations certain resonant frequencies exist (generally, dependent on time, because the system is nonlinear). The frequency spectrum of pure sinusoidal loading function consists of only one frequency and thus it is very difficult to produce the complex resonant phenomena that, in our opinion, might be responsible for the stress serration effect. By adding a small constant strain rate, we effectively expand the frequency spectrum to infinity, enabling the system to "pick up" the frequencies which are close to those at which it resonates. Thus, the stress serration effect can be explained as a non-linear resonance phenomenon in a system of ordinary differential equations. Analogous effects were observed and studied, mostly via computer experiments, in chemical engineering (in the theory of periodically controlled chemical reactors), [9].

We also made an attempt to reproduce the Neumann bursts in a ramp-loaded metallic system. The bursts occur during the gradual increase of the stress amplitude from zero, Fig. 1c. Comparing our solution to the experimental result of Neumann, Fig.3, one can see how the dislocation density bursts obtained in numerical simulations correlate to the characteristic stress-strain response observed in experiment, [6].

Concluding this brief preliminary discussion of our results which will be presented in detail elsewhere, [7], one can nevertheless say that the modified Ananthakrishna-Valsakumar model that we used gives a possibility to reproduce and explain the experimentally observed stress serrations under the conditions of pure harmonic loading, creep fatigue and ramp loading, thus establishing for the first time the fundamental connection between the micro- and macro-mechanics of cyclic deformation.

LITERATURE

1. G.Nicolis and I.Prigogine, Exploring Complexity. (Freeman and Co.,New York, 1989).
2. M.V.Glazov, L.Llanes and C.Laird, Phys. Stat. Sol. (B), 149, 297-320 (1995).
3. M.V.Glazov and C.Laird, Acta Metall. Mater., 43, 2849-2857 (1995).
4. S. I. Hong and C. Laird, Mat. Sci. Eng., A124, 183 (1990).
5. W.Rauchle, O.Vohringer, E.Macherauch, Mat. Sci. Eng., 12, 147 (1973).
6. P.Neumann, Acta Metall., 17, 1219 (1969).
7. M.V.Glazov, D.R.Williams and C.Laird, submitted to Acta Metall. Mater.
8. G.Ananthakrishna and M.Valsakumar, Phys. Letts., 95A, 69 (1983).
9. P.Rehmus, J.Ross, in: Oscillations and Traveling waves in Chemical Systems, edited by R.Field and M.Burger, (Wiley Interscience, NY, 1985) pp.287-331.

Part VI

Fractals and Scaling in Fracture

SCALING AND UNIVERSALITY
IN REAL CRACKS

Pascal DAGUIER (daguier@onera.fr), Elisabeth BOUCHAUD (bouchaud@onera.fr)
O.N.E.R.A. (OM), 29 Av. de la Division Leclerc
B.P. 72, 92322 Châtillon Cedex, FRANCE

Abstract

The morphology of fracture surfaces in complex metallic alloys is analysed. The simultaneous use of Atomic Force Microscopy (AFM) and Scanning Electron Microscopy (SEM) allows the measurement of the universal roughness exponent $\zeta_\perp = 0.78$ over five decades of lengthscales (0.5 nm - 0.5 mm). Furthermore, a small lengthscales regime (1nm - 1 μm) is shown to be characterised by a roughness index $\zeta_{\perp_{QS}} \simeq 0.5$.
On the other hand, cracks fronts stopped during their propagation at pinning microstructural obstacles in two different metallic alloys is analysed, and their "in-plane" roughness index is determined for the first time. In the case of the 8090 Al-Li alloy, which has a very anisotropic microstructure, the roughness of the tensile crack front propagating along the grains length is equal to $\zeta \simeq 0.60$. On the other hand, the roughness of the purely three-dimensional fatigue crack front in the Superα_2 Ti$_3$Al-based alloy is equal to $\zeta \simeq 0.54$.

Introduction

There is an upsurge of interest in the problem of crack advance through locally heterogeneous materials since a complete understanding could be a key to tough materials design. As a matter of fact, microstructural elements as grain boundaries or sub-boundaries, second phase precipitates, dislocation assemblies, are able not only to force crack deflection by modifying locally the stress intensity factors, but also to "trap" the front which then is constrained to advance in a nonuniform way. On the other hand, progress has been made in statistical physics in the understanding of line pinning by randomly distributed impurities. It was recently proposed that the two problems could be related [1], and crack advance was described within a framework proposed for very different purposes by Ertas and Kardar [2, 3, 4, 5]. Although this first model suffers from a few (important) weak points, it provides a rich phase diagram containing part of the observed characteristics of real fracture surfaces.

As a matter of fact, out-of-plane roughness indices (see Fig 1) have been measured on various materials, using different experimental techniques. In the case where fracture profiles perpendicular to the direction of crack propagation (profiles along the x axis) are concerned [6, 7, 8], there always seems to exist a lengthscales domain where it is close to the value $\zeta_\perp \simeq 0.8$, which we conjectured to be universal, i.e. independant of the fracture mode, mechanical properties and of the material [9]. Results are presented

343

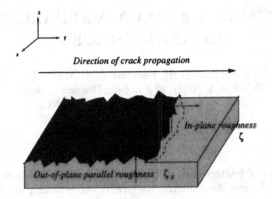

Figure 1: *The crack propagates along the y direction, in the xy plane, while the tensile axis is parallel to z (mode I). Out-of-plane roughness indices refer to profiles contained in the xz ($\zeta_\perp$) or in the yz ($\zeta_{//}$) planes, while in-plane exponent (ζ) refer to the roughness of the projection of the front on the xy fracture plane.*

in Section 1, where it is shown that the simultaneous use of a standard scanning electron microscope (SEM) and of an atomic force microscope (AFM) allows to observe this regime over *five decades of lengthscales* [10]. Note that in this universal regime, Schmittbuhl *et al.* [11] have recently shown that the self-affine correlation length $\xi_\perp$ is varying with the distance y to the initial notch as a power law, $\xi_\perp \propto y^\alpha$, with $1/\alpha \simeq 1.2$. This implies in turn that the parallel roughness index should be slightly different from the perpendicular one: $\zeta_{//} = \alpha\zeta_\perp \simeq 0.7$. This seems to be confirmed by direct measurements on profiles along the y axis [12]. Note that the model refered to above [1] predicts in some particular cases $\zeta_\perp = 0.75$ and $\zeta_{//} \simeq 0.5$, i.e. $1/\alpha = 1.5$, which is not that far from experimental results [6, 7, 9, 11, 12].

For crack velocities tending to zero, i.e. in the vicinity of the so-called "depinning transition" [5] in the line trapping problem, the roughness index $\zeta_\perp$ perpendicular to the direction of propagation of the crack is predicted to be close to 0.5. More generally, for a given velocity, this model also predicts the existence of a crossover $\xi_{\perp_{QS}}$ at some length-scale from the "quasi-static" to the "dynamic" behaviour ; the cross-over length *decreases* quite rapidly with the crack velocity (velocity to the power $\phi = -3$). Correlatively, it is expected, within the framework of this model, that the crossover length decreases with increasing stress intensity factor K_I.

Another interesting model was proposed by S. Roux *et al.* [13, 14] for the fracture of plastic materials: the fracture surface is expected then to be a *minimum energy surface* [15], the roughness index of which is lying between 0.4 and 0.5 [16, 17, 18, 19]. Note that this mechanism, although different in its physical content than the "line depinning" one described above, is also "quasi-static", since it is based upon an equilibrium model. In this case, however, the crossover between the small and the large lengthscales regimes should lie around the plastic zone size, and hence, *increase* with K_I. We are faced then with two different models, which both predict a small lengthscales regime characterised by a roughness index close to 0.5, but in one case the crossover length is predicted to decrease with increasing the stress intensity factor, and in the second case, it increases with that parameter as the plastic zone size.

Here again, a simultaneous use of SEM and AFM allowed us to observe this regime over approximately three decades of lengthscales (see section 1). Although the crack velocities could not be measured, we have clearly shown that the crossover length ξ_{Qs} which separates the two regimes depends on the crack length, and hence on the stress intensity factor K_I.

It will be emphasized that the existence of two exponents may lead to *apparent* roughness indices continuously varying with the stress intensity factor and the crack velocity (see fig 2).

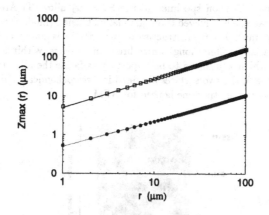

Figure 2: *How to fit a crossover with a continuously variable exponent over two decades of lengthscales.*
  ⋯ *power law regression : slope 0.66*
  — *power law regression : slope 0.75*
  • *real function : $Zmax(r) = 1.12 \times ((\frac{r}{10})^{0.5} + (\frac{r}{10})^{0.8})$*
  □ *real function : $Zmax(r) = 0.56 \times ((\frac{r}{0.1})^{0.5} + (\frac{r}{0.1})^{0.8})$*

In contrast with the out-of-plane properties, the in-plane crack roughness ζ (see Fig. 1) has never, to our knowledge, been characterised in the past, although it should provide interesting elements for modeling. In section 2, such results are presented [28], in the case where crack propagation is slow enough to be stopped before complete fracture
of the sample.

As for the case of perpendicular "out-of-plane" properties, it is expected that, for intermediate values of the crack velocity, one should observe the quasi-static exponents at small lengthscales and the dynamic ones at large lengthscales, the two domains being separated by a cut-off lengthscale depending on crack velocity. S. Roux and A. Hansen [20] have recently developped a bidimensional model based on a linear Langevin equation [21]. It describes the quasi-static movement (the velocity of the front tends to zero) of an elastic front through randomly distributed impurities. The line is supposed to respond locally (it has a line tension) to elastic perturbations.

Since the impurities are supposed to describe fixed microstructural obstacles, the disorder is quenched. In this case, the front is not self-affine, but self-similar, with roughness index: $\zeta \simeq 1.2$. This quasi-static case corresponds, in the Ertas and Kardar's model [4] to the vicinity of the "depinning transition". More recently, still in the framework of the linear bidimensional model developed by Roux et al. [20], Schmittbuhl et al. [22] have shown that ζ can decrease down to 0.35 if non local elastic effects are taken into account. For that purpose, they use a non-local elastic kernel determined to first order by Gao and Rice [23].

1- Out-of-plane perpendicular properties of fracture surfaces

A notched Compact Tension specimen of the Super-α_2 alloy (Ti$_3$Al-based) is pre-cracked in fatigue. Fracture is achieved through uniaxial tension (mode I) with a constant opening rate (0.2 mm/mn). The microstructure of our material is mainly constituted of α_2 needles ($\simeq 1\mu$m thick and $\simeq 20\mu$m long) which break in cleavage, within a β matrix, the plastic behaviour of which was shown to be important as far as the alloy fracture toughness is concerned. One of the two surfaces obtained is electrochemically Ni-Pd plated for SEM observations, while the other one is used for AFM.

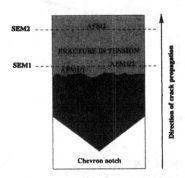

Figure 3: *Localisation of the various observed zones*

Two profiles (SEM1 and SEM2, see Fig.3) located respectively just behind the fatigue zone and closer to the edge of the specimen, are obtained by subsequently cutting and polishing the sample perpendicularly to the direction of crack propagation. These profiles are observed with a scanning electron microscope Zeiss DSM 960 at various magnifications, ranging from x50 to x10 000, with a backscattered electron contrast. Images in 256 grey levels are registred and the profiles are extracted by image analysis (Visilog 4.1.1). The length of the images is 1024 pixels, and adjacent fields (overlapping over fifty pixels with each other) have been explored in order to build up profiles of 6 000 and 7 000 points.

Ten profiles are registred in each of the three different regions, AFM1/1 and AFM1/2, comparable to SEM1, and AFM2, comparable to SEM2 (see Fig. 3). The lengths the profiles are 2.5, 20 and 1 μm, in regions AFM1/1, AFM1/2 and AFM2 respectively, with 10 000 points for profiles registered in zones AFM1/1 and AFM2, and 20 000 points for those in AFM1/2.

In order to determine the roughness exponent $\zeta_\perp$ and the cross-over length ξ_{QS}, three methods are used: the 'variable band width' method, the return probability and the spectral method. The results of the latter two methods are perfectly compatible with those of the 'variable band width' method, which, being much less noisy, are the only ones presented here. In this method, the following quantity is computed :

$$< Zmax(r) >_{r_o} = < \max\{z(r')\}_{r_o<r'<r_o+r} - \min\{z(r')\}_{r_o<r'<r_o+r} >_{r_o} \propto r^\zeta \qquad (1)$$

where r is the width of the window. $Zmax(r)$ is the difference between the maximum and the minimum heights z within this window, averaged over all possible origins r_o of the window belonging to the profile.

In the case of AFM2+SEM2 (see Fig. 4), the best fit is obtained with one power law regression, with exponent $\zeta_\perp = 0.78$. For the first time to our knowledge, the universal exponent $\zeta_\perp \simeq 0.8$ is observed on *five and a half decades* of lengthscales .

In the case of the AFM1/1+SEM1 zone, a non linear curve fit is needeed. Three power-law regimes can be observed on the plot of Fig. 4. The roughness exponent which can be determined at short lengthscales is close to 1, which is characteristic of a flat surface. This regime will be briefly discussed just below.
At larger lengthscales, two regimes can be observed, which correspond to the "quasi-static" (with $\zeta_{\perp_{QS}} \simeq 0.5$) and "dynamic" (with $\zeta_\perp \simeq 0.8$) regimes already observed [8] on this material. As in ref.[8], the simplest form of the crossover function is chosen, i.e. $Zmax(r)$ is fitted with the sum of two power laws:

$$Zmax(r) = a0 * ((\frac{r}{\xi_{\perp_{QS}}})^{0.5} + (\frac{r}{\xi_{\perp_{QS}}})^{0.78}) \qquad (2)$$

which allows to define the crossover length $\xi_{\perp_{QS}}$. Exponent $\zeta_{\perp_{QS}} \simeq 0.5$ is measured for lengthscales approximately ranging between 1nm and 1μm, i.e. on roughly three decades. Finally, at lengthscales larger than $\xi_{\perp_{QS}} \simeq 1\mu$m, the "universal" exponent $\zeta_\perp \simeq 0.8$ is recovered. If the same fit (Eq. (2)) is applied to AFM2+SEM2, $\xi_{\perp_{QS}}$ is shown to decrease down to 5 nm.

In the case of AFM1/2+SEM1 (see Fig. 4), the same behaviour is observed. As could be expected, $\xi_{\perp_{QS}}$ is the same as in the AFM1/1+SEM1 zone. However, the crossover length separating the "flat" and the quasi-static regimes is significantly larger (ranging from 50 nm to 0.1 μm) than in the previous case. This discrepancy is due to the fact that the flat 'very low lengthscales' regime is mixed with a non physical signal linked to the experimental limitation of the AFM used.

These results confirm those obtained in [8], where the intermediate and large length-scales regimes were interpreted respectively as "quasi-static" and "dynamic". They are

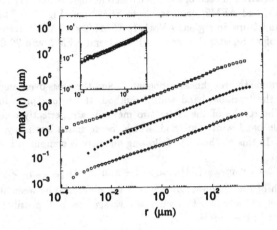

Figure 4: *Zmax(r) is plotted versus r on a log-log plot for three zones.*
    o : *AFM1/1+SEM1*
    * : *AFM1/2+SEM1 (Zmax(r) × 100)*
    □ : *AFM2+SEM2 (Zmax(r) × 10000)*
*Note that the experimental points obtained with the two techniques gently collapse on the
same curve (the region of overlap of the two techniques extends over 2 decades - see insert
wich correspond to AFM1/1 (*) and SEM1 (o)).*
The fits (-) correspond to : $Zmax(r) \propto r^{0.78}$ for AFM2+SEM2.
$$Zmax(r) = a0\left(\left(\frac{r}{\xi_{\perp_{QS}}}\right)^{0.5} + \left(\frac{r}{\xi_{\perp_{QS}}}\right)^{0.78}\right) \text{ with } \xi_{\perp_{QS}} \simeq 0.1\mu m \text{ for}$$
AFM1/1+SEM1 and AFM1/2+SEM1 zones.

also in agreement with the experimental results obtained by Milman and coworkers [25, 26]
and by Mc Anulty et al. [27], and with previously quoted theoretical models [1-4,13-19,24].
However, the "moving line" models [1-4,24] predict the short lengthscales regimes to be
independant of the crack velocity[29]. This is clear for the first series of experiments [8],
and also compatible with the results presented here.

Finally, it has to be noted that the "quasi-static" regime has only been observed on
metallic materials [30]. Hence, plasticity might be an important factor for the onset of
this regime, either because different fracture mechanisms are indeed involved within the
plastic zone, or because plastic dissipation may slow down crack propagation at small
lengthscales.

On the other hand, quantitative experiments for fracture in fatigue, where the crack
velocity and the load are measured should allow to clarify the meaning of this regime
and are currently being performed. In particular, it is not clear from this first set of
experiments that the crack velocity is the relevant parameter to be taken into account.

Since fracture does not occur at constant load, it is not obvious that region SEM1 corresponds to a smaller stress intensity factor than region SEM2, and our attempts to perform measurements of the load and of the crack velocity in that region failed, because fracture occurs too quickly (approximately in 200 ms, with a sharp variation of the velocity within 10ms). However, it can be noted that an upper bound of the maximum crack velocity could be estimated to be less than 1% of the speed of sound in the material: hence, no inertial effect has to be taken into account.

It would also be of great interest to relate $\xi_{\perp_{QS}}$ to characteristic lengths of the microstructure. Note that $\xi_{\perp_{QS}}$ obtained for AFM1/1+SEM1 is of the order of the average thickness of the α_2 needles. However, in a previous study of the same material [8], a cross-over length of $10\mu m$ could be determined for fracture in fatigue. The distribution of the stresses felt by the crack front during its propagation depending both on the microstructural disorder and on the loading conditions, it is expected that $\xi_{\perp_{QS}}$ is linked both to the microstructure, and to the local stress intensity factor. Further experiments on an aluminium alloy will also be performed to help clarifying that point.

2- In-plane roughness properties of crack fronts

Two different materials are considered here. The first one (specimen 1) is a sample made of the 8090 aluminium-lithium alloy, extruded, and heat treated in a way which leads to an intergranular fracture mode. Metallurgical grains are very elongated in the direction of extrusion: they are a few millimeters long, while they are only approximately 100 micrometers wide (along x) and 20-50 micrometers thick (along the $z \equiv$ short transverse direction). Cracks are grown in tension from a chevron-notch bar specimen, with the tensile axis z parallel to the short-transverse direction, while the crack propagates along $y \equiv$ long direction. The sample is 12mm high (parallel to z), 27.5mm long (along y), and 13mm wide (along x). It can be seen on the fracture surface that fracture occurs by successive jumps of the fracture front [28]: when the front is arrested at microstructural obstacles, enhanced plastification at the tip occurs before propagation is allowed back by a further increase of the load. After each jump, the crack having no more elastic energy to release propagates at a nearly null velocity, and is pinned by microstructural obstacles, the efficiency of which controls the local shape of the front at various lengthscales [31]. The specimen is unloaded before complete breaking.

Specimen 2 is made of the Superα_2 Ti$_3$Al-based material, which is forged, and hence has a far less anisotropic microstructure than specimen 1. A compact tension specimen 31 mm high, 31 mm long and 12.5 mm wide is broken during fatigue testing (cyclic tension-tension loading). As in the previous case, the specimen is unloaded before complete breaking, and the crack front stops at microstructural obstacles. In this case, however, the pinning sites are not as efficient on large lengthscales as the grain boundaries in the 8090. On the other hand, one can note the presence of many overhangs and secondary cracks on the fracture surface.

Many attempts to visualize the crack fronts were made in both cases. The only method

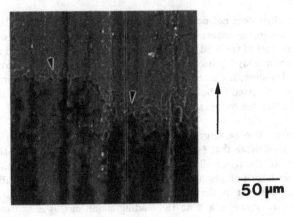

Figure 5: *Fracture surface of specimen 1 (8090), observed with secondary electrons. The direction of crack propagation is indicated by the arrow on the right). Characteristic dimples (indicated by small arrows) are visible just beyond the front marked with the ink.*

Figure 6: *Fracture front marked with China ink in the 8090 (specimen 1).*

which provides interesting observations is an impregnation of China ink under a moderate vacuum. In the case of the Superα_2, oxydation of the sample (340°C for 2 hours) leads to a visible mark, but does not allow a quantitative study. It is used however to check that the China ink wets the whole crack.

In the case of specimen 1, the front marked with black ink (see Fig. 6) has exactly the same structure as the arrested fronts observed subsequently, in particular with characteristic dimples (see Fig. 5).

Crack fronts are observed with the scanning electron microscope Zeiss DSM 960 using a backscattered electron contrast at magnifications ranging from x50 to x500 for specimen 1, and from x1500 to x500 for specimen 2. Images of size 1024x1024 pixels with 256 grey levels are directly sent from the microscope to a workstation where the image analysis is performed using Visilog 4.1.4. In the case of sample 2, for which the highly perturbed relief forbids intermediate magnifications, some images are directly registered from a CCD camera, digitized to 768x512 pixels images with 256 grey levels, and subsequently analysed with Visilog. For specimen 1, a 5000 pixels long profile is constructed from adjacent micrographs at magnification x200 in order to improve the statistics.

$z_{max}(r)$ is computed for each image, and averaged over all the images relative to one front, taking into account the real scale in each case. Error bars on z_{max} are estimated from the dispersion of the results over the various profiles. Fig. 7 shows the evolution of $z_{max}(r)$ with r, for specimens 1 and 2. The power law regimes extend from 1 to 630 micrometers in the case of the 8090, and from 1 to 1500 micrometers in the case of the Superα_2 (note however that although intermediate lengthscales are missing in the latter case, because they cannot be observed, there is no change in roughness with the scale). The measured exponents are the following:

$$\zeta_1 = 0.60 \pm 0.04 \tag{3}$$

$$\zeta_2 = 0.54 \pm 0.03 \tag{4}$$

for specimens 1 and 2 respectively.

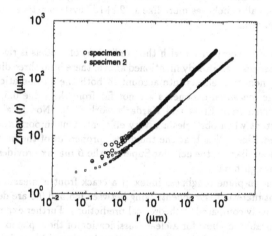

Figure 7: $z_{max}(r)$ as a function of r on a log-log plot:
(1)- Specimen 1. A power-law fit provides : $\zeta_1 = 0.60 \pm 0.04$.
(2)- Specimen 2: A power-law fit provides : $\zeta_2 = 0.54 \pm 0.03$.

Error bars on the values of ζ are estimated from the scattering of the measured slopes of $\log(z_{max}(r))$ as a function of $\log(r)$ for all the analysed profiles. One can see in Fig.7 that there might be a different regime at very small lengthscales; measurements however are not accurate enough to allow for its characterization.

Because the function $z(r)$ is not univalued in this case (due locally to the presence of overhangs, see Fig.6), we cannot use spectral methods. However, it was shown [7] that the return probability $P_0(r)$, i.e. the probability that z comes back to its initial value after a distance r along the axis perpendicular to crack propagation can be successfully determined on such objects, giving results consistent with those obtained from the computation of $z_{max}(r)$, although $P_0(r)$ is far noisier than $z_{max}(r)$. Furthermore, the linear drift of all considered profiles, which is always present at high enough magnification, has to be suppressed to compute $P_0(r)$, whereas it hardly has any influence on the power-law behaviour of $z_{max}(r)$ (it usually affects the saturated "plateau" regime, which disappears in case of a linear trend). Nevertheless, values of the roughness indices ζ determined from $z_{max}(r)$ are compatible with the decrease of $P_0(r)$ as $r^{-\zeta}$.

ζ_1 and ζ_2 both determined over approximately three orders of magnitude in length-scales (1 micron to 1 millimeter), are not significantly different, and the slight difference could very well arise from the difference in crack velocity in the two cases, and hence from the difference in unknown crossover lengthscales between the two regimes (quasi-static and dynamic) which should be observed. Upper self-affine correlation lengths are likely to be different as well, thus introducing a further source of uncertainty in the measured exponents.However, one cannot exclude that the two exponents could be different due to an essential difference in microstructures: due to crack propagation along very anisotropic grains, the aluminium alloy behaves more like a "2d+1d" system rather than like a plain three-dimensional material.

On the other hand, a comparison with the predictions of models is rather difficult in both cases, mainly because, as already mentioned above, there is no three dimensional non linear model taking non-local effects into account in both the quasi-static and dynamic cases ! However, the measured exponents are not far from the value 0.5, which results from model [1], derived from Ertas and Kardar's work [2-3]. Non local elastic effects could be less important when obstacles are very efficient. One important assumption in the work of Gao and Rice [23] is that the toughness properties of the obstacles and the "matrix" have to be similar. In the acicular Superα_2, the β matrix provides a toughening effect, especially through blunting.

In conclusion, the in-plane roughness index of a crack front is measured for the first time on two different materials. Exponents lying between 0.5 and 0.6 are determined, but cannot be quantitatively compared to theoretical predictions. Further experiments on the topic are highly desirable to allow for a clear classification of these phenomena. Marking the crack front in non-porous brittle materials as granite should be an interesting possibility. In the quasi-static case, more likely to be modelled in a near future, observations might be made on transparent materials if the relevant lengthscales are not too small to be quantitatively observed. Other metallic materials broken with different procedures will also be explored.

Conclusion

It has been shown in this paper that :

- The universal exponent $\zeta_\perp \simeq 0.8$ could be measured over five decades of lengthscales (5 nm to 0.5 mm) by using simultaneously an AFM and a SEM.

- There is a "small lengthscales" exponent close to 0.5 which could be characteristic of plasticity. Let us emphasize that the existence of two exponents can lead to the apparent continously variable indices measured over two decades of lengthscales wich are reported in the literature.

- The in-plane roughness exponent, close to 0.5 - 0.6, has been measured for the fisrt time.

Aknowledgements: We are indebted to S. Navéos, P. Josso, D. Boivin, J.-L. Pouchou and J.-M. Dorvaux for their help. AFM experiments were achieved in collaboration with S. Hénaux and F. Creuzet at Saint Gobain. Finally, we are particularly grateful to J.-P. Bouchaud, D. Ertas and S. Roux for interesting discussions.

References

[1] J.-P. Bouchaud, E. Bouchaud, G. Lapasset, J. Planès, Phys. Rev. Lett. **71**, 2240 (1993).

[2] D. Ertas, M. Kardar, Phys. Rev. Lett. **69**, 929 (1992).

[3] D. Ertas, M. Kardar, Phys. Rev. E, **48**, 1228 (1993).

[4] D. Ertas, M. Kardar, Phys. Rev. Lett. **73**, 1703 (1994).

[5] O. Narayan, D.S. Fisher, Phys. Rev. B **48**, 7030 (1993)

[6] K. J. Maloy, A. Hansen, E. L. Hinrichsen, S. Roux, Phys. Rev. Lett. **68**, 213 (1992).

[7] E. Bouchaud, G. Lapasset, J. Planès, S. Navéos, Phys. Rev. B **48**, 2917 (1993).

[8] E. Bouchaud, S. Navéos, J. de Physique I France, **5**, *Brèves Communications*, 547 (1995).

[9] E. Bouchaud, G. Lapasset, J. Planès, Europhys. Lett. **13**, 73 (1990).

[10] P. Daguier, S. Hénaux, E. Bouchaud, F. Creuzet, submitted to Phys. Rev. E, (nov. 1995).

[11] J. Schmittbuhl, S. Roux, Y. Berthaud, Europhys. Lett., **28**, 585 (1994).

[12] E. Bouchaud, P. Daguier, G. Lapasset, ASM Int. Conf. on Metallography (Colmar, France, May 1995).

[13] S. Roux, D. François, Scripta Metall. **25**, 1092 (1991).

[14] A. Hansen, E.L. Hinrichsen, S. Roux, Phys. Rev. Lett., **66**, 2476 (1991).

[15] M. Kardar, Nucl. Phys. **B290**, [FS20], 582 (1987).

[16] M. Mézard, G. Parisi, J. Phys. France I **1**, 809 (1991).

[17] T. Halpin-Healy, Phys. Rev. A **42**, 711 (1990).

[18] G. Batrouni, S. Roux, preprint (1995).

[19] A.A. Middleton, Phys. Rev. E, **52**, R3337 (1995).

[20] S. Roux, A. Hansen, J. de Physique I (Paris) **4**, 515 (1994).

[21] S.F. Edwards, D.R. Wilkinson, Proc. R. Soc. London A **381**, 17 (1982).

[22] J. Schmittbuhl, S. Roux, J.-P. Villotte, K. J. Maloy, Phys. Rev. Lett. **74**, 1787 (1995).

[23] H. Gao, J.R. Rice, J. of Applied Mechanics **56**, 828 (1989).

[24] D. Ertas, M. Kardar, preprint (1995).

[25] V.Y. Milman, R. Blumenfeld, N.A. Stelmashenko, R.C. Ball, Phys. Rev. Lett. **71** 204 (1993).

[26] V.Y. Milman, N.A. Stelmashenko, R. Blumenfeld, Prog. Mater. Sci. **38** 425 (1994).

[27] P. McAnulty, L.V. Meisel, P.J. Cote, Phys. Rev. A **46**, 3523 (1992).

[28] P. Daguier, E. Bouchaud, G. Lapasset, Europhys. Lett. , **31**, 367 (1995).

[29] S. Roux, private communication (1995).

[30] F. Plouraboué, K.W. Winkler, L. Petitjean, J.P. Hulin, S. Roux, preprint 1995.

[31] H. Cai, J.T. Evans, N. J. H. Holroyd, Acta Metall. Mater., **39**, 2243 (1991).

MODELING ACOUSTIC EMISSION IN MICROFRACTURING PHENOMENA

S. ZAPPERI[†], A. VESPIGNANI[‡] AND H.E. STANLEY[†]

[†] Center for Polymer Studies and Department of Physics, Boston University, Boston, MA 02215 ,USA

[‡] Instituut-Lorentz, University of Leiden, P.O. Box 9506 The Netherlands.

ABSTRACT

It has been recently observed that synthetic materials subjected to an external elastic stress give rise to scaling phenomena in the acoustic emission signal. Motivated by this experimental finding we develop a mesoscopic model in order to clarify the nature of this phenomenon. We model the synthetic material by an array of resistors with random failure thresholds. The failure of a resistor produces an decrease in the conductivity and a redistribution of the disorder. By increasing the applied voltage the system organizes itself in a stationary state. The acoustic emission signal is associated with the failure events. We find scaling behavior in the amplitude of these events and in the times between different events. The model allows us to study the geometrical and topological properties of the micro-fracturing process that drives the system to the self-organized stationary state.

INTRODUCTION

Acoustic Emission (AE) is produced by sudden movements in stressed systems. Several experiments have recently observed this phenomenon on very different length scales (from the largest scale of an earthquake to the smallest one of dislocation motions) [1, 2, 3]. Unfortunately, the AE analysis is a rather delicate technique since each external stress is unique and tests the whole sample. For instance, it is very difficult to obtain in this way insight on the microscopic dynamics of the fracturing phenomena. The statistical analysis, however, gives rise to the hypothesis that AE is generated by fracturing phenomena which are similar to critical points. Correlations develops leading to cascade events which drive the systems into a critical stationary state. For this reason, as a working hypothesis, the mechanism of Self-Organized Criticality (SOC) [4] has been invoked.

The understanding of this statistical behavior calls for a model that can simulate the fracturing phenomenon. Unfortunately, the models usually considered describe the formation of a macroscopic crack [5]. These can therefore model AE of fracturing phenomena on mesoscopic scale culminating in a large event that change the system's properties dramatically [6]. This is very different from the stationary state generated by stressing the sample below the breaking threshold of the system. In fact, in this case the fracturing phenomenon produces an energy release that changes the physical properties in a non destructive way (like the passage to a different metastable state).

Here we show a statistical model for fracturing phenomena, where the rupture burst and the following energy release changes the properties but does not destroy the system. This allows us to obtain a stationary state for AE of which we can investigate the statistical properties in space, time and magnitude.

355

Mat. Res. Soc. Symp. Proc. Vol. 409 ® 1996 Materials Research Society

THE MODEL

The mesoscopic description of an elastic disordered medium is obtained discretizing macroscopic elastic equations. In the theory of linear elasticity, these equations relate the stress tensor $s_{\alpha\beta}$ to the strain tensor $\epsilon_{\gamma\delta}$ via the Hooke tensor $C_{\alpha\beta\gamma\delta}$. The full tensorial formalism is quite heavy to handle numerically. A compromise is obtained by considering scalar elastic equations. In fact the phenomenology of fractures in scalar models captures many essential features of more complex tensorial models. Scalar elasticity is formally equivalent to electricity, provided one identifies the current I with the stress, the voltage V with the strain and the conductivity σ with the Hooke tensor.

The discretization scheme we use corresponds to the study a resistor network. For symmetry reason we will consider a rotated square lattice. The disorder, due to the inhomogeneity in the synthetic material, is introduced in the model in the failure thresholds I_c of the resistors. For simplicity we will use an uniform distribution. The crucial part of the model is the breaking criterion, which describe the dynamics of the micro-fracturing process. Typically the breaking criterion is chosen so that if the current flowing in a resistor exceed the failure threshold the conductivity of the bond drops abruptly to zero. In this way the system develops a macroscopic crack and the lattice breaks apart. To describe the micro-fracturing phenomena in the stage preceding the onset of the macroscopic crack it is useful to consider the concept of damage. For a macroscopic elastic material, in which micro-fracturing processes are taking place, the damage D is a tensor relating the effective Hooke tensor $\hat{C}$ to the Hook tensor of the undamaged material. The damage is defined as $D = I - C\hat{C}^{-1}$. For scalar elasticity the damage is just a constant relating the effective resistance of the damaged material to that of the undamaged one. We generalize the concept of damage from the macroscopic to the mesoscopic description, using it in the breaking criterion. When the current in a bond exceeds the threshold we impose a permanent damage to the bond. In other words, the conductivity of the bond drops by a factor $a = (1 - D)$.

In the synthetic materials we are describing, after a micro-fracturing event, local rearrangements take place. We model this effect by changing at random the breaking threshold of the damaged bond as well as those of the neighboring bonds. This rearrangement of the disorder emphasize the probability of breaking successively neighboring bonds. In crack models this process is enforced by imposing the connectivity of the crack or by similar rules.

SIMULATION RESULTS

To simulate the model we start from an undamaged lattice where the conductivities are equal to one for all the bonds. The breaking threshold are chosen at random between zero and one. We then impose an external voltage between two edges of the lattice and we use periodic boundary conditions in the other direction.

The voltage is increased until the current in some bond exceed the threshold. When this happens we apply the breaking rule and we check if subsequent failures occurs. In fact due to the long range elastic interactions combined with the redistribution of the disorder, a single failure can be followed by other similar events, thus generating an avalanche. We consider the number of bonds that break in an avalanche to be proportional to the amplitude of the emitted acoustic signal. In the early stage of the process only small avalanches occur and the current carried by the material steadily increases. In this stage the damage is scattered through the lattice in a random uniform way. After some time when some damage has been accumulated into the system the activity starts to increase. Eventually the system reaches a stationary state where the current does not increase anymore. In other words the increase of the voltage is exactly balanced by the damage, in such a way that the current is kept constant. In this state the damage is no longer homogeneously scattered, but tends to be localized along lines. We are modeling an ideal case in which a bond can suffer a very big damage without breaking completely. In fact we can slightly modify the model by introducing a threshold in the conductivity after which the bond is consider broken (i.e. the conductivity drops to zero). One can then easily understand that the regions of localized damage are those where the macroscopic crack will eventually form and this is indeed what we observe in our simulations.

The activity in this stationary state is highly fluctuating and the distribution of amplitude follows a power law. The scaling region increases with the system size and the fact is a signature of an underlying critical state (see figure 1).

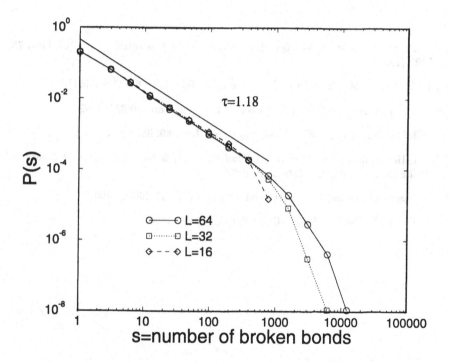

Figure 1: The avalanche distribution in the stationary state for different system sizes.

Another sign of the criticality of the system is provided by the distribution of time intervals between subsequent avalanches. Since the voltage is increased linearly in time, this corresponds to consider the distribution of voltage increases ΔV. The power law with slope close to $x = -1$ is reminiscent of the Omori [7] law for fore-shocks in earthquakes.

CONCLUSIONS

In summary we have introduced a statistical model for fracturing phenomena. This model describes on a mesoscopic scale the energy release and the following rearrangements in the material produced by fracturing events. The simulations on the model show that the system organizes itself in a stationary state where we can relate the AE signal with the rupture events. We investigate the statistical properties of rupture sequences during fracturing and their correlation properties. We find that the stationary state develops critical correlations and scaling behavior through a self-organization process. This kind of analysis is usually considered in the study of experimentally detected AE signals and the proposed model could give important clues in the understanding of the general scale invariance of the fracturing phenomena. The Center for Polymer studies is supported by NSF.

[1] A. Petri, G. Paparo, A. Vespignani, A. Alippi and M. Costantini, Phys. Rev. Lett. **73**, 3423 (1994)

[2] P. Diodati, F. Marchesoni and S. Piazza, Phys. Rev. Lett. **67** 2239 (1991)

[3] G. Cannelli, R. Cantelli and F. Cordero, Phys. Rev. Lett. **70** 3923 (1993)

[4] P.Bak ,C.Tang and K. Wiesenfeld, Phys. Rev. Lett. **59**, 381 (1987).

[5] H.J. Herrmann and S. Roux (eds.), *Statistical Models for the Fracture of Disordered Media* (North Holland, Amsterdam, 1990)

[6] F. Tzschichholtz and H. J. Herrmann, Phys. Rev. E, **51** 1961 (1995).

[7] F. Omori, J. Coll. Sci. Imper. Univ. Tokyo, **7**, 111, (1894)

FRACTURE SURFACE FRACTAL DIMENSION DETERMINED THROUGH AN EXTENSIVE 3-D RECONSTRUCTION

J.J. AMMANN AND A.M. NAZAR
Materials Engineering Department, Faculty of Mechanical Engineering, State University of Campinas, Campinas, Brazil

ABSTRACT

An efficient image processing method is developed to determine the fractal dimension from a high resolution 3-D reconstruction of fracture surface. The third dimension of the fracture surface is obtained from a stereo pair. The elevation is locally determined through a recursive window matching algorithm centered on the cross-correlation operation. The method allows to reach high vertical and horizontal resolutions of the elevation map.

The fractal dimension is obtained by an implementation of the slit-island method. A series of contour lines is extracted from the reconstructed surface. The images of the contour lines are progressively reduced in size and at every step the curve length is estimated. The result is then normalized according to the scale factor to build a Kolmogorov plot.

The performance of this method is demonstrated on a synthetic image generated by a fracture surface mathematical model.

INTRODUCTION

Quantitative fractography has become an essential tool in materials science to understand fracture processes [1-3]. Recently, increasing efforts have been done to develop more accurate methods for quantitative characterization of the fracture surfaces and their specific features [4-7]. In this way, several parameters have been proposed to characterize unflat, rough fracture surfaces and, between them, the fractal dimension has been recognized as one of the most suitable and most accurate way to characterize such surfaces [1, 3-6]. Indeed, various works have already correlated this parameter with mechanical properties [8-10].

However, in order to quantify the true aspect of the surface features, the main requirement is to access the elevation (3-D) information of the surface. From this point of view, the scanning electron microscope (SEM), particularly indicated for fracture surface observation because of its large depth of focus and its high spatial resolution, is very limited in its ability to acquire full 3-D data of the surface [11-13]. In spite of the development of various new methods, the 3-D data acquisition step is still limited: 1) to partial surface topography acquisition, for example through 2-D vertical or horizontal sections, 2) to low resolution reconstruction or 3) by destructive observation techniques, for example through mechanical sectioning or polishing of the specimen.

In order to investigate in a more reliable and accurate way, the fractal dimension of fracture surfaces, a new, powerful method [14] has been used to reconstruct the surface topography up to a high level of spatial and vertical resolution. The method produces a dependable 3-D digital replica of the surface allowing the determination of independent parameters by complementary and otherwise exclusive methods.

In this work, the topography of a synthetic surface is reconstructed and the fractal dimension is determined, based on the slit-island method [15].

EXPERIMENT

Equipment and software

The procedure presented here requires powerful image processing tools. In this field, the KHOROS system (University of New Mexico & KRI Inc.) [16] provides a wide range of image processing operations and routines, attached to a powerful and flexible programming graphic interface (GUI). KHOROS is run on a Sun SPARC 20 workstation.

3-D Surface reconstruction.

The three dimensional reconstruction method is based on the principle of stereoscopy to build an elevation map of the surface [14]. The process extracts locally the third dimension information from the stereo pair. To achieve this goal, the cross-correlation operation is performed between small corresponding windows located at the same relative position in the two images. The position of the maximum in the cross-correlation function is directly related to the local parallax between the images and thus, to the elevation of the surface at this location. A complete elevation map of the region of interest is built through an x-y scan of the image.

The method has been improved to reach a high level of spatial resolution by progressively reducing the cc-window size in successive run of the program. This can be done by taking advantage of the 3-D information previously obtained in the first runs of the program, allowing to pre-align the two corresponding windows. This method has already shown to be powerful to reconstruct very uneven surfaces [14].

Images

The images used in this work have been produced by E. Bouchaud at ONERA (France) [17]. Figure 1 shows the original image created by a fractional brownian model for fracture surfaces (Hurst exponent of 0.8), corresponding to a theoretical fractal dimension of 2.20. In this image, the elevation of the surface is represented as a gray level: high regions of the fracture surface appear as bright regions in the synthetic image. From this synthetic image, a stereo pair has been generated [17] for 2.5 and –2.5 degrees of inclination of the surface (figure 2). The images have 256x256 pixels with 256 levels of gray.

Figure 1- Original synthetic images generated from a mathematical model of fracture surface and used to produce the stereo pair of this work [17].

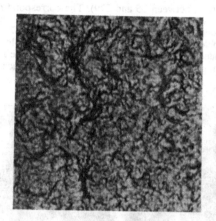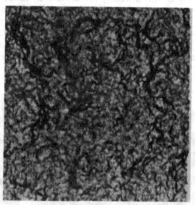

Figure 2-   Stereo pair produced from image of Figure 1 [17], used in the reconstruction process.

Fractal dimension - method implementation

The principle of the slit-island method is to measure at several scales, the length of a contour line of the surface [15]. The log-log plot of the curve length versus the scale factor (Kolmogorov plot) presents a straight line which slope (negative) is related to the fractal dimension ($\mathfrak{D}$) through the following relationship: $\mathfrak{D} = 2$-slope.

In the implementation of the method, contour lines are obtained directly by threshold of the gray-level representation of the elevation map. Regions higher than the threshold level appear in white on a black background. In order to get a 1-pixel wide curve (8-connectivity), a new image is produced by the erosion of the bright regions by a cross (3x3) structural element and an "exclusive-Or" operation is performed between the two images.

The length of contour lines is determined by a simple box counting method. This considers that the number of pixels constituting the curve (curve area) in the digital image corresponds to the length of the curve.

To build the Kolmogorov plot, the elevation map is progressively reduced by increasing scale factors. At each step, the true curve length is determined and normalized by multiplication by the corresponding scale factor. The slope of the curve lengths against the scale factors is obtained by linear regression and the fractal dimension is then determined.

RESULTS

In a first stage, the stereo pair presented in figure 2 is processed by the reconstruction program. Seven successive runs of the program have produced the surface shown in figure 3a. The following cross-correlation window sizes have been successively used: 60, 42, 21, 29, 14, 10 and 7 pixels respectively. In the perspective view (figure 3b), the "z" axis has been normalized to show the real aspect of the synthetic surface.

In the last run of the program, the position of the cross-correlation maximum has been accurately determined to better than a pixel by using a quadratic interpolation of the cross-correlation function. In this case, the vertical resolution can be estimated to half of the image spatial resolution. For further processing, borders of the reconstruction have been rejected.

The slit-island method has been performed on nine contour lines. The elevation map has been normalized to [0, 255] and nine equally spaced threshold levels have been chosen between 10%

and 90% of the intensity (or elevation) range (i.e., between 25 and 229). The corresponding contour lines are shown in figure 4 for the reconstructed and the original surfaces. For each curve, the image is reduced in nine successive steps with scale factors ranging from 1 to 37. At each step, the box counting method is applied to determine the curve length and a log-log plot is built (figure 5). The slopes of the plots are estimated by linear regression for each contour line and the fractal dimensions are determined. The average $\mathfrak{D}$ value of the reconstructed 3-D surface is 2.214 (Std. dev.: 0.07).

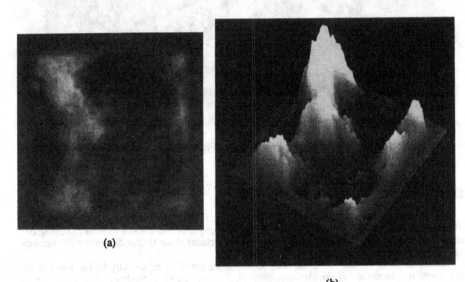

(a)

(b)

Figure 3- Reconstructed surface produced from the stereo pair of figure 2. In (a), the elevation map is presented as a gray level image, in (b) as a perspective view.

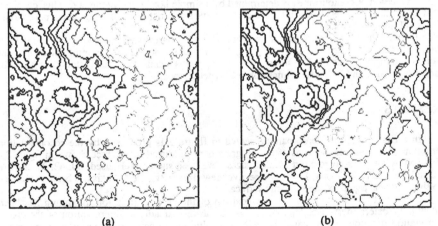

(a)   (b)

Figure 4- Image of the contour lines obtained by thresholding the elevation map: (a) for the original elevation map, (b) for the reconstructed surface. The elevation of corresponding contour lines can be slightly different for the two images.

The fractal dimension has been also measured on the original synthetic image (figure 1). The estimated value of $\mathfrak{D}$ = 2.230 (Std. dev.: 0.07) shows a very good agreement between the fractal dimension of the 3-D reconstructed surface and the original one.

The fractal dimension obtained on the original image (2.23) is close from the theoretical value (2.20) for which the surface model has been build. This confirms the validity of the slit island implementation to investigate the fractal characteristics of the fracture surface.

Moreover, the good agreement between the fractal dimension of the original and reconstructed surfaces shows clearly the efficiency and accuracy of the reconstruction method.

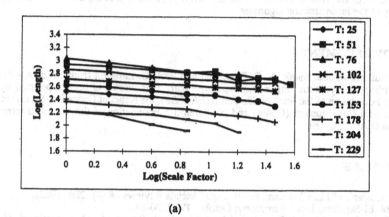

(a)

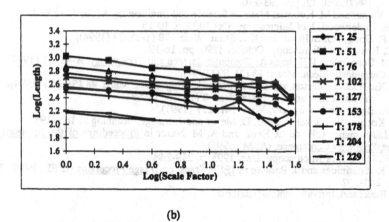

(b)

Figure 5- Log-log plot of the contour line length versus the scale factor for the reconstructed elevation map (a) and for the original elevation map (b). The values of the threshold levels used for each contour line are given on the graph.

CONCLUSION

The iterative 3-D reconstruction method provides a high quality representation of the fracture surfaces. The resulting elevation map has a high vertical and a high horizontal resolution and shows to be a powerful and accurate tool to determine the fractal dimension.

The method has been applied successfully to a synthetic fracture surface model of known fractal dimension. The fractal dimension measured on the original elevation map agrees very well with the theoretical one, validating the implementation of the slit-island method.

Moreover, the fractal dimension measured on the reconstructed surface presents a very good agreement with the fractal dimension of the original surface, proving also the validity and efficiency of the reconstruction algorithm.

ACKNOWLEDGMENT

The authors wish to thank Dr. E. Bouchaud and Prof. G. Lapasset from ONERA - France for providing the synthetic image and the stereo pair. This work has been supported by the Scientific Foundation of the State of São Paulo (FAPESP), the National Council of Scientific and Technological Development (CNPq) and the Scientific Foundation of the State University of Campinas (FAEP), Brazil.

REFERENCES

1. M. Coster and J.L. Chermant, International Metals Reviews 28 (4), 228 (1983).
2. S.M. El-Soudani, J. of Microscopy; October 1990, 20-27.
3. E.E. Underwood and K. Banerji, in Metals Handbook, 9th ed. (ASM International, Metals Park, 1987), vol. 12, pp. 193-210.
4. L. Wojnar and M. Kumosa, Material Science and Engineering A128, 45 (1990).
5. E.E. Underwood, J. of Microscopy; Oct 1990, p. 10-15.
6. W.P. Dong, P.J. Sullivan and K.J. Stout, Wear 178 (1-2), 29 (1994).
7. J.C. Russ, J. of Microscopy; October 1990, pp. 16-19.
8. F.M. Borodich in IFIP trans A. Computer Science and Technology, A-41, 61 (1994).
9. A. Carpinteri, Mech. Mater. 18 (2), 89 (1994).
10. H. Xie, D.J. Sanderson and D.C.P. Peacock, Eng. Fract. Mech. 48 (5), 655 (1994).
11. J.J. Friel and C.S. Pande, J. Mater. Res. 8 (1), 100 (1993).
12. C.A. Brown and G. Savary, Wear 141, 211 (1991).
13. G. Koenig, W. Nickel, J. Storl, D. Meyer and J. Stange, Scanning 9, 185 (1987).
14. J. J. Ammann, L.R. de O. Hein and A. M. Nazar in Proceedings of the International Metallography Conference, (ASM , 1995).
15. J. C. Russ, J. of Microscopy; Dec. 1993; p. 239-248.
16. K. Konstatinides and J. Rassure in IEEE Trans. on Image Processing (IEEE, 1994), 3 (3) pp. 243-252.
17. E. Bouchaud, (private communication).

FRACTOGRAPHY OF GLASS AT THE NANOMETER SCALE

E. GUILLOTEAU, H. ARRIBART and F. CREUZET
Laboratoire C.N.R.S./Saint-Gobain, "Surface du Verre et Interfaces"
39 Quai Lucien Lefranc, 93303 Aubervilliers, France.

ABSTRACT

We present a nanometer scale description of the fracture surface of soda-lime glass. This is achieved by the use of Atomic Force Microscopy. The mirror zone is shown to be built with elementary entities, the density of which increases continuously while the mist and hackle zones are approached. Moreover, the overall picture leads to some kind of self-similarity, in the sense that small regions of the hackle zone exhibit the full set of mirror, mist and hackle areas.

INTRODUCTION

In the field of brittle fracture, it is well known that the analysis of the fracture surface can provide fruitful information on the fracture dynamics [1]. This so-called fractography is based on the observation of three different zones (namely mirror, mist and hackle) ; optical criteria are used to define the frontier between these zones. In other words, this is the observation of the surface which introduces a cut-off at a scale corresponding to the wavelength of light. The purpose of this work is to investigate the topography of these zones at a much shorter scale, in the nanometer range.

EXPERIMENT

A piece of soda-lime silicate glass has been broken in the four points bending geometry after a defect had been created by Vickers indentation of the surface. The fracture surface revealed the three different zones described above. Atomic Force Microscopy [2] has been used to study these zones. Details on the AFM experimental technique can be found elsewhere [3]. Basically, a micro-fabricated cantilever is terminated by a Si_3N_4 tip which is brought close to the surface of interest. The low spring constant of the cantilever ensures that its deflexion can be measured even if the tip-sample interaction is weak. 256×256 imaging is performed by scanning the cantilever over the surface. All the results presented here have been obtained in the contact mode ; taking into account the force involved in this case, we can infer that the elastic deformation (in the context of the Hertz contact) of the glass is much smaller than the size of the features that have been observed. Therefore, all the images described below are an exact representation of the surface topography, unless tip artifacts occur. Such problems may arise when the shape of the tip is such that the contact point is not the apex of the tip.

With our samples, the crack propagation is known to be radial, starting from the surface defect which initiates the fracture. We have taken care to acquire the images in such a way that the direction of propagation is locally at an angle of 45 degrees with respect to the x-axis of the image.

RESULTS

Figure 1 displays the results obtained at the beginning of the mirror zone, close to the defect which initiated the fracture. The z-scale (perpendicular to the surface) is shown as a color scale. The main striking feature is that the surface is not smooth ; precisely, the RMS roughness on a $1.5\mu m \times 1.5\mu m$ area (Figure 1a) is 0.7 nm, to be compared with the value of 0.2 nm for bare soda-lime float glass [4]. In fact, this deviation from flatness is due to the existence of rather well defined entities which can be clearly seen at high magnification (Figure 1b). Their height can be reliably measured in the range of a few nanometers. However, the apparent width (tens of nanometers) may be the consequence of artifacts if we consider the radius of curvature of the tip (typically 20 nm). For instance, we cannot exclude that the entities have a spherical shape which

Mat. Res. Soc. Symp. Proc. Vol. 409 ° 1996 Materials Research Society

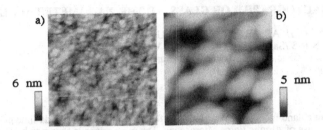

Figure 1 : AFM images of the beginning of the mirror zone in soda-lime float glass
a) 1.5μm × 1.5μm b) 375nm × 375nm

cannot be accurately probed by the tip. In any case, the real diameter of these entities will remain smaller than the apparent one. We want to emphasize that this possible effect will not significantly modify the RMS roughness measurements for large areas (typically above 1 μm).

When the end of the mirror zone is imaged, the topography is largely modified, as it can be seen in Figure 2. At large scale (Figure 2a, 5μm × 5μm), the RMS roughness goes up to 11.4 nm. It is important to point out that the slight difference in size (compared to Figure 1a) cannot account for this very large increase ; this is particularly obvious if we notice that the color scale corresponds to an order of magnitude change. Actually, systematic roughness measurements of 2 μm long profiles have shown that the RMS value varies <u>continuously</u> from roughly 0.4 nm to more than 15 nm between the start and the end of the mirror zone. At short scale (Figure 2b, 375nm × 375nm), elementary entities are still observed, the characteristics of which are very similar to those of Figure 1b. Therefore, the following description can be given : the entities simply build the texture of the mirror zone, the only noticeable change being that they are more and more numerous as the distance to the defect increases.

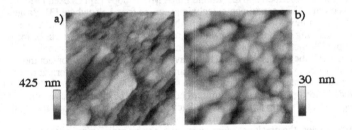

Figure 2 : AFM images of the end of the mirror zone in soda-lime float glass
a) 5μm × 5μm b) 375nm × 375nm

In the mist zone (Figure 3), the RMS roughness is still higher, namely 20.7 nm for an area of 5μm × 5μm (Figure 3a). The fracture surface exhibits bumps and holes, the size of which is in the micron range which corresponds to the wavelength of visible light. Moreover, these features clearly show the direction of crack propagation at 45 degrees. This helps to have a better look at the Figure 2a where we can actually discern the same direction, although the features are much less pronounced. Again, the elementary entities are observed at high magnification (Figure 3b) ; it is important to emphasize that Figures 1b, 2b and 3b do not reveal any meaningful change between mirror and mist regions at a length scale of 375 nm. At this stage, the entities aggregate to form the rather large objects of Figure 3a.

Imaging the hackle zone completes the continuous description of the fracture surface. Figure 4a shows that the RMS roughness is still higher (55.6 nm over an area of 5μm × 5μm).

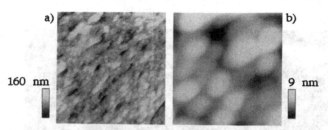

Figure 3 : AFM images of the mist zone in soda-lime float glass
a) 5μm × 5μm b) 375nm × 375nm

Moreover, the direction of crack propagation is now unambiguously settled. At short scale (Figure 4b), the elementary entities are still present and remain identical to those in the mirror and mist regions. As far as roughness is concerned, it is important to note that no detectable jump is observed when the mirror-mist and mist-hackle frontiers are crossed.

Figure 4 : AFM images of the hackle zone in soda-lime float glass
a) 5μm × 5μm b) 550nm × 550nm

Large areas of Figure 4a can be described as a fracture surface at the microscopic scale. To make this statement clear, we have used a Scanning Electron Microscope (SEM) with a field emission source. This permits to image insulators without metallization of the surface since the SEM can be operated at low voltage. Figure 5 displays a typical image of a detail of the hackle zone ; in that particular image, the direction of propagation is not aligned along the 45 degrees direction. Since the resolution of SEM is not as good as the one of AFM, a mirror zone can be defined in the bottom left corner of the image. The starting point of this mirror seems to correspond to some kind of micro-branching. Far from this point, the mirror transforms into a mist zone, again at the microscopic scale. This description corresponds to a self-similar character of the fracture surface.

AFM has been used to image these newly identified zones. The results are summarized in Figure 6 which displays three 1.5μm × 1.5μm images of these mirror and mist zones (belonging to the μm scale fracture surface of the macroscopic hackle zone). Figure 6a has been taken in a region close to the initiation due to micro-branching. The RMS roughness has been drastically reduced and is now 1.4 nm. The elementary entities are clearly evidenced and cannot be distinguished from those observed in the macroscopic mirror region. In the μm scale mist zone (Figure 6c), the entities gather in such a way that the direction of crack propagation appears along the 45 degrees line. Correlatively, the RMS roughness goes up to 7.0 nm. In between (Figure 6b), the direction of propagation is barely noticeable while the RMS roughness is 1.7 nm. In all cases, the entities are present and appear to be the building bricks of the fracture surface.

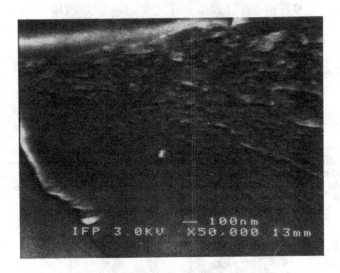

Figure5 : SEM image of a small area (2.3μm × 1.8 μm) in the hackle zone of soda-lime float glass.

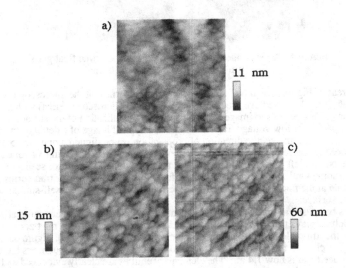

Figure 6 : 1.5μm × 1.5μm AFM images of the a) mirror b) mirror-mist and c) mist zones of the μm scale fracture surface in the macroscopic hackle region.

DISCUSSION AND CONCLUSION

For clarity, we have restricted the presentation of our experimental results to the case of soda-lime silicate glass. More investigations have been carried out with silica and various glasses like fluoride (non oxide) glass and lead oxide silicate glass. So far, all the results described above are found to be independent of the chemical nature of the glass. Moreover, as far as the existence of the elementary entities is concerned, we have checked that the water does not play a major role. Explicitly, we have broken glass at liquid nitrogen temperature which prevents water to move with the crack front ; no difference has been detected.

Therefore, the above description of the fracture surface can be used to discuss the characteristics of brittle fracture. Close to the defect responsible for fracture, the crack velocity (or stress intensity factor) is low and the behavior is expected to be almost quasi-static. At this stage, the elementary entities are revealed without any apparent order related to the direction of crack propagation. The characteristic length scale is far below the wavelength of visible light and the fracture surface looks like a mirror. When the crack velocity increases, the entities gather and form rather large objects (the size of which is in the micron range) which can diffuse light. This is the genesis of the mist zone. The dimension of these objects remain small in the direction perpendicular to the crack propagation, so that no alignment can be optically detected, although this is clearly evidenced with AFM at the nanometer scale. In the hackle zone, the dimensions are so large that optical means are sufficient to follow the propagation.

Finally, the self-similarity indicates that the crack velocity (and/or the stress intensity factor, whichever of the parameter is relevant) is not uniform in the hackle region : locally, the existence of a fracture surface (with mirror and mist regions) at the micron scale suggests that the crack velocity might be significantly reduced at the location of micro-branching. It is well known that the fractal approach [5] can be used to analyze this self-similarity. This must be performed at a length scale larger than the size of the elementary entities. Preliminary measurements have been carried out to evaluate the roughness exponent ζ of fracture surfaces prepared at constant velocity. The measurements have been carried out by recording the height profile in a direction perpendicular to the fault. The results obtained at two different crack velocities (1 m/s and 1μm/s) do not reveal any remarkable difference : the roughness exponent ζ is found to be close to 0.87, with a possible appearance of $\zeta=0.5$ at short length scale. As suggested in [6], this might be the indication that the relevant parameter is not the roughness exponent itself, but possibly the length scale which defines the crossover between these two regimes. More work is now under progress to analyze the height profile in the regime of very low crack velocity.

The elementary entities that have been disclosed by this AFM work seem to be characteristic of the vitreous state. The nature of this new structure at the nanometer scale is still an open question. Two main directions may be proposed on the basis of the glass structure which is expected to exhibit some heterogeneity in this range [7]. First, a possible explanation is that the crack follows a pathway of low energy barrier which should be related to the structural network. This is supported by recent molecular dynamics studies on silica [8,9] which predict that the breaking energy depends on the precise molecular arrangement in the material. Second, a structural relaxation may occur after the fracture itself, even if the crack growth takes place in a well-defined plane. It has to be noted that the heterogeneity of glass has been recently investigated by NMR [10]. For the glass studied in this latter work, medium range order is interpreted as the existence of clusters, the diameter of which is typically 20 nm and depends upon the temperature cycling. From this point of view, the elementary entities observed with AFM would correspond to these dense clusters. Of course, other interpretations may be proposed, like dynamic instability in the nanometer range. Finally, ageing proceeses have to be taken into account, as pointed out by Watanabe et al [11].

More details cand be found in ref [12].

ACKNOWLEDGEMENTS

We would like to thank S. Valladeau for technical assistance and the Institut Français du Pétrole for SEM imaging. E.G. acknowledges the financial support of Isover Saint-Gobain.

REFERENCES

1. see for example Fractography of glasses and ceramics Ed. J.R. Varner and V.D. Frechette, Advances in Ceramics, Vol.22, 1986.

2. G. Binnig, C.F. Quate and Ch. Gerber, Phys. Rev. Lett. **56**, 930 (1986).

3. Scanning Tunneling Microscopy Ed. R. Wiesendanger and H.J. Güntherodt, Springer Series in Surface Science, 1992.

4. F. Creuzet, D. Abriou and H. Arribart, in Fundamentals of the Glass Manufacturing Process 1991 Proc. of the First Conference of the European Society of Glass Science and Technology, Sheffield (1992), p.111.

5. J.J. Mecholsky Jr. and J.R. Plaia, J. of Non Cryst. Solids **146**, 249 (1992).

6. E. Bouchaud and J.P. Bouchaud, Phys. Rev. **B50**, 17752 (1994).

7. P.H. Gaskell, Materials Science and Technology, Vol. 9 (VCH, New-York, 1991) pp175-278.

8. J.H. Simmons, T.P. Swiler and R. Ochoa, J. of Non-Cryst. Solids **134**, 179 (1991).

9. J.K. West and L.L. Hench, J. of Materials Science **29**, 3601 (1994).

10. S. Sen and J.F. Stebbins, Phys. Rev. **B50**, 822 (1994).

11. Y. Watanabe, Y. Nazamura, J.T. Dickinson and S.C. Langford, J. of Non Cryst. Solids **177**, 9 (1994).

12. E. Guilloteau, PhD thesis, Université Paris-XI, 1995.

SCALING OF FRACTURE IN QUASIBRITTLE MATERIALS AND THE QUESTION OF POSSIBLE INFLUENCE OF FRACTAL MORPHOLOGY

ZDENĚK P. BAŽANT
Walter P. Murphy Professor of Civil Engineering and Materials Science
Northwestern University, Evanston, Illinois 60208, USA, z-bazant@nwu.edu

ABSTRACT

The paper presents a review of recent results on the problem of size effect (or the scaling problem) in nonlinear fracture mechanics of quasibrittle materials and on the validity or recent claims that the observed size effect may be caused by the fractal nature of crack surfaces. The problem of scaling is approached through dimensional analysis and asymptotic matching. Large-size and small-size asymptotic expansions of the size effect on the nominal strength of structures are presented, considering not only specimens with large notches (or traction-free cracks) but also structures with no notches. Simple size effect formulas matching the required asymptotic properties are given. Regarding the fractal nature of crack surfaces, it is concluded that it cannot be the cause of the observed size effect.

INTRODUCTION

Scaling is a salient aspect of all physical theories. Nevertheless, little attention has been paid to the problem of scaling or size effect in solid mechanics. Up to the middle 1980's, observations of the size effect on the nominal strength of a structure have generally been explained by Weibull-type theory of random strength. However, recent in-depth analysis (cf. Bažant, 1995a) has shown that this Weibull-type theory does not capture the essential cause of size effect for quasibrittle materials such as rocks, toughened ceramics, concretes, mortars, brittle fiber composites, ice (especially sea ice), wood particle board and paper, in which the fracture process zone is not small compared to structural dimensions and large stable crack growth occurs prior to failure. The dominant source of size effect in these materials is not statistical but consists in the release of stored energy from the structure engendered by fracture propagation.

By approximate analysis of energy release from the structure, a simple size effect law has been derived in 1983 for quasibrittle fracture. This law subsequently received extensive justifications, based on: (1) comparisons with tests of notched fracture specimens of concretes, mortars, rocks, ceramics, fiber composites as well as unnotched reinforced concrete structures, (2) similitude in energy release and dimensional analysis, (3) comparison with discrete element (random particle) numerical modeling of fracture, (4) derivation as a deterministic limit of a nonlocal generalization of Weibull statistical theory of strength, and (5) comparison with finite element solutions based on nonlocal model of damage (see ref. in Bažant et al., 1994; Bažant, 1995a,b). The simple size effect law has been shown useful for evaluation of material fracture characteristics from tests. Recently, the fractal nature of crack surfaces in quasibrittle materials has been studied intensively by Mandelbrot, Mecholsky and Mackin, Molosov and Borodich, Xie and others (see ref. in Bažant, 1955a,b). It has been proposed that the crack surface fractality might be an alternative source of the observed size effect (Carpinteri et al. 1955, Lange, Saouma and others). This paper, which summarizes a lecture presented at the IUTAM Symposium on Nonlinear Fracture Mechanics at Cambridge University (1995), tersely outlines a generalized asymptotic theory of scaling of quasibrittle fracture and at the same time explores the possible role of the crack surface fractality in the

371

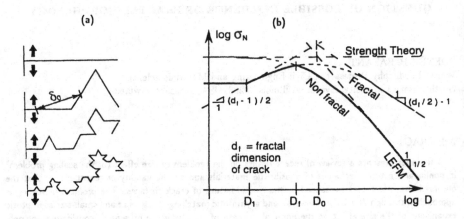

Figure 1: (a) Von Koch curves as examples of fractal crack at progressive refinement. (b) Size effect curves obtained for geometrically similar specimens with nonfractal and fractal cracks and finite size of fracture process zone (possible transition to horizontal line for nonfractal behavior is shown for $D < D_1$)

size effect.

LARGE SIZE ASYMPTOTIC EXPANSION OF NONFRACTAL AND FRACTAL SIZE EFFECT

For the sake of brevity, the analysis will be made in general for fractal cracks and the nonfractal case will then simply be obtained as a limit case. Consider a crack representing a fractal curve (Fig. 1a) whose length is defined as $a_\delta = \delta_0 (a/\delta_0)^{d_f}$ where d_f = fractal dimension of the crack curve (≥ 1) and δ_0 = lower limit of fractality implied by material microstructure, which may be regarded as the length of a ruler by which the crack length is measured (Mandelbrot). Unlike the case of classical, nonfractal fracture mechanics, the energy W_f dissipated per unit length of a fractal crack cannot be a material constant because the length of a fractal curve is infinite. Rather, it must be defined as

$$W_f/b = G_{fl} a^{d_f} \tag{1}$$

where b = thickness of the structures (considered to be two-dimensional), G_{fl} = fractal fracture energy, of dimension Jm^{-d_f-1}. A nonfractal crack is the special case for $d_f = 1$, and them $G_{fl} = G_f$, representing the standard fracture energy, of dimension Jm^{-2}).

The rate of macroscopic energy dissipation $\mathcal{G}_{cr}$ with respect to the 'smooth' (projected, Euclidean) crack length a is:

$$\mathcal{G}_{cr} = \frac{1}{b}\frac{\partial W_f}{\partial a} = G_{fl} d_f a^{d_f-1} \tag{2}$$

(e.g., Borodich, 1992). To characterize the size effect in geometrically similar structures of different sizes D, we introduce the usual nominal stress $\sigma_N = P/bD$ where D = characteristic size (dimension) of the structure, P = applied load If $P = P_{max}$ = maximum load, σ_N is called the nominal strength.

The problem will be analyzed under the following three hypotheses: (1) Within a certain range of sufficiently small scales, the failure is caused by propagation of a single fractal crack. (2) The fractal fracture energy, G_{fl} is a material constant correctly defining energy dissipation. (3) The material may (but need not) exhibit a material length, c_f.

The material length, c_f, may be regarded as the size (smooth, or projected) of the fractal fracture process zone in an infinitely large specimen (in which the structure geometry effects on the process zone disappear). The special case $c_f = 0$ represents fractal generalization of linear elastic fracture mechanics (LEFM). Alternatively, if we imagine the fracture process zone to be described by smeared cracking or continuum damage mechanics, we may define $c_f = (G_{fl}/W_d)^{1/(2-d_f)}$ in which W_d = energy dissipated per unit volume of the continuum representing in a smeared way the fracture process zone (area under the complete stress-strain curve with strain softening). As still another alternative, in view of nonlinear fracture mechanics such as the cohesive crack model, we may define $c_f = (EG_{fl}/f_t^2)^{1/(2-d_f)}$ in which E = Young's modulus and f_t = material tensile strength.

We have two basic variables, a and c_f, both of the dimension of length. Let us now introduce two dimensionless variables: $\alpha = a/D$ and $\theta = c_f/D$. In view of Buckingham's theorem of dimensional analysis, the complementary energy Π^* of the structure with a fractal crack may be expressed in the form:

$$\Pi^* = \frac{\sigma_N^2}{E} bD^2 f(\alpha, \theta) \tag{3}$$

in which f is a dimensionless continuous function characterizing structure geometry.

The energy balance during crack propagation (first law of thermodynamics) must be satisfied by nonfractal as well as fractal cracks. The energy release from the structure as a whole is a global characteristic of the state of the structure and must be calculated on the basis of the smooth (projected, Euclidean) crack length a rather than the fractal curve length a_δ, i.e.

$$\frac{\partial \Pi^*}{\partial a} = \frac{\partial W_f}{\partial a}. \tag{4}$$

Substituting (3) and differentiating, we obtain an equation (see Bažant, 1995a,b) containing the derivative $g(\alpha, \theta) = \partial f(\alpha, \theta)/(\partial \alpha)$, which represents the dimensionless energy release rate. The derivative of (3) must be calculated at constant load (or constant σ_N) because, as known from fracture mechanics, the energy release rate of a crack is the derivative of the complementary energy at constant load, i.e. $\sigma_N/\partial a = 0$. In this manner (Bažant, 1995a,b) one obtains the equation $\sigma_N = \sqrt{E\mathcal{G}_{cr}/Dg(\alpha_0, \theta)}$ where α_0 = relative crack length α at maximum load.

Because $g(\alpha_0, \theta)$ ought to be a smooth function, we may expand it into Taylor series about the point $(\alpha, \theta) \equiv (\alpha_0, 0)$. This leads to the result (Bažant, 1995a,b):

$$\sigma_N = \sqrt{\frac{E\mathcal{G}_{cr}}{D}} \left[g(\alpha_0, 0) + g_1(\alpha_0, 0)\frac{c_f}{D} + \frac{1}{2!}g_2(\alpha_0, 0)\left(\frac{c_f}{D}\right)^2 + ... \right]^{-1/2} \tag{5}$$

in which $g_1(\alpha_0, 0) = \partial g(\alpha_0, \theta)/\partial\theta$, $g_2(\alpha_0, 0) = \partial^2 g(\alpha_0, \theta)/\partial\theta^2, ...$, all evaluated at $\theta = 0$. This equation represents the large-size asymptotic series expansion of size effect. To obtain a simplified approximation, one may truncate the asymptotic series after the linear term, i.e.

$$\sigma_N = Bf_t' D^{(d_f - 1)/2} \left(1 + \frac{D}{D_0}\right)^{-1/2} \tag{6}$$

in which D_0 and B are certain constants depending on both material and structure properties, expressed in terms of function $g(\alpha_0, 0)$ and its derivative. For the nonfractal case, $d_f \to 1$, this reduces to the size effect law deduced in 1983 by Bažant (see Bažant et al., 1994), which reads $\sigma_N = Bf_t'/\sqrt{1+\beta}$, $\beta = D/D_0$ in which β is called the brittleness number.

If only geometrically similar fracture test specimens are considered, α_0 is constant (independent of D), and so is D_0. For brittle failures of geometrically similar quasibrittle structures without notches, it is often observed that the crack lengths at maximum load are approximately geometrically similar. For concrete structures, the geometric similarity of cracks at maximum load has been experimentally demonstrated for diagonal shear of beams, punching of slabs, torsion, anchor pullout or bar pullout, and bar splice failure, and is also supported by finite element solutions (e.g. Bažant et al., 1994) and discrete element (random particle) simulations, albeit for only a limited size range of D. Thus, k, c_0, D_0, σ_N^0 and $B f_t'$ are all constant. In these typical cases, (6) describes the dependence of σ_N on size D only, that is, the size effect. Fig. 1b shows the size effect plot of $\log \sigma_N$ versus $\log D$ at constant α_0. Two size effect curves are seen: (1) the fractal curve and (2) the nonfractal curve (for the latter, the possibility termination of fractality at the left end is considered in the plot).

The curve of fractal scaling obtained in Fig. 1b disagrees with the bulk of experimental evidence (for concrete, see e.g. the review in Bažant et al. 1994); for carbon fiber epoxy composites used in aerospace industry, as shown by Bažant, Daniel and Li. It follows that crack fractality cannot be the cause of the observed size effect.

What aspect of the fracture process causes that crack fractality has no significant effect on scaling of failure? The fracture front in quasibrittle materials does not consist of a single crack, but a wide band of microcracks, which all must form and dissipate energy before the fracture can propagate. Only very few of the microcracks and slip planes eventually coalesce into a single continuous crack, which forms the final crack surface with fractal characteristics. Thus, even though the final crack surface may be to a large extent fractal, the fractality cannot be relevant for the fracture process zone advance. Most of the energy is dissipated in the fracture process zone by microcracks (as well as plastic-frictional slips) that do not become part of the final crack surface and thus can have nothing to do with the fractality of the final crack surface.

GENERALIZATIONS

Material length c_f can be defined as the LEFM-effective length of the fracture process zone, measured in the direction of propagation in a specimen of infinite size. In that case, $\theta = c_f/D = (a - a_0)/D = \alpha - \alpha_0$, and so $g(\alpha, \theta)$ reduces to the LEFM function of one variable, $g(\alpha)$. Thus Eq. (6) yields (Bažant, 1995a,b):

$$D_0 = c_f \frac{g'(\alpha_0)}{g(\alpha_0)}, \qquad B f_t' = \sqrt{\frac{EG_f}{c_f g'(\alpha_0)}}, \qquad \sigma_N^0 = \sqrt{\frac{EG_{fl} d_f \alpha_0^{d_f-1}}{c_f g'(\alpha_0)}} \qquad (7)$$

and so Eq. (6) takes the form:

$$\sigma_N = \sqrt{\frac{EG_{fl} d_f \alpha_0^{d_f-1}}{g'(\alpha_0)c_f + g(\alpha_0)D}} \qquad (8)$$

The advantage of this equation is that its parameters are directly the material fracture parameters. For $d_f = 1$, Eq. (8) reduces to the form of size effect law derived in 1990 in a different manner by Bažant and Kazemi (see Bažant, 1995b). Fitting this equation to size effect data, which can be done easily by rearranging the equation to a linear regression plot, one can determine G_f or G_{fl} and c_f. This serves as the basis of the size effect method for measuring the material fracture parameters, which has been adopted by RILEM as an international standard for concrete.

More generally, one may introduce general dimensionless variables $\xi = \theta^r = (c_f/D)^r, h(\alpha_0, \xi) = [g(\alpha_0, \theta)]^r$, with any $r > 0$. Then, expanding in Taylor series function $h(\alpha_0, \xi)$ with respect to ξ, one obtains by a similar procedure as before a more general large-size asymptotic series expansion (whose nonfractal special case was derived by Bažant in 1985):

$$\sigma_N = \sigma_P \left[\beta^r + 1 + \kappa_1 \beta^{-r} + \kappa_2 \beta^{-2r} + \kappa_3 \beta^{-3r} + \ldots \right]^{-1/2r} \qquad (9)$$

in which $\beta = D/D_0$ and $\kappa_1, \kappa_2, \ldots$ are certain constants. However, based on experiments as well some limit properties, it seems that $r = 1$ is the appropriate value for most cases.

The large-size asymptotic expansion (9) diverges for $D \to 0$. For small sizes, one needs a small-size asymptotic expansion. The previous energy release rate equation $(\sigma_N^2/E)Dg(\alpha, \vartheta) = \mathcal{G}_{cr}$ is not meaningful for the small size limit because the zone of distributed cracking is relatively large. In that case, the material failure must be characterized by W_f rather than G_f. In that case, the energy balance equation (first law) for $\partial\sigma_N/\partial a = 0$ (second law) must be written in the form $\sigma_N^2[\psi(\alpha, \eta)]^r/E = W_f$ where $\psi(\alpha, \eta) =$ dimensionless function of dimensionless variables $\alpha = a/D$ and $\eta = (D/c_f)^r = \vartheta^{-r}$ (variable ϑ is now unsuitable because $\vartheta \to \infty$ for $D \to 0$), and exponent $r > 0$ is introduced for the sake of generality, same as before. Because, for very small D, there is a diffuse failure zone, a must now be interpreted as the characteristic size of the failure zone, e.g., the length of cracking band. The same procedure as before now leads to the result (Bažant, 1995a):

$$\sigma_N = \sigma_P \left[1 + \beta^r + b_2\beta^{2r} + b_3\beta^{3r} + \ldots\right]^{-1/2r} \tag{10}$$

in $\sigma_P, D_0, b_2, b_3, \ldots$ are certain constants depending on both material and structure properties and expressed in terms of function $\psi(\alpha_0, 0)$ and its derivatives. Eq. (10) represents the small-size asymptotic series expansion.

An important common characteristic of the large-size and small-size asymptotic series expansions in Eqs. (9) and (10) is that they have the first two terms in common. Therefore, if either series is truncated after the second term, it reduces to the same generalized size effect law (Bažant, 1985a):

$$\sigma_N = \sigma_P(1 + \beta^r)^{-1/2r} \qquad (\beta = D/D_0) \tag{11}$$

Because this law is anchored to the asymptotic cases on both sides and shares with both expansions the first two terms, it may be regarded as a matched asymptotic, that is, an intermediate approximation of uniform applicability for any size. The value $r = 1$ appears, for several reasons, most appropriate.

A different approach is needed for unnotched quasibrittle structures that reach the maximum load when the crack initiates from a smooth surface, as exemplified by the standardized bending test of modulus of rupture f_r of a plain concrete beam. Applying the size effect law in Eq. (6 for the case $\alpha_0 \to 0$ is impossible because $g(\alpha_0, 0)$ vanishes as $\alpha_0 \to 0$. To deal with this case, one must truncate the large-size asymptotic series expansion only after the third term. Then, considering that $r = 1$ and $g(\alpha_0, 0) = 0$, and restricting attention to the nonfractal case only, a similar procedure as that which led to Eq. (8) some further asymptotic approximations lead (Bažant, 1995a) to the following size effect law (Fig. 2a) for failures at crack initiation from a smooth surface:

$$\sigma_N = Bf_r^\infty \left(1 + \frac{D_b}{D}\right) = f_r^\infty \left[1 - 0.0634g''(0)\frac{\bar{c}_f}{D}\right] \tag{12}$$

(the first part of this equation was derived by Bažant and Li in a different manner; see ref. in Bažant, 1995a,b). Here f_r^∞ is the modulus of rupture for and infinitely large beam (but not so large that Weibull statistical size effect would become significant), and B is a dimensionless parameter. This equation can be arranged as a linear regression plot of σ_N versus $1/D$, which is again helpful for easy identification of the constants from tests.

Asymptotic matching of the three asymptotic expansions, namely: (1) the large-size expansion for large α_0, (2) the large-size expansion for varnishing α_0, and (3) the small-size expansion for large α_0, leads (Bažant, 1995a) to the following approximated universal size effect law (Fig. 2b) valid for failures at both large cracks and crack initiation from a smooth surface:

$$\sigma_N = \sigma_0 \left(1 + \frac{D}{D_0}\right)^{-1/2} \left\{1 + \frac{1}{s}\left[\left(\eta + \frac{D}{D_b}\right)\left(1 + \frac{D}{D_0}\right)\right]^{-1}\right\}^s \tag{13}$$

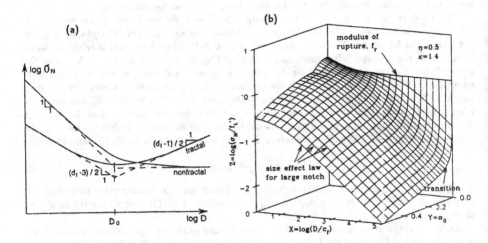

Figure 2: (a) Size effect curves obtained for unnotched specimens, nonfractal and fractal. (b) The surface of universal size effect law for notched as well as unnotched fracture specimens

in which σ_0, D_0, D_b and $\bar{c}_f$ are constants expressed in terms of $g(\alpha_0)$ and its first and second derivatives, and of EG_f, and η and κ are additional empirical constants.

ACKNOWLEDGEMENT: Partial financial support under NSF grant MSS-911447-6 to Northwestern University and additional support from the ACBM Center at Northwestern University are gratefully acknowledged.

REFERENCES

Bažant, Z.P. (1995a). "Scaling theories for quasibrittle fracture: Recent advances and new directions." in *Fracture Mechanics of Concrete Structures* (Proc., 2nd Int. Conf. on Fracture Mech. of Concrete and Concrete Structures (FraMCoS-2), held at ETH, Zürich), ed. by F.H. Wittmann, Aedificatio Publishers, Freiburg, Germany, 515–534.

Bažant, Z.P. (1995b). "Scaling of Quasibrittle Fracture and the Fractal Question", *ASME J. of Materials and Technology* (Special issue of Diamond Jubilee Symp. of Materials Division) 117, 361–367.

Bažant, Z.P., Ožbolt, J., and Eligehausen, R. (1994). "Fracture size effect: review of evidence for concrete structures." *J. of Struct. Engrg., ASCE*, 120 (8), 2377–2398.

Borodich, F. (1992). "Fracture energy of fractal crack, propagation in concrete and rock" (in Russian). *Doklady Akademii Nauk* 325 (6), 1138–1141.

Carpinteri, A., Chiaia, B., and Ferro, G. (1995). "Multifractal nature of material microstructure and size effects on nominal tensile strength." *Fracture of Brittle Disordered materials: Concrete, Rock and Ceramics* (Proc., IUTAM Symp., Univ. of Qeensland, Brisbane, Sept. 1993), eds G. Baker and B.L. Karihaloo, E. & F.N. Spon, London, 21–50.

Fracture Surfaces in 3D Fuse Networks

V. I. Räisänen[1], M. J. Alava[2,3], and R. M. Nieminen[1,3]

[1] Centre for Scientific Computing, P. O. Box 405, FIN-02101 Espoo, Finland

[2] Michigan State University, Department of Physics and Astronomy, E. Lansing, MI 48824-1116, U.S.A*

[3] Laboratory of Physics, Helsinki U. of Technology, Otakaari 1M, SF-02150 Espoo, Finland

(December 5, 1995)

We study a 3D random fuse network model with computer simulations. The breaking thresholds are distributed randomly, corresponding to quenched disorder. We find for the roughness exponent of the final fracture surface $\zeta = 0.47 \pm 0.19$, which is close both the minimum energy surface value and the directed percolation depinning model value in 2+1 dimensions. It is also similar to results from measurements of fracture surfaces at nanometer scale, and from experiments in which the fracture process occurs slowly as in fatique. The traditional measure of damage, the number of broken bonds grows faster than the area effect ($n_b \sim L^{2.28}$), with no signs of a trivially brittle regime.

I. INTRODUCTION

The characterization of crack surfaces is a problem with both engineering and theoretical interest. The question of how 'rough' a crack becomes is connected with the microstructure of the material, and with the mode of failure. It has also to do with practically important qualities such as fracture toughness. The roughness of the surfaces, defined via the fractal dimension D or in case of self-affine ones the roughness exponent ζ, may yield fundamental information about the processes active both at the microscale and globally, as in the case of crack formation that proceeds by searching for a minimum energy configuration.

In experiments performed with a wide range of materials, it has been found that the crack roughness exponent seems to be rather non-universal, in contrast to early claims [1]. For 'fast' failure at large enough length scales ($d \gg 1\mu m$), there exists a spectrum of results centered around $\zeta = 0.8$. On the other hand, for quasi-static failure the observed roughness is smaller and one reported exponent $\approx 0.4(5)$ [1–5]. At nanometer scales the crack roughness seems to be systematically smaller, e.g. for graphite 0.43 [6]. The mode of fracture, especially speed of crack propagation, has in general a complicated effect on the roughness of the final crack [7], as the branching of rapidly advancing cracks tends to increase the observed roughness.

The '3D' roughness of crack surfaces in metals was first analyzed by Mandelbrot and coworkers [8] the indication being that high crack roughness corresponds to low fracture toughness. The later experimental evidence is, however, inconclusive [9,10]. Theoretically, it has been shown that the wandering of the crack should increase the free surface to be created projected along the macroscopic direction of the crack surface [2,11].

Thus the fracture toughness, as defined by Griffith's criterion, should increase with D as $K \sim \sqrt{D-1}$ for fractal cracks

An electric analogue of mechanical failure, the random fuse network (RFN) model, has been studied by Hansen et al. [12] as a simplified model for crack surfaces in disordered systems. The results obtained are fairly close to the surface variation expected from a directed polymer in a random energy landscape [13], whose conformations can be exactly mapped to the Kardar-Parisi-Zhang universality class in 1+1 dimensions [14]. For ideal plasticity, being another version of the minimum energy problem, the correspondence has been claimed to be exact [15], giving a surface roughness of $\zeta = 2/3$ in 1+1D. The roughness of *three-dimensional* fracture/yield models has not been studied, although in 2+1D the directed surface problem gives a roughness of $\zeta = 0.5 \pm 0.1$, as calculated by Kardar and Zhang with a transfer matrix method [16]. Another type of universality class, possibly relevant for the scaling of fracture surfaces, is given by the directed percolation depinning models (DPD) [17,18] depending on whether the surface is considered at the pinning threshold or above it. Bouchaud *et al.* have developed a random Langevin-type 2D crack model with DPD-type scaling properties [19].

The general features of fracture of RFN's with a breaking threshold probability defined with $P(i_c) = 1/w$, $i_c \in [1 - w/2, 1 + w/2]$ have been studied by Kahng et al. [20] in 2D already a decade ago. The number of broken bonds n_b scales with L, the linear system size, trivially with a decreasing finite size correction term for all $w < 1.2$. Only in the case of $w \approx 1.5$ does the finite size scaling contribution become important in the thermodynamic limit. For $w = 2$ the scaling of the breakdown voltage, which equals in RFN's fracture strain, was found to follow roughly the expected result from the dilute system limit ($v_b \sim 1/(\ln L)^{0.8}$, in contrast to $v_b \sim 1/\sqrt{\ln L}$ [21]).

The largest difference to be expected when moving from 2D models to 3D is due to the lesser current enhancement near the crack tip, as in the 2D case $\Delta_i = 4/\pi \approx 1.2732$ for a single missing bond/fuse and in 3D $\Delta_i \approx 1.0926$ [21,20]. The existing studies of 3D fuse models have, however, either not concentrated on the threshold case [22], or been restricted to rather small systems [23]. Note that Sahimi and Arbabi, after reanalyzing their data, claim that the VI-curves from various system sizes can be collapsed by $F \sim \left[L^{d-1}/(\ln L)^{\Psi}\right] h\left(U/L^{d-1}\right)$, with a quite small Ψ (0.2) for $d = 3$ [24].

Next the roughness of crack surfaces is studied numerical simulations of 3D RFN lattice models. The roughness exponent is acquired using the standard procedure of calculating $w^2 = \langle (h^2) - h^2 \rangle$, where h is the local crack interface height, excluding overhangs and secondary branches. In addition, we calculate macroscopic quantities like the total number of broken bonds, n_b, breakdown current and voltage, v_b and i_b, and compare them to known 2D and 3D results.

II. MODEL

We study RFNs in a cubic three-dimensional lattice with a random distribution of breaking thresholds i_c described above. Periodic boundary conditions are applied in the y-direction, free B.C.'s in z-direction and an external voltage in the x-direction. In this study, we mainly have used the value $w = 0.5$ for the width of the breaking limits distribution. A larger width increases the diffuse damage prior to failure, and therefore also the time required to break the system apart rendering systematic studies with varying distribution width w impossible. Note that the 2D results of Hansen et al. seem to indicate that the actual distribution used is not of importance in determining the crack roughness.

We solve the electrical equilibrium state using conjugate gradient method to the accuracy of $\Delta_{res} = 10^{-8}$. Convergence checks with smaller Δ_{res} showed no difference in any of the fracture quantities. Once the equilibrium has been found, the most stressed bond is removed (corresponding to fast relaxation time or the 'single hot bond algorithm' in algorithmic parlance). This procedure of calculating the equlibrium and breaking the most stressed bond is continued until the sample is split into two parts by a fracture surface. The Conjugate Method was parallelized with the PVM (Parallel Virtual Machine) library [25] for use with Cray T3D distributed memory parallel computer, and was also run on a C90 as an ordinary shared memory parallel/vectorized program. System sizes range from 8^3 to 40^3. The number of realizations per system size are as follows: 80 (L=8), 50 (L=12), 50 (L=14), 50 (L=16), 35 (L=24), 30 (L=32) and 20 (L=40). For the results reported here, the total CPU-time expenditure is of the order of 200 equivalent Cray C90 CPU-hours.

The fracture surface (excluding sidebranches) is extracted from the data of broken bonds. An example case is shown in Fig. 1.

AN EXAMPLE OF A FRACTURE SURFACE IN 3D

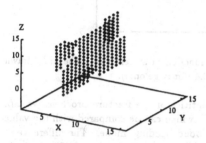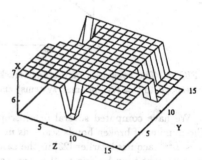

FIG. 1. An example of a fracture surface in 3D. The system size is 16^3 and $w = 1$. Left, placement of the fracture surface in the lattice; right, surface plot of the same data.

III. RESULTS

The scaling of the interface width of the fracture surface can be divided into two regimes. As Fig. 2 shows, for large enough systems ($L = 16 \cdots 40$) we obtain $\zeta = 0.47 \pm 0.19$ [26], which indicates a relatively smooth surface (compare with the example in Fig. 1). For smaller systems there seems to be a cross-over from a flat crack to a rough one, when disorder begins to dominate entropy effects (remember the periodic boundary conditions used, which also tend to increase the 'surface tension' of the crack). Thus the roughness exponent is much larger, $\zeta \approx 1.3 \pm 0.4$.

The scaling of the interface width is, as can be expected, characterized with a smaller exponent as compared to the 2D case ($\zeta_{2D} \simeq 0.7$). Our result is, although somewhat smaller, close to both the DPD and minimum energy surface estimates for the roughness exponent ($\zeta = 0.48 \pm 0.01$ and 0.5 ± 0.1, respectively). These models would imply different reasons for the microscopic dynamics of crack propagation: in the DPD case the surface would be pinned/restricted by clusters of resistent bonds, and in the latter one the crack would grow by choosing the energetically most favourable configuration over *all* possible ones. In any case, the numerical value obtained for the roughness exponent is close to the results acquired in nanoscale/static growth experiments, which would imply universality of crack growth processes in these cases.

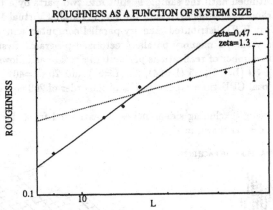

ROUGHNESS AS A FUNCTION OF SYSTEM SIZE

FIG. 2. The roughness of the final crack as a function of system size. The lines drawn correspond to $\zeta = 1.3$ (for small systems) and $\zeta = 0.4$ (for large ones). $w = 1$.

We have computed several macroscopic quantities of the fracture processes vs. L. The number of broken bonds scales as $n_b \sim L^{2.28}$, which can be compared with the value $n_b \sim L^{2.65}$ acquired earlier [23] in the case of a 'bond bending' model. The difference in the exponent may be due to the larger system sizes in our study (in ref. [23] $L \le 12$) or the difference between a vectorial model and a scalar one. It is also conceiveable that the different threshold distributions may affect the scaling results, although, as earlier pointed out, for the RFN-model the roughness exponent seems to be universal.

The maximum/breakdown current scales as $i_{max} \sim L^{2.0}$ and the voltage corresponding to the maximum current as $v_{max} \sim L^{0.9}$. The i_{max} result is similar to those acquired in 2D, conveying the fact that maximum current scaling is mainly due to the trivial area effect. The voltage corresponding to the maximum current, on the other hand, scales as $v_{max}/L \sim 1/(\ln L)^{0.34}$, using a Kahng/Duxbury-type ansatz [20] (Fig. 3). The fit is not an exact one, but an extra curvature is seen in the same way as in the results of Kahng et al..

We have also computed results for broad distribution of fracture limits, $w = 2$. Simulations with this distribution width extending to zero require much more CPU time, as more bonds are broken per case. For example, in the 16^3 case, $n_b = 565$ for $w = 1$ whereas for $w = 2$ the corresponding $n_b = 2159$ (note that a trivial planar crack would imply $n_b = 256$). Preliminary results show, that v_{max} and i_{max} behave similarly as in the

$w = 1$ case. Due to the broad fracture limit distribution the number of broken bonds grows almost with volume, $n_b \sim L^{2.9}$.

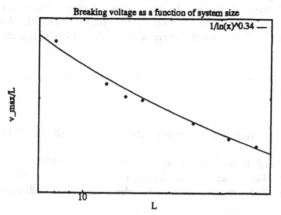

FIG. 3. The voltage corresponding to the maximum current as a function of system size on log-log scale. $w = 1$. The solid line corresponds to an ansatz of the form $v_{max}/L \sim 1/(\ln L)^\gamma$ with $\gamma = 0.34$.

IV. CONCLUSIONS

The roughness exponent of the final fracture surface in a 3D random fuse network model, $\zeta = 0.47$, is roughly the same as that observed in experiments with a quasi-static time scale experiments [1,2] and at nanometer scale [6]. This observation suggests that the microscopic processes governing the formation of the crack and leading to the final roughness of the interface may be universal in all these cases.

The total number of broken bonds scales with system size as $n_b \sim L^{2.28}$ for the width of distribution of the breaking limits ($w = 1$) for which $n_b \sim L$ in 2D case [20]. For the broader distribution ($w = 2$), the number of broken bonds seems to be almost a volume effect, $n_b \sim L^{2.9}$. The maximum current of the fracture process scales as $i_{max} \sim L^{2.02}$ and the maximum voltage as $v_{max}/L \sim 1/(\ln L)^{0.34}$. The last two quantities are rather accurately independent of the fracture threshold distribution.

Our results are interesting in several ways. Firstly, universality is observed with respect to the width of the threshold distribution. Secondly, the calculated roughness exponent is quite close to theoretical values for depinning-type models and minimum energy surfaces leaving room for future investigation. Definite improvements in the estimates of the interface width would currently seem prohibitively difficult in terms of CPU-time. There are still several questions one may ask about the development and final properties of the cracks as branching, critical crack formation, distributions of local currents (which in 2D have been shown to be multifractal [27]), and comparisons to scaling arguments. Work in this direction is in progress.

Acknowledgements

We thank Edinburgh Parallel Computing Centre (EPCC) and Center for Scientific Computing (CSC) in Otaniemi, Finland for very generous computing resources. This work has been supported by the European Union TRACS scheme, the Technology Development Centre of Finland (TEKES), the Academy of Finland, and by DOE grant DE-FG02-090-ER45418.

---

[1] E. Bouchaud, G. Lapasset and J. Planès, Europhys. Lett. **13**, 73 (1990) (and references therein).

[2] E. Bouchaud and J.-P. Bouchaud, Phys. Rev. B **50**, 17 752 (1994).

[3] K. J. Måløy, A. Hansen and E. L. Hinrichsen, Phys. Rev. Lett. **68**, 213 (1992).

[4] V. Yu. Milman, R. Blumenfeld, N. A. Stelmashenko and R. C. Ball, Phys. Rev. Lett. **71**, 204 (1993).

[5] E. Bouchaud and S. Navéos, J. Phys. I (France) **5**, 547 (1995).

[6] S. Miller and R. Reifenberger, J. Vac. Sci. Tech. **B10**, 1203 (1992).

[7] J. Schmittbuhl, S. Roux and Y. Berthaud, Europhys. Lett. **28**, 585 (1994).

[8] B. B. Mandelbrot, D. E. Passoja and A. J. Paullay, Nature (London) **308**, 721 (1984).

[9] Z. Q. Mu and C. W. Lung, J. Phys. D **21**, 848 (1988).

[10] Q. Y. Long, L. Suqin and C. W. Lung, J. Phys. D **24**, 602 (1991).

[11] A. B. Mosolov, Europhys. Lett. **24**, 673, 1993.

[12] A. Hansen, E. L. Hinrichsen and S. Roux, Phys. Rev. Lett. **66**, 2476 (1991).

[13] M. Kardar and Y.-C. Zhang, Phys. Rev. Lett. **58**, 2087 (1987); D. A. Huse and C. L. Henley, Phys. Rev. Lett. **54**, 2708 (1985).

[14] M. Kardar, G. Parisi and Y.-C. Zhang, Phys. Rev. Lett. **56**, 889 (1986).

[15] S. Roux, A. Hansen and E. Guyon, J. Phys. (Paris) **48**, 2125 (1987).

[16] M. Kardar and Y.-C. Zhang, Europhys. Lett. **8**, 233 (1989).

[17] cf. e.g. Ch. 26 in A.-L. Barabási, H. E. Stanley, *Fractal Concepts in Surface Growth*, Cambridge University Press, 1995, Cambridge, UK.

[18] L. A. N. Amaral et al., Phys. Rev. E **51**, 4655 (1995).

[19] J. P. Bouchaud et al., Phys. Rev. Lett. **71**, 2240 (1993).

[20] B. Kahng et al., Phys. Rev. B **37**, 7625 (1988).

[21] P. M. Duxbury, P. L. Leath and P. D. Beale, Phys. Rev. B **36**, 367 (1987); P. M. Duxbury, P. L. Leath and P. D. Beale, Phys. Rev. Lett. **57**, 1053 (1986).

[22] P. M. Duxbury, P. D. Beale and C. Moukarzel, Phys. Rev. B **51**, 3476 (1995); C. Moukarzel and P. M. Duxbury, J. Appl. Phys. **76**, 4086 (1994).

[23] S. Arbabi and M. Sahimi, Phys. Rev. B **41**, 772 (1990).

[24] M. Sahimi and S. Arbabi, Phys. Rev. B **47**, 713 (1993).

[25] A. Geist et al., *PVM 3 User's Guide and Reference Manual*, ORNL/TM-12187, Oak Ridge National Laboratory, Oak Ridge, Tennessee, U.S.A, 1994.

[26] V. I. Räisänen and M. J. Alava, submitted for publication.

[27] L. de Arcangelis and H. J. Herrmann, Phys. Rev. B **39**, 2678 (1989).

Part VII

Friction, Wear, and Peel Processes

EXPERIMENTAL ANALYSIS OF INSTABILITY EFFECTS ON THE SURFACE OF SCHALLAMACH'S WAVES PRODUCED BY SLIDING FRICTION OF RUBBER-LIKE MATERIALS

A. KOUDINE and M. BARQUINS
PMMH/ESPCI, 10, rue Vauquelin, 75231 Paris Cedex 05, France, koudine@pmmh.espci.fr

ABSTRACT

It is shown that a small ridge exists on the surface of the detachment fold or Schallamach's wave (1971). Such waves take place in the contact area between a moving transparent hemispherical asperity and the smooth, flat surface of a soft elastomer sample. The space-time evolution diagram, composed by juxtaposition of contact area cross-sections, recorded one after the other, proves that there is a correlation between the appearance of these ridges and microdynamic shifting of forward and backward detachment fold limits. A rolling of a rubber band, compressed between two rigid plates, enables us to reproduce the ridges in the macroscale approach.

INTRODUCTION

We describe a new experimental approach of superficial phenomena which take place in the contact area between a moving glass ball and the flat, smooth surface of a natural rubber sample (Fig. 1). It is well known that at low sliding speeds, the slider forms a viscoelastic bulge in front of it that propagates at the same velocity. Thus, there is a fold between the forward limit of the contact area and the bulge, as seen in Fig. 1b and 1c. Simultaneously, at the rear of the contact zone, the rubber begins to become unstuck from the glass ball, peeling off at a low angle. A continuous displacement of the backward limit, represented by points N, N' and N", towards the centre of a motionless contact area, is produced.

As soon as the critical sliding speed is reached, the viscoelastic bulge is overtaken by the slider and sticks to it near point M (Fig. 1c). From this moment, the fold is formed and begins to propagate across the area of contact as a detachment wave. This phenomenon was first observed by Schallamach [1] in 1971. These are small regular folds which cross the contact area from a compressive zone to a tensile zone, at rates greater than the imposed velocity. In these circumstances, true sliding does not occur, waves are formed in the rubber and these cause the relative motion of the two surfaces, in adhesive contact, without slipping between two successive waves.

The formation mechanism of the detachment fold is wholly governed by viscoelastic effects and results from the surface instability near the leading edge of the contact area. Each

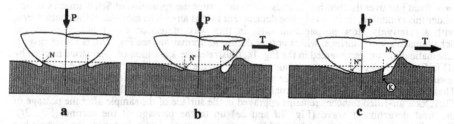

Fig. 1. Schematic diagram showing the formation of a detachment fold between a glass ball and a smooth, flat surface of natural rubber: (a) equilibrium adhesive contact; [(b) and (c)] formation of a viscoelastic bulge at the leading edge. Low angle peeling at the contact area rear limit is represented by the points N, N' and N". A microscale ridge is created at the leading edge near point K.

Mat. Res. Soc. Symp. Proc. Vol. 409 © 1996 Materials Research Society

Schallamach's wave can be schematised by two crack tips which propagate in the same direction and at the same velocity. There is an opening crack in the frontal region, where peeling at an angle of $\pi/2$ is produced, and a closing crack, or readhesion, at the rear, where the junction between the rigid slider and the elastomer is tangential [2].

The question that provides the backdrop for this study is this: Why, under some circumstances, do the dynamics of Schallamach's waves look like a stick-slip movement ? To explain this phenomenon, in this study, we looked at fine structure instability, as small ridges on the surface of detachment folds.

EXPERIMENTAL EQUIPMENT

All experiments were carried out using a mechanical apparatus to study the sliding friction of rubber-like materials [2]. This apparatus is composed of a high sensibility balance attached to the support of the microscope. A hemispherical glass lens on the tip of the balance, adjusted in the optical axis of the microscope, is in contact with the flat, smooth surface of a rubber sample, placed on the plate. This plate is dragged by a micro-motor through flat springs, which enables us to measure the advancing resistance force. The contact area and its immediate vicinity, which remain continually centred on the optical axis, are recorded by a video camera located at the top of the microscope. The images are then processed by computer to obtain a space-time representation of the Schallamach wave propagation. The space-time diagram is composed by juxtaposition of contact area cross-sections, recorded one after the other. The thickness of each digital line is equal to one pixel which corresponds approximately to 1 μm. Each cross-section coincides with the horizontal symmetry axis of the contact area in the direction of the slider displacement. A typical space-time evolution diagram is shown in Fig. 3, illustrating the detachment fold passage, with "grooves", represented by AD. In order to obtain the best resolution with respect to time, the period between two consecutive cross-sections is chosen so that it is equal to 0,06 s, i.e. 17 cross-sections per second. This is the maximum frequency that our computer program allows us.

The mean velocity of the detachment fold propagation can be obtained by simply measuring the slope of the space-time evolution diagram. Also, the microdynamic shifting of Schallamach's wave limits, analogous to stick-slip motion with a small amplitude, is easily seen.

In some cases, an ordinary camera was used to obtain high resolution photographs.

RESULTS OF THE DIRECT CONTACT AREA OBSERVATIONS

A series of contact area snapshots, showing the formation of the first detachment fold between a spherical asperity and a moving rubber sample, is presented in Fig. 2. Fig. 2a shows the situation just after the appplication of a tangential force The round limit of an initial equilibrium adhesive contact is clearly visible. In order to obtain better image resolution, necessary to visualise fine structure instability elements, the values of load and dragging velocity were fixed just over the threshold values which determine the apparition of Schallamach's waves. Under this condition, there is only one detachment fold at any given moment in the contact area, with a relatively slow propagation rate. In this way, detection of irregularities on the Schallamach's wave surface, which appear during its formation (see Fig. 2b), is rather easy. A schematic diagram, represented in the Fig. 1c, illustrates the apparition of clear striae marked by (1) in Fig. 2b. Indeed, a microscale ridge is created in the vicinity of the viscoelastic bulge at the leading edge, near point K. In Fig. 2c we can see the striae corresponding to the irregularities. These striae show up as trails of small ridges that fracture the Schallamach's wave surface. Therefore, air-filled trenches remains engraved in the surface of the sample after the passage of the first detachment wave (Fig. 2d and 2e) up to the passage of the second (Fig. 2f). Nevertheless, all deformations of the rubber surface are quasi-elastic. In consequence, there is a rapid surface re-covering above the asperity and no trails are observed in the rubber outside the considered contact zone.

We use space-time analysis to explain the correlation between the appearance of ridges and microdynamic shifting of forward and backward Schallamach's waves limits.

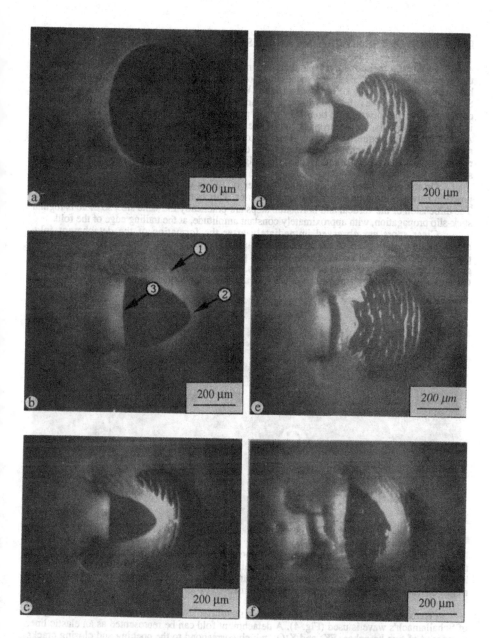

Fig. 2. First detachment fold forming (normal load P = 10 mN, sliding speed V_e = 10 μm.s⁻¹) in the contact area between a rigid spherical asperity (R = 2.19 mm) and a smooth surface of a natural rubber sample (Young's modulus E = 2.0 MPa). Ridge (1) on the surface of the forming detachment fold (2); backward limit (3) of contact area resulting from low angle peeling.

DISCUSSION OF THE SPACE-TIME ANALYSIS RESULTS

As already mentioned, a typical space-time diagram is shown in Fig. 3. Curves such as that represented by AD, show the successive positions of the same detachment fold along the horizontal symmetry axis of the contact area. We are interested in the propagation phase of Schallamach's wave evolution which is given by position BC. Its dark left-hand border corresponds to the forward fold boundary which propagates as an opening crack. This borderline is almost continuous, so the propagation of the forward fold boundary is practically regular. Arrows mark clear equidistant spots (Fig. 3) corresponding to the cross-section of the small ridges on the Schallamach's wave surface. The dimensions of vertical spots in relation to time remain approximately constant. Hence, the ridges are motionless with respect to the rubber surface.

On the other hand, the right-hand border of BC is very light, owing to the tangential junction between the rigid slider and the elastomer. This borderline has a stairway-like profile. The dimension of the vertical and horizontal steps are practically the same size, so there is regular stick-slip propagation, with approximately constant amplitude, at the trailing edge of the fold.

Clear spots are observed immediately after the formation of the detachment fold, represented by section AB of the space-time diagram. At this moment there is no stairway-like pattern on the right-hand border. During the fold propagation phase (section BC), each clear spot goes before the next right-hand border step. Thus, the small ridges act as subsequent motivator of the stick-slip propagation mechanism of the forward fold boundary.

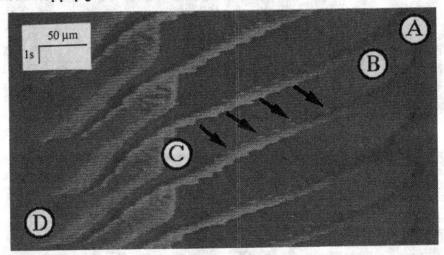

Fig. 3. Space-time evolution diagram with "grooves", such as AD, illustrating the detachment fold passage. Formation, propagation and vanishing phases of the fold evolution are represented by positions AB, BC and CD respectively. Clear, equidistant spots corresponding to the small ridges on the surface of Schallamach's wave are marked by arrows.

To explain the generation of the ridges, a simplified, two-dimensional schematic diagram of Schallamach's wave is used (Fig. 4). A detachment fold can be represented as an elastic line, composed of two branches, FK_i and K_iQ_i, which correspond to the opening and closing cracks respectively. Regular glass ball displacement causes an advance of the forward branch and, hence, an increase in its length. Thus, a continuous decrease of the local curvature radius in the vicinity of the two branches' intersection point, K, is produced. Owing to rubber's incompressibility, as soon as this local radius reaches a threshold value, the ridge is formed in a

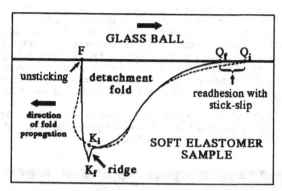

Fig. 4. Schematic view of a detachment fold composed of two branches, FK_i and K_iQ_i, corresponding to its forward and backward boundary-lines. The ridge is formed in a singularity zone of boundary-line branch intersection marked by points K_i and K_f. Two consecutive positions of rubber adhesion to the glass sphere, illustrating the advancing of the fold backward limit, are represented by points Q_i and Q_f.

singularity zone near point K_f (Fig. 4, see curve FK_fQ_i). Still engraved in the rubber surface, it traverses this strong compression zone. At this time, the ridges, or more precisely, the corresponding clear spots on the grooves of the space-time diagram in Fig. 3, are still visible. After traversing the strong compression zone, the ridge is conserved for a moment along the closing crack's branch. Now the clear spots are unobservable because a groove's background becomes light. The ridge relaxes in the tangential part of K_fQ_i branch and produces a forward jerk of the fold's backward limit, illustrated by positions Q_i and Q_f in Fig. 4, and the cycle repeats itself.

In order to obtain a criterion of ridge formation on the detachment fold surface, a simple experimental macroscale approach is proposed (Fig. 5). A band of the rubber such as a shoe

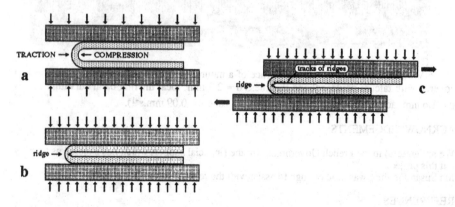

Fig. 5. Schematic diagram showing the rolling of a rubber band placed between two rigid plates: (a) regular compression of the plates in the absence of shifting, (b) ridge apparition phenomenon when the local internal band curvature radius exceeds a threshold value, (c) formation and vanishing of ridges on the surface of the compressed, rolling rubber band.

elastomer sole, is compressed between two rigid plates as shown in Fig. 5a and 5b. When the local internal radius of curvature reaches a critical value, the macroscale ridge can be seen in the band (Fig. 5b). Experiments carried out using different bands show that this critical value depends mainly upon the viscoelastic properties and the thickness of rubber sample. If a regular displacement is imposed on the rigid plates as seen in Fig. 5c, band rolling is produced. Equidistant ridges, engraved in the rubber surface, are formed one after the other. At the end of the band curvature, rapidly vanishing tracks of relaxed ridges can be observed. In order to measure the distance between consecutive ridges, the internal face of the rubber band was covered with talcum powder (Fig. 6).

On-going studies using this experimental macroscale approach should enable us to determine criteria concerning the formation of the small ridges on the surface of Schallamach's waves as a function of its local internal curvature radius.

Fig. 6. View of the ridges (1) on the surface of a natural rubber band (E = 0.89 MPa, ν = 1/2) covered with talcum powder (band thickness d ≈ 2.3 mm; local internal band curvature radius ρ ≈ 1.6 mm; advancing speed of the band frontal zone v = 0.09 mm.s^{-1}).

ACKNOWLEDGEMENTS

We are grateful to the French Government for the financial support which has enabled us to carry out this project.
Ms Susan Fielding was kind enough to assist with the reading of the proofs.

REFERENCES

1. M. Barquins, Wear **158**, 87 (1992)
2. A. Schallamach, Wear **17**, 301 (1971)

TRIBOEMISSION AND WEAR OF HYDROGENATED CARBON FILMS

Keiji Nakayama
Mechanical Engineering Laboratory , Namiki 1-2, Tsukuba, Ibaraki 305, Japan

ABSTRACT

It is suggested that perfluoropolyether lubricating oil coatings applied to the carbon overcoat film of magnetic recording layers become decomposed by electrons emitted from frictional surfaces. However, no work has at yet been reported as to triboemission of electrons from frictional carbon films.

This paper describes the behavior of triboemission of electrons and the friction coefficient during wear of sputtered hydrogenated carbon films (with various hydrogen contents on the glass substrate). The triboemission of electrons, together with friction coeficient, was measured in a frictional system of Al_2O_3 sliding on carbon films in a reduced dry air atmosphere. The worn surfaces of the carbon films were then observed using both a SEM and an AFM. The results showed that intense triboemission of electrons were observed during wear of hydrogenated carbon films. The electron emission intensity and friction coefficient transit from low to high with hydrogen content in the film. These results are discussed including physical properties of the carbon films such as internal stress and surface wettability.

1. INTRODUCTION

Hydrogenated amorphous carbon films are coated on the thin film magnetic recording layer to protect them from wear which occurs when the flying head is landing on and taking off the disk in the recording media. To reduce wear of the carbon film, perfluoropolyether (PFPE) lubricating oil is coated on the carbon film. The PFPE molecules have good characteristics as a lubricant for magnetic recording disks, such as chemical inertness, low vapor pressure, low thermal decomposition and good lubricity properties. However, the stable PFPE lubricating oil molecule is decomposed at the point of head/disk sliding contacts to evolve gases [1]. This is serious, because the PFPE coating has only only a thickness of approximately monomolecular layer and the decomposition of the lubricant leads to the wear of carbon films and in turn to the failure of the important magnetic recording layer.

The cause of the decomposition of PFPE molecules was explained with thermal reactions due to frictional heating [2] and catalytic reactions by head materials [3-5]. Recently, however, it has been reported that the PFPE molecules are decomposed not by thermal and catalytic reactions but through triboelectrical reactions by electrons which may be emitted from wearing carbon films [6-7]. Triboemission of electrons, ions and photons have been observed in various solids in various gas atmospheres, in ambient air and even under hydrocarbon oil lubricated conditions [8-15]. However, no work has been done on triboemission from wearing hydrogenated carbon films.

This paper describes the triboemission characteristics of electrons from wearing carbon films having various hydrogen content in reduced dry air atmospheres in the frictional system of Al_2O_3 ball sliding on hydrogenated amorphous carbon films. The Al_2O_3 ball was selected as a model head material. Triboemission characteristics were discussed with wear mechanisms which are explained by physical properties of the carbon films such as wettability and internal compressive stress.

2. EXPERIMENTALS

Amorphous and hydrogenated amorphous carbon films have been deposited on K^+ ion exchanged glass substrate using the DC magnetron sputtering method. The films were deposited using an amorphous carbon plate as a target in Ar or gas mixture of Ar and H_2 under a gas pressure of 0.33 Pa changing H_2 gas content in spattering gas ranging from 0 to 30 %. The surface roughness of the glass substrate was Ra < 1 nm. The materials and sputtering conditions are summarized in Table 1.

The hydrogen content in the hydrogenated carbon films are determined using the

Table 1 Materials, sputtering conditions and film thickness of hydrogenated carbon films

Hydrogenated carbon film	Film thickness	20 nm
Substrate	Soda lime glass	K$^+$ ion exchanged (mirror polished, Ra < 1 nm)
Sputtering condition	Sputtering gas	Ar + H$_2$ (H$_2$ content: 0 ~ 30 vol.%)
	Target	Amorphous carbon
	Gas pressure	0.33 Pa
	Sputter power	0.55 kW

hydrogen forward scattering (HFS) method incidenting He$^+$ ion in an incident energy of 2.3 MeV, the beam current of 7.0 nA, the incident angle of 75 degrees, the reflective angle of 30.0 degrees with the diameter of the ion beam being 1 mm. Table 2 shows the hydrogen content in the carbon films prepared in the gases having various hydrogen contents.

Table 2 Hydrogen content in sputtering gases and in carbon films

Hydrogen content in gas V_{H2} (vol.%)	Hydrogen content in carbon film C_H (at.%)
5	15
10	27
15	30
20	37
30	40

Figure 1 shows the triboemission and friction measuring apparatus. The number of electrons emitted and friction coefficient were measured *in situ* during sliding of a normal sintered alumina (Al$_2$O$_3$) ball. This wqas carried out on amorphous carbon film and hydrogenated amorphus carbon films coated on glass substrate in reduced dry air atmosphere under a normal force of 440 mN and a sliding velocity of 22 mm/s for 10 minutes. The gas pressure of the dry air was kept constant to be 5 mPa by balancing the evacuation and introduction rate of the dry air in the vacuum chamber. The Al$_2$O$_3$ ball had a diameter of 4 mm and a Vickers hardness of 11.5 GPa. The ball specimen surface was finished with #200 abrasive paper and washed with petroleum benzine and then acetone in an ultrasonic bath.

The electrons emitted from wearing surfaces were detected using a channeling type electron multiplier, the Ceratron–EMW(6081B), amplified by a pre–amplifier and linear amplifier and then counted by a counter and a rate meter. The friction coefficients were detected with strain gauges and the signals were then amplified with a strain amplifier. The electron emission intensity was measured by a rate meter as a counts per second (cps) and recorded simultaneously with friction coefficient on a strip chart recorder.

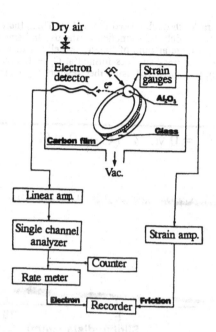

Fig. 1 Block diagram of electron measuring system during sliding

3. RESULTS AND DISCUSSION

Before measuring electron emission from carbon films, the electron emission from the glass substrate was measured as a reference. The electron emission intensity from the glass substrate was extremely low, almost noise level, while the friction coefficient was high with a value of around 0.7. However, as seen in Fig. 2, electron emission was definitely observed from amorphous carbon film of $C_H = 0$ at.%, while the friction coefficient was much smaller with values of 0.1 to 0.2 than that of the glass substrate. In the region of C_H < 25 at.%, the electron emission intensity and friction coefficient were low having approximately the same values as observed at $C_H = 0$. However, 25 at.% < C_H < 33 at.%, both electron emission intensity and friction coefficient increased sharply with the increase of the hydrogen content, C_H in the carbon film. In C_H > 33 at.%, the electron emission intensity and friction coefficient was high as seen in Fig. 3 for $C_H = 37$ at.%, where triboemission intensity of electrons reaches to the order of 10^4 cps with high fluctuation.

Figure 4 shows the dependence of the average electron emission intensity in counts /min (a) and average friction coefficient (b) on hydrogen content in the carbon film. As seen in Fig. 4, both the electron emission intensity and friction coefficient was low in C_H < 25 at.% (State I). The low electron emission intensity and low friction coefficient sharply increases in the region of 25 at.% < C_H < 33 at.% (Transition state). At a hydrogen content of C_H > 33 at.%, both electron emission intensity and friction coefficient retained high values (State II). Note that friction coefficient increases only in doubles, while electron emission intensity increases by about 10^4 times as C_H increases.

To find out the origin of the electron emission and the transition of electron emission intensity of low to high values, the wear track surfaces of the carbon film were observed using a SEM and an AFM. In State I, many parallel cracks together with a small amount of film flaking was found. Previously, it has been reported that triboemission takes place during propagation of cracks through coated films [16]. Then, the electron emission in the region of State I is caused mainly by formation of the parallel cracks.

At transition state, a large amount of flaking or delamination of the carbon films occurred as seen on the wear track surfaces formed at C_H = 27 at.% in Fig. 5(a). Internal stress of the carbon film was measured using the Newton ring method and showed that the

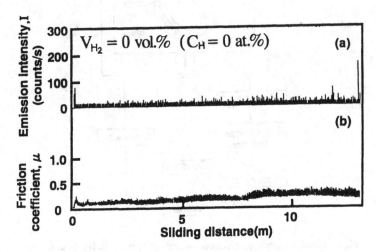

Fig. 2 Electron emission intensity (a) and friction coefficient (b) during sliding of Al_2O_3 ball sliding on carbon film with hydrogen content C_H = 0 at.%

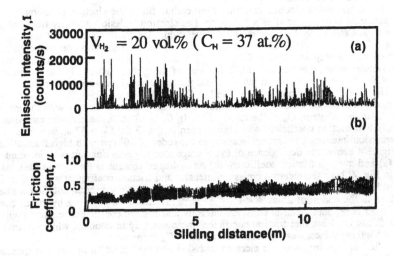

Fig.3 Emission intensity of electrons (a) and friction coefficient (b) during sliding of Al_2O_3 ball on carbon film with hydrogen content of C_H = 37 at.%.

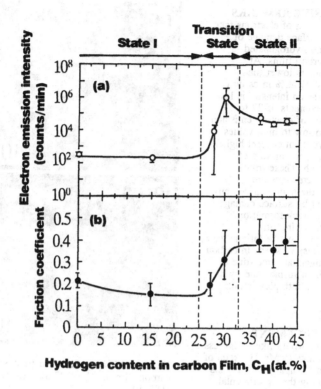

Fig. 4 Dependence of emission intensity and friction coefficient on hydrogen content in the carbon film, C_H: F_N = 440 mN; V = 22 mm/s in dry air (gas pressure, p = 5 mPa).

films have compressive stresses, having a maximum at C_H = 27 at.%. The increase of film flaking is thought to be caused by a combined effect of high internal compressive stress and the increased tangential force due to a high friction coefficient. At C_H = 30 at.%, heavy removal of flaked film occurred. Then the increase of the triboemission intensity is attributed to the increase of the film flaking and the subsequent breakdown of the flaked films.

At State II, the flaked film is all removed off as seen in the SEM images of wear track surfaces formed at C_H = 37 at.% in Fig. 5(b). This breakdown of the flaked film is thought to be caused by high tangential force due to high friction . This removing of the flaked film generated increased formation of a fresh carbon film surface which is responsible for the intense triboemission of electrons. The contact angle, measured by water drop, decreased with the decrease in hydrogen content in the film. Namely, with decreasing C_H, wettability increased. This means that with decreasing hydrogen content, surface energy increased to give good boundary lubrication by adsorbed water molecules to reduce friction, surface damage and triboemission intensity.

4. CONCLUSIVE REMARKS

Triboemission of electrons take place during surface damage of amorphous and hydrogenated amorphous carbon films. At hydrogen contents lower than approximately 25 at.% in carbon film, triboemission intensity is low. At hydrogen contents of 27 to 33 at.%, transition of electron emission intensity from low to high values occurs. At hydrogen content higher than 33 at.%, triboemission of electrons is high. These tribo-emission characteristics are dependent on the hydrogen content and are caused by surface damage due to fatigue crack formation, flaking and removing off the flaked film. For the fatigue wear, internal compressive stress in carbon films, and tangential force (due to friction in which surface energy has important role), both play important parts.

ACKNOWLEDGEMENT

The author would like to express his sincere thanks to Mr. H. Ikeda of Nippon Sheet Company Ltd. for his help to obtaining the experimental results.

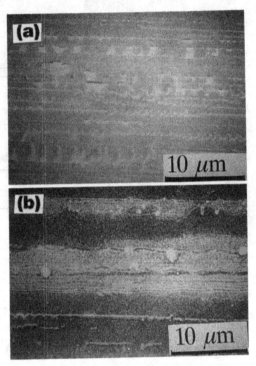

Fig. 5 SEM images of wear track surfaces of carbon film in $C_H = 27$ at.% (a) and $C_H = 37$ at.% (b).

REFERENCES

[1] S. Mori and W. Morares, Tribology Trans., **33**, p. 325–332(1990).
[2] G. Vurens, R. Zehringer, D. Saperstein and S. Jose, Surface Science Investigation in Tribology edited by Y. Chung, et al (ACS Symposium, 1992), 485, p.169–180.
[3] M.J. Zehe and O.D. Faut, NASA TM, **101962**, p.1–15(1989).
[4] P.H. Kasai, T. Tang and P. Wheeler, Appl. Surf. Sci., **51**, p.201–211(1991).
[5] P.H. Kasai, Adv. Info. Storage Syst., **4**, p. 291–314(1992).
[6] J. Packansky and R.J. Waltman, J. Phys. Chem., **95**, p.1512–1518(1991).
[7] B.D. Strom, D.B. Bogy, R.G. Walmsey, J. Brandt and C.S. Bhatia, IEEE Trans., **29**, p.253–258(1993).
[8] K. Nakayama, J.A. Leiva and Y. Enomoto, Tribology Int. in press.
[9] K. Nakayama, N. Suzuki and H. Hashimoto, J. Appl. Phys., **25**, p.303–308(1992).
[10] K. Nakayama and H. Hashimoto, Tribology Trans., **35**, p.643–650(1992).
[11] K. Nakayama and H. Hashimoto, Tribology Trans., **38**, p. 35–42(1995).
[12] K. Nakayama and H. Hashimoto, Wear 185, p.183–188(1995).
[13] K. Nakayama, Wear, No.178, p.61–67(1994).
[14] K. Nakayama and H. Hashimoto, Tribology Trans., **38**, p.541–548.
[15] K. Nakayama and H. Hashimoto, Tribology Int. in press.
[16] K. Nakayama, H. Hashimoto and Y. Fukuda, Proccedings of the Japan Int. Tribology Conference, Nagoya, Oct. 29–Nov. 1, 1990, p. 1141–1146.

FRACTURE PROCESS CONTROL
FOR A PEEL-APART IMAGING FILM

H.-C. CHOI, A. KNIAZZEH, F. HABBAL
Polaroid Corporation, W4-4, 1265 Main St., Waltham, MA 02154

ABSTRACT

Peeling delamination and image-layer fracture of a multilayer imaging film is analyzed by a simple beam approximation. Criteria for propagation of the delamination cracks and tensile failure of the imaging layer are established. The critical role of substrate strain in controlling tensile strain in the imaging layer is explored. For the case of a thin imaging layer with a fracture strain below a critical level, image layer fracture occurs while the leading-edge delaminations length is comparable to the layer thickness. Experimental Helios film image quality is optimized at a small level of substrate tensile strain where delamination length at leading and trailing edges of small holes is balanced and small and where hole size is minimized.

INTRODUCTION

Recently, a multi-layered peel-apart imaging film has been developed for high resolution imaging. The imaging film has a micro-thin carbon black imaging layer (image layer in Fig. 1) sandwiched between thick transparent layers (peel-off and substrate layers in Fig. 1). The interface adhesion fracture energies on both sides of the imaging layer are precisely maintained over a large area. In the imaging process laser light heats the front surface of the carbon layer (imaged area) to change its adhesion energy. Subsequently the two transparent layers are peeled apart causing the carbon layer to fracture and form transparent pels.

The quality of a peeled image depends on the quality of an individual pel which is created by its formation mechanism – initiation and termination. The involved mechanisms for pel initiation and termination will be quite different because a crack directly meets a discontinuity of an interface adhesion on its way for the pel initiation, while, for the pel termination, it only feels the adhesion discontinuity from the other interface. Both pel initiation and termination are involved in the crack jumping from one interface to the other. The crack propagation path (or jumping condition) depends on many parameters such as relative adhesion properties between the two sides of the imaging layer, the stress field near a crack tip (mode mixity and T-stress), material properties and thicknesses of a structure, and size of imaged area. Depending on these parameters, a crack may (1) stay on the same interface or kink and propagate inside the image layer [Fig. 1(a)], or jump to the other interface (2) by directly kinking through the imaging layer [Fig. 1(b)], (3) by delaminating a small area on the other interface and then directly kinking to connect the two cracks [Fig. 1(c)], or (4) by delaminating an extensive area on the other interface which can cause a bending failure of the imaging layer when the two cracks connect [Fig. 1(d)]. By investigating these mechanisms, not only the minimum pel size produced by peeling an imaging film but also the variation of pel sizes can be estimated.

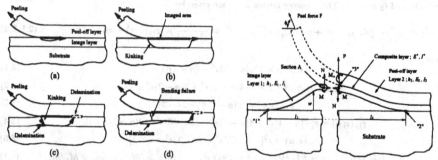

Fig. 1 Pel formation mechanisms   Fig. 2 The growth of a delaminated crack

A pel is a transparent region on the substrate.

In this paper, only the pel initiation mechanism by extensive delamination is considered, even though both pel initiation and termination mechanism should be well balanced in order to have a good quality image. This extensive delamination crack jumping mechanism will be dominant when both interface adhesions are

397

relatively weak compared to other material failure strengths. In the following, a crack delamination during the peeling process is analytically modeled, and its parametric study is presented. At the end, experimental results for various types of pel formation and the effect of substrate bending on the quality of the image are also presented.

DELAMINATION OF AN IMAGING LAYER

In this section, we consider the pel initiation mechanism caused by the extensive delamination process between an image layer and a substrate during the peeling process as shown in Fig. 2. Delamination will occur if the image/substrate interface is relatively weaker than the image/peel-off layer is. The delamination crack may grow. As it grows a bending stress in the image layer may reach its critical value to cause a bending failure if it is the weakest failure mechanism of the system. Because this sequence happens the crack jumps to the image/substrate interface from the peel-off/image interface. In the following, the growth of a crack delamination of the image/substrate interface is analytically modeled using a simple beam theory, and then parametrically studied. At the end the fracture of the image layer is also discussed.

Analysis of the delamination during peeling

Let's assume that the image layer is delaminated from a semi-infinite substrate during peeling of a peel-off layer as shown in Fig. 2. It is also assumed that the delamination length is large enough such that a simple beam theory is valid (beam length is larger than at least 2 times of the beam height).

From the free body diagram in Fig. 2, the energy release rates at crack tip 1 and 2 are given by (Suo and Hutchinson [1])

$$\sqrt{2E_1 I_1 \mathcal{G}_1} = N l_1 - M, \quad \text{and} \quad \sqrt{2E^* I^* \mathcal{G}_2} = (P - N)l_2 + M_o - M \qquad (1.1, 1.2)$$

where h_1 and E_1 are the thickness and Young's modulus of layer 1, and h_2 and E_2 are those for layer 2. In the above, $E^* I^*$ is the bending stiffness for the composite layer (layer 1 and 2), and l_1 and l_2 are the crack lengths measured from the crack tip 3 to the crack tip 1 and crack tip 2, respectively. The continuity conditions for the deflection and slope across section A give

$$\frac{1}{E_1 I_1}\left(\frac{1}{2}N l_1^2 - M l_1\right) = -\frac{1}{E^* I^*}\left[\frac{1}{2}(P - N)l_2^2 - (M - M_o)l_2\right], \qquad (1.3)$$

$$\frac{1}{E_1 I_1}\left(\frac{1}{3}N l_1^3 - \frac{1}{2}M l_1^2\right) = \frac{1}{E^* I^*}\left[\frac{1}{3}(P - N)l_2^3 - \frac{1}{2}(M - M_o)l_2^2\right]. \qquad (1.4)$$

For a given delaminated crack geometry (l_1 and l_2) and loading conditions (P and M_o), the energy release rates at both crack tip 1 and 2 can be easily calculated by solving for N and M from Eqs. (1.3) and (1.4), and then by substituting them into Eqs. (1.1) and (1.2). To do that, let $\bar{l}_o = \bar{l}_1 + \bar{l}_2$ and $s = \bar{l}_1/\bar{l}_o$ where $\bar{l}_1 = l_1/h_1$ and $\bar{l}_2 = l_2/h_1$. Then, energy release rates are given by

$$\bar{\mathcal{G}}_1 = \frac{1}{g_o^2(s;\alpha)}[p\bar{l}_o g_{1p}(s;\alpha) + m_o g_{1m}(s;\alpha)]^2, \qquad \bar{\mathcal{G}}_2 = \frac{\alpha}{g_o^2(s;\alpha)}[p\bar{l}_o g_{2p}(s;\alpha) + m_o g_{2m}(s;\alpha)]^2 \qquad (2.1, 2.2)$$

where

$$g_o(s;\alpha) = \alpha + \alpha(\alpha - 1)s^4 - (\alpha - 1)(1 - s)^4, \qquad (3.1)$$

$$g_{1m}(s;\alpha) = (\alpha - 1)s^4 - 4(\alpha - 1)s^3 + 3(\alpha - 2)s^2 + 4s - 1, \qquad (3.2)$$

$$g_{2m}(s;\alpha) = s[(\alpha - 1)s^3 + 3s - 2], \qquad (3.3)$$

$$g_{1p}(s;\alpha) = s[-(\alpha - 1)s^4 + 4(\alpha - 1)s^3 - (5\alpha - 6)s^2 + 2(\alpha - 2)s + 1], \qquad (3.4)$$

$$g_{2p}(s;\alpha) = s^2[-(\alpha - 1)s^3 + (\alpha - 1)s^2 - s + 1] \qquad (3.5)$$

$$n = N h_1/M_1, \qquad p = P h_1/M_1, \qquad t = T h_1/M_1, \qquad m = M/M_1, \qquad m_o = M_o/M_1, \qquad (4.1)$$

$$\bar{\mathcal{G}}_1 = \mathcal{G}_1/\mathcal{G}_c, \qquad \bar{\mathcal{G}}_2 = \mathcal{G}_2/\mathcal{G}_c, \qquad \alpha = E^* I^*/(E_1 I_1) \qquad (4.2)$$

where $M_1 = \sqrt{2E_1 I_1 \mathcal{G}_c}$ and $\mathcal{G}_c$ is the critical energy release rate per unit width for the image/substrate interface crack at both crack tip 1 and 2. The function g's in Eqs. (3) depend only on the ratio of crack lengths, s, and the bending stiffness ratio of a composite layer to an image layer, α.

Growth of a delaminated crack

From the peel analysis (Aravas et al. [2]), a moment m_o and shear force p can be related to a peel force F by

$$p = \sqrt{6\mathcal{G}_c/(E_1 h_1)}\,\bar{F}\sin\phi \quad\text{and}\quad m_o = \sqrt{\bar{E}_2\bar{h}_2^3 \bar{F}(1-\cos\phi)} \tag{5}$$

where $\bar{E}_2 = E_2/E_1$, $\bar{h}_2 = h_2/h_1$, $\bar{F} = F/\mathcal{G}_c$, ϕ is the peel angle and F is the peel force per unit width as shown in Fig. 2. In this paper only a 90° peel is considered. Then, by dimensional analysis, the energy release rates at both crack tips can be considered as functions of the crack configuration, thicknesses and material properties of each layer; that is,

$$\bar{\mathcal{G}}_{1,2} = f_{1,2}(\bar{F}, E_1 h_1/\mathcal{G}_c, \bar{E}_2, \bar{h}_2, \bar{l}_1, \bar{l}_2). \tag{6}$$

In the following, $(h_1 E_1/\mathcal{G}_c) = 500$, $\bar{E}_2 = 1$; all other parameters are varied.

The energy release rates at both delaminated crack tips are plotted as a function of the crack configuration s in Fig. 3 for various parameters. Energy release rates strongly depend on the crack length, l_o, when the peel-off layer is very thin compared to the image layer (small $\bar{h}_2$) as shown in Fig. 3(a). But, they are very insensitive to l_o when $\bar{h}_2$ is large [Fig. 3(c)]. Even though the linear model based on a simple beam theory may cause a large error for large $\bar{h}_2$ and small $\bar{l}_2$, it is purposely used here to investigate the phenomenological delamination process.

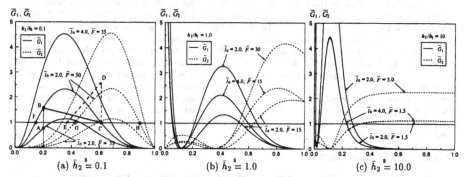

Fig. 3 Energy Release rates during peeling

Using these plots, the growth of a delaminated crack can be easily predicted. A crack grows if it reaches a critical point where the energy release rate reaches its critical value ($\bar{\mathcal{G}}_1 = 1$ for crack tip 1 and/or $\bar{\mathcal{G}}_2 = 1$ for crack tip 2). Multiple critical points are possible. For example, when $\bar{F} = 50$, point F and C are two critical points for crack tip 1 as shown in Fig. 3(a). Point F is *unstable* because the energy release rate increases as crack tip 1 grows at that point, while point C is *stable* because the energy release rate decreases as crack tip 1 grows. Point G is *stable* while point H is *unstable* for the growth of crack tip 2. When the initial crack configuration is between the stable and unstable point, a crack should grow to reach its stable point. To show this situation, let $\bar{h}_2 = 0.1$, $\bar{l}_1 = 0.4$ and $\bar{l}_2 = 1.6$, initially. A crack will not grow for $\bar{F} = 35$ [point A in Fig. 3(a)] because $\bar{\mathcal{G}}_1, \bar{\mathcal{G}}_2 < 1$. When $\bar{F} = 50$ (point B), $\bar{\mathcal{G}}_1 > 1$, but $\bar{\mathcal{G}}_2 < 1$. In this case crack tip 1 grows unstably until it reaches point C. Note that a point C is not accurately located in Fig. 3(a) because $\bar{\mathcal{G}}_1$ is changing due to crack growth ($\bar{l}_o$ increases). As crack tip 1 moves toward a stable position, crack tip 2 may also grow depending on the value of $\bar{\mathcal{G}}_2$ at that crack configuration. If $\bar{\mathcal{G}}_2 < 1$, crack 2 will not grow. But, if $\bar{\mathcal{G}}_2 > 1$, crack tip 2 should grow further until it reaches a stable position [say point E in Fig. 3(a)]. The delaminated crack will grow completely without bounds if the stable configuration can not be found such that $\bar{\mathcal{G}}_1 = 1$ and $\bar{\mathcal{G}}_2 = 1$. There is one more stable point for crack tip 1 when $\bar{h}_2$ is large as shown in Figs. 3 (b) and (c)], and the corresponding $\bar{l}_1$ may be too small for the simple beam theory to be valid.

In order to more clearly see the growth of a crack, crack growth maps are plotted in Fig. 4 as a function of loading $\bar{F}$. The energy release rate is greater than one in the *active region* which is bounded by *unstable* and *stable* boundaries. If the initial crack configuration is inside the active region for a given peel force $\bar{F}$, a crack should grow up to the corresponding stable boundary; that is, if the initial configuration is inside the active region for crack tip 1, $\bar{l}_1$ grows (point A to point B in Fig. 4), while, if the initial configuration is inside the active region for crack tip 2, $\bar{l}_2$ grows instead (point C to point D). But, if either point B or point D are beyond point E at which $\bar{\mathcal{G}}_1$ =1 and $\bar{\mathcal{G}}_2$ = 1 as shown in Fig. 4(a), the crack will be completely delaminated for a given loading.

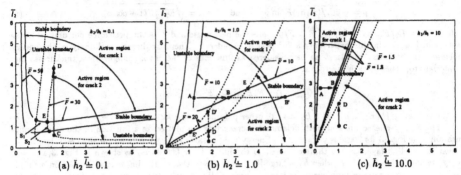

Fig. 4 Crack growth maps for a delaminated crack during peeling

For $\bar{h}_2 = 0.1$ and $\bar{F} = 30$, stable growth can only be possible on the stable boundary of either S_1 ($\bar{\mathcal{G}}_1 = 1$, $\partial \mathcal{G}_1/\partial l_1 < 0$ and $\bar{\mathcal{G}}_2 < 1$) or S_2 ($\bar{\mathcal{G}}_2 = 1$, $\partial \mathcal{G}_2/\partial l_2 < 0$ and $\bar{\mathcal{G}}_1 < 1$) as shown in Fig. 4(a). Note that both crack tips are stably active at point E (*dual active point*) because $\bar{\mathcal{G}}_1 = 1$, $\bar{\mathcal{G}}_2 = 1$, $\partial \mathcal{G}_1/\partial l_1 < 0$ and $\partial \mathcal{G}_2/\partial l_2 < 0$. All other initial crack configurations create complete delaminations. For $\bar{h}_2 = 1.0$ and $\bar{F} = 10$, if an initial crack configuration is at point A in Fig. 4(b), crack tip 1 grows to point B, while if it is at point C, crack tip 2 grows to point D. When initial crack lengths are larger than those at point E ($\bar{l}_1 = 3.53$ and $\bar{l}_2 = 2.74$), both crack tips are active and a crack will be completely delaminated. For $\bar{F} = 20$, either the initial crack configuration of point A or point C are now beyond the corresponding *dual active point* E'. Therefore, both crack configuration create complete delaminations. When $\bar{h}_2 = 10.0$, only one crack tip can grow depending on the initial crack configuration, and both crack tips can not be simultaneously active as shown in Fig. 4(c) at relative low $\bar{F}$. But, for higher $\bar{F}$ a crack will be completely delaminated regardless of its initial crack geometry.

Bending failure of an image layer

In the previous section, the growth mechanism of a delaminated crack was discussed. This section explains how an image layer breaks during peeling.

As the peel force $\bar{F}$ increases during peeling, the strength of the stress concentration near crack tip 3 changes and the tensile surface stress in the delaminated image layer due to bending moment M (see Fig. 2) also increases. Three possible situations can occur as follows: (1) a crack tip 3 propagates along an image/peel-off layer interface without further delaminating the image/substrate interface (if $\mathcal{G}_3$ is larger than the critical energy release rate for that interface, where $\mathcal{G}_3$ is the energy release rate for crack tip 3); (2) crack tip 3 may kink and break the image layer (if the necessary kinking energy is the lowest one in the system); (3) the image layer may break due to a high tensile bending stress in the image layer (top surface at section A in Fig. 2). The actual process depends on relative initiation strengths of each mechanism. In this paper, only the bending failure is considered. The kinking condition of an interface crack can be found in He and Hutchinson [3] and Wang *et al.* [4].

In Fig. 5, the crack growth ($\bar{l}_1$ and $\bar{l}_2$) of a delaminated crack, whose initial lengths are $\bar{l}_1 = 1$ and $\bar{l}_2 = 1$, is plotted as the peel force $\bar{F}$ increases. When $\bar{h}_2 = 0.1$, a crack grows unstably for $\bar{F} > 37$, but it will not grow at all below that loading. When $\bar{h}_2 = 1.0$, as $\bar{F}$ exceeds 13.1 only crack tip 1 grows stably up to its maximum length $\bar{l}_1 = 1.42$. Then as $\bar{F}$ exceeds 17.3 both crack tips become active and delamination becomes unbounded. When $\bar{h}_2 = 10.0$, only crack tip 2 grows for $1.36 < \bar{F} < 1.9$ up to $\bar{l}_2 = 5.06$. For $\bar{F} > 1.9$, delamination again becomes unbounded.

If the tensile failure of an image layer is the weakest failure mechanism in the system, the failure of an image layer could be predicted during the peeling process by calculating the maximum tensile strain in the image layer. The maximum tensile strain ϵ_t appears at the top surface of section A (Fig. 2), and is given by

$$\epsilon_t = \epsilon_m + \epsilon_a \tag{7}$$

where $\epsilon_m = 6M/(E_1h_1^2)$ is the maximum bending strain due to M, $\epsilon_a = T/(E_1h_1)$ is the strain due to the horizontal (axial) force T.

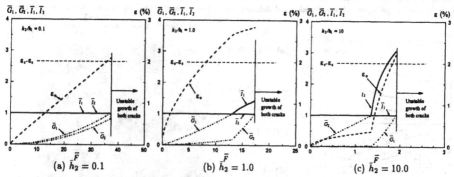

(a) $\bar{h}_2 = 0.1$ (b) $\bar{h}_2 = 1.0$ (c) $\bar{h}_2 = 10.0$

Fig. 5 Bending failure of an image layer due to the crack growth

In this section we approximate ϵ_a by ϵ_s, the strain at the surface of the substrate produced by stretching or bending of the substrate. Fracture of the image layer occurs when ϵ_t exceeds the tensile fracture strain ϵ_c. In Fig. 5 the maximum bending strain is plotted as $\bar{F}$ increases showing how control of the substrate bending (ϵ_s) can accelerate or delay the bending fracture. A compressive substrate bending moment will prevent tensile failure of the image layer, but an excessive compressive bending moment may buckle the image layer. As an example, $\epsilon_c - \epsilon_s = 2\%$ is chosen. In this case, The image layer will fail when $\bar{F} = 35.6$ for $\bar{h}_2 = 0.1$, $\bar{F} = 8.1$ for $\bar{h}_2 = 1.0$, and $\bar{F} = 1.83$ for $\bar{h}_2 = 10$. For $\bar{h}_2 = 10$, $\bar{l}_2 = 2.93$ when the image layer breaks.

FRACTURE OF THIN IMAGING LAYER

When $\bar{h}_2$ is very large as is required to form fine pels in imaging media, the crack length l_2 becomes much smaller than the thickness of the peel-off layer h_2 and a new approach is required to describe the conditions of crack tip 2. Now crack tip 2 is like a crack in an infinite solid in regard to the relation of the crack opening displacement and crack shear displacement to K_1 and K_2, the stress intensity factors for crack 2 (Broek [5]). Also in regard to the contributions of n, m and t to K_1 and K_2, use is made of the M-integral (Freund [6]) considering these as point forces and moments. The beam contribution of m_o and ϵ_s to K_1 and K_2 is treated by the method of Thouless et al. [7].

In order to estimate the delamination length $\bar{l}_1$ after the fracture of the image layer due to the peeling process, Equations (1) are modified as follows; both crack tip 1 and 2 are stably active, $\epsilon_t = \epsilon_c$ and the deflection w is small. Then, the resulting set of equations are

$$(m - \bar{l}_1 n)^2 + \frac{(t - \beta_s)^2}{12} = 1, \quad k_1^2 + k_2^2 = 1, \quad \bar{l}_1 n - 2m = -\frac{2}{\bar{l}_1\sqrt{3\pi l_3}}\,k_1, \quad 2\bar{l}_1 n - 3m = \frac{4\sqrt{3l_3}}{\bar{l}_1^2\sqrt{\pi}}\,k_1, \quad (8.1-4)$$

$$6m + t = \beta_c, \qquad \bar{l}_1 t = \beta_s \bar{l}_o + 8\sqrt{3l_3/\pi}\,k_2 \qquad (8.5, 8.6)$$

where

$$k_1 = m_o \sin\omega/\sqrt{\alpha} + \beta_s \cos\omega\,\sqrt{\bar{h}_2/12} + -n(\bar{l}_1/2 + \bar{l}_3)/\sqrt{3\pi \bar{l}_3^3} + m/\sqrt{3\pi \bar{l}_3^3}$$

$$k_2 = -m_o \cos\omega/\sqrt{\alpha} + \beta_s \sin\omega\,\sqrt{\bar{h}_2/12} - t/\sqrt{3\pi \bar{l}_3}$$

$$\beta_s = \epsilon_s \sqrt{6E_1h_1/\mathcal{G}_c}, \quad \beta_c = \epsilon_c \sqrt{6E_1h_1/\mathcal{G}_c}, \quad \bar{l}_3 = 4(1/\sqrt{\bar{l}_2} + 1/\sqrt{\bar{l}_o})^{-2}$$

and $k_1 = K_1/\sqrt{E_1\mathcal{G}_c}$, $k_2 = K_2/\sqrt{E_1\mathcal{G}_c}$. Equation (8.6) results from conservation of length $\bar{l}_o$ in the cracking process. The numerical solution is shown in Fig. 6.

Interpretation of Eqs. (8) shows that these results are unstable. As the cracks start at small $\bar{l}_1$ and $\bar{l}_2$ and m_o is increases, $\bar{l}_1$ will start to grow because Eq. (8.1) is satisfied before Eq. (8.2). Eqs. (10.1 to 4) give for $l_2 \ll l_1$, $2\bar{l}_1 n = 3m$, $m - \bar{l}_1 n = -1$ so $m = 2$. Then Eq. (8.5) is satisfied for $\beta_c < 12$, but the equality of Eq. (8.5) is established as $\bar{l}_1$ grows only after $\beta_c < 6m + t$ at smaller $\bar{l}_1$. However for small $\bar{l}_1 < 2$ crack

1 does not yet approximate a beam and so is not yet governed by Eqs. (8). We conclude that for $\beta_c < 12$ and $\beta_s > 0$ (moderate tensile substrate bending) the image layer breaks before extensive delamination. For $\beta_c > 12$ the image layer will not break until delamination is so extensive that the beams become elasticae.

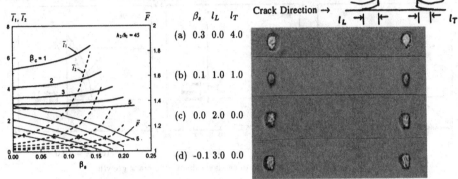

	β_s	l_L	l_T
(a)	0.3	0.0	4.0
(b)	0.1	1.0	1.0
(c)	0.0	2.0	0.0
(d)	-0.1	3.0	0.0

Fig. 6 Delamination lengths and
required peel force.

Fig. 7 Small holes in carbon-black imaging layer
at various substrate strains ϵ_s. Peeling direction is left to right.

EXPERIMENTS

Control of substrate strain β_s ($= 122\ \epsilon_s$ using $\mathcal{G}_c = 2J/m^2$) by substrate bending changes the shape of the leading and trailing edges of the peeled holes in the carbon-black image layer of our Helios film. As β_s decreases the length l_L of the delaminated leading edge of the small imaged holes in Fig. 7 (units of l_L and l_T are μm) progressively increases. Simultaneously the delamination length of the trailing edge decreases. The opening width of the hole goes through a minimum and the delamination lengths of the leading and trailing edges are equal at an optimal small tensile value of β_s.

CONCLUSIONS

Beam analysis of peeling delamination provides a means of describing the fracture of interfaces and an intervening carbon-black image layer in terms of forces and moments which are easily understood. The analysis suggests that the strain in the substrate controls the strain in the image layer and this controls the structure of the holes in the image. High image quality requires precise control of the substrate strain.

ACKNOWLEDGMENTS

Authors gratefully thank Prof. K.-S. Kim of Brown University for his helpful suggestions and discussions.

REFERENCES

1. Z. Suo and J.W. Hutchinson, Int. J. Fracture. **43**, p. 1-18 (1990).

2. N. Aravas, K.-S. Kim and M.J. Loukis, Mat. Sci. and Eng. **A107**, p. 159-168 (1989).

3. M.Y. He and J.W. Hutchinson, J. Appl. Mech. **56**, p. 270-278 (1989).

4. T.C. Wang, C.F. Shih and Z. Suo, Int. J. Solids Structures **29**, p. 327-344 (1992).

5. D. Broek, Elementary Engineering Fracture Mechanics, Nijhoff, Boston, 1986.

6. L.B. Freund, Int. J. Solids Structures **14**, p. 241-250 (1978).

7. M.D. Thouless, A.G. Evans, M.F. Ashby and J.W. Hutchinson, Acta metall. **35**, p. 1333-1341 (1987).

AUTHOR INDEX

SUBJECT INDEX

Printed in the United States
By Bookmasters